AAPS Introductions in the Pharmaceutical Sciences

The *AAPS Introductions in the Pharmaceutical Sciences* book series is designed to support pharmaceutical scientists at the point of knowledge transition. Springer and the American Association of Pharmaceutical Scientists (AAPS) have partnered again to produce a second series that juxtaposes the *AAPS Advances in the Pharmaceutical Sciences* series. Whether shifting between positions, business models, research project objectives, or at a crossroad in professional development, scientists need to retool to meet the needs of the new scientific challenges ahead of them. These educational pivot points require the learner to develop new vocabulary in order to effectively communicate across disciplines, appreciate historical evolution within the knowledge area with the aim of appreciating the current limitations and potential for growth, learn new skills and evaluation metrics so that project planning and subsequent evolution are evidence-based, as well as to simply "dust the rust off" content learned in previous educational or employment settings, or utilized during former scientific explorations. The *Introductions* book series will meet these needs and serve as a quick and easy-to-digest resource for contemporary science.

Márcia C. Breitkreitz • Hector C. Goicoechea
Editors

Introduction to Quality by Design in Pharmaceutical Manufacturing and Analytical Development

Springer

aaps

Editors

Márcia C. Breitkreitz
Laboratory of Pharmaceutical Research and Chemometrics (LABFARQUI), Institute of Chemistry
University of Campinas (UNICAMP)
Campinas, SP, Brazil

Hector C. Goicoechea
Laboratorio de Desarrollo Analítico y Quimiometría (LADAQ), Cátedra de Química Analítica I, Facultad de Bioquímica y Ciencias Biológicas
Universidad Nacional del Litoral, Ciudad Universitaria, Santa Fe
Santa Fe, Argentina

Consejo Nacional de Investigaciones Científicas y Técnicas (CONICET)
Ciudad Autónoma de Buenos Aires, Buenos Aires, Argentina

Jointly published with American Association of Pharmaceutical Scientists

ISSN 2522-834X ISSN 2522-8358 (electronic)
AAPS Introductions in the Pharmaceutical Sciences
ISBN 978-3-031-31504-6 ISBN 978-3-031-31505-3 (eBook)
https://doi.org/10.1007/978-3-031-31505-3

This Springer imprint is published by the registered company Springer Nature Switzerland AG
The registered company address is: Gewerbestrasse 11, 6330 Cham, Switzerland

Preface

This book was inspired by the demand of modern pharmaceutical development to incorporate quality into their products, processes and methods in the early stages of development. This is the goal of Quality by Design and Analytical Quality by Design approaches, in which Design of Experiments plays a central role, together with the risk assessment. At this moment we are facing a paradigm change regarding quality – from final product testing to building the quality into the products. As any paradigm change, it requires studying, training and learning from examples. Therefore, solid materials are needed for the present and future generations and this is the major goal of this book: to organize in a single volume the most important topics, including both theory and examples, on QbD and AQbD, with special attention to our favorite part: DOE. We can say this material is the result of our passion for applied research and the use of DOE to fully understand the systems we study. Over the years we have become enthusiastics about it and therefore we are really glad to contribute to the spread of this magnificent tool within a modern applied scenario.

Another important aspect we would like to highlight is the so-needed bound of Academia, Industry and Pharmacopeias we have reached when building the content of this book. More than ever, the academic knowledge is very welcome into the industrial scenario in order to promote quality to products and methods. Pharmacopeias play an important role in providing guidance to this process.

We do hope this book helps the readers to walk towards the understanding of the importance of QbD, AQbD and the implementation of DOE in routine for product and method development. We hope you enjoy reading as much as we enjoyed organizing it.

Campinas, São Paulo, Brazil — Márcia C. Breitkreitz
Santa Fe, Santa Fé, Argentina — Hector C. Goicoechea

Acknowledgments

To Prof. Roy Edward Bruns for the careful revision of the text, for his willingness to teach DOE and for being an inspiring example to all of us.

To all the authors who, in spite of their hectic routine of work, went on board with us on this knowledge endeavour and contributed to the final result of this book.

Contents

Part 1
Quality by Design (QbD)

Chapter 1
Pharmaceutical Quality by Design (QbD) Overview and Major Elements

Ana Vanessa Dutra Silva and Márcia C. Breitkreitz

Abstract The risk-based approach initiative for pharmaceutical development initiated in 2004 by the Food and Drug Administration (FDA, USA) paved the way for the Quality by Design framework to be implemented in the pharmaceutical industry. Ever since, much work has been done; however, there is still a long way to go for its complete understanding and implementation. This chapter intends to discuss the need for the QbD methodology as well as provide an overview of its major elements. An illustrative example of direct compression has been created to elucidate the concepts of quality target product profile (QTTP), critical quality attributes (CQAs), and risk analysis (RA). The benefits of using design of experiments (DOE) within the context of QbD will be discussed, including the concept of *design space*. At the end, a regulatory overview with practical references will be provided.

Keywords Quality by design · Regulatory flexibility · Lifecycle approach · QbD framework · Risk-based approach

1.1 Introduction

The launching of FDA innovative guideline "Pharmaceutical CGMPs for the twenty-first century – a risk-based approach" [1], in 2004, encouraged, for the first time, the pharmaceutical industry to present a rational, proactive, and flexible attitude toward pharmaceutical development. Therefore, this was a historical mark starting a new era for product development. A "risk-based approach" represents a new way of thinking guided by the previous identification of risky elements that could cause failure of the final product in achieving its quality performance. With this approach, FDA gave to the pharma companies the opportunity to develop and

A. V. D. Silva · M. C. Breitkreitz (✉)
Laboratory of Pharmaceutical Research and Chemometrics (LABFARQUI), Institute of Chemistry, University of Campinas (UNICAMP), Campinas, SP, Brazil
e-mail: marciacb@unicamp.br

M. C. Breitkreitz, H. C. Goicoechea (eds.), *Introduction to Quality by Design in Pharmaceutical Manufacturing and Analytical Development*, AAPS Introductions in the Pharmaceutical Sciences 10, https://doi.org/10.1007/978-3-031-31505-3_1

present their rational thinking and research results describing what is important to monitor in their products in order to achieve quality.

The Quality by Design (QbD) concept was mentioned within the risk-based approach presented by the FDA and described in more details by the International Council for Harmonisation of Technical Requirements for Pharmaceuticals for Human Use (ICH) guideline ICH Q8 [2] and was intended to provide guidance on the contents of Section 3.2.P.2 (Pharmaceutical Development) for drug products. This guideline, together with ICH Q9 (Quality Risk Management) and ICH Q10 (Pharmaceutical Quality System), served as the basis for the new quality concepts expected for pharmaceutical development in the twenty-first century [3, 4]. The ICH Q8 (R2) defines QbD formally as "a systematic approach for product development that begins with predefined objectives and emphasizes product and process understanding and process control, based on sound science and quality risk management."

The idea of QbD by itself has been known years before the ICH guideline in many other different areas. The concept was introduced and developed by Joseph Juran and can be found in a series of publications [5–7]. Perhaps the major additional aspect in pharmaceutical QbD is to also take into account the safety and efficiency to patients, more than process improvements as sought by other areas. In a pharmaceutical QbD framework, it is important to understand how input parameters (material attributes, composition, and process variables) affect the quality attributes of the product, decrease out-of-specification results on the production line, and make the transfer to other manufacturing sites easier, saving time and decreasing losses while also guaranteeing safety and efficacy to the patient.

The pre-QbD era in the pharmaceutical field was guided by testing the final product (Quality by Testing, QbT), justification of choices based on general monographs or supplier specifications, and univariate thinking for experimentation. The QbD proposal was to replace the QbT concept by the attitude of building the quality during product development and replace general justifications based on monographs and supplier specifications for rational thinking based on the chosen dosage forms and manufacturing processes. This allows identifying the critical material attributes (CMAs) and critical process parameters (CPPs) in a more trustworthy way.

QbD also highlighted – in time – the importance of multivariate thinking, which has been established in academic research for many years. The one-factor-at-a-time (OFAT) approach of studying the influence of experimental variables on a given response is faulty in detecting interactions between factors – very common in many different systems. One factor interacts with another if the influence of the first one depends on the level of the second one. Furthermore, the OFAT approach does not allow the generation of models to describe the cause-and-effect relationships required by the QbD approach. DOE methods, besides enabling the detection of interactions, allow mapping the experimental domain by model equations, which is highly desirable to fully understand the system and avoid scale-up problems. These models specify the design space, "the multidimensional combination and interaction of input variables (e.g., material attributes) and process parameters that have demonstrated to provide assurance of quality." If the company demonstrates knowledge about the root causes of failure and the relationship between inputs and the quality of the final product by specifying a design space, flexibility regarding materials and process parameters can be assured by regulatory agencies. Therefore, the whole development shifts the focus from a strict, highly controlled environment to a prospective and proactive approach, where the companies have the opportunity to demonstrate a deep understanding of their products to the regulatory bodies. For this and many other reasons, design of experiments (DOE) is very welcome in the QbD scenario and will be described in detail in other chapters of this book.

ICH Q8 Table of Appendix 1 compares the actual "Minimum Approaches" and "Enhanced Quality by Design Approaches." The major aspects to be highlighted are:

- The overall pharmaceutical development is currently mostly empirical and conducted using the abovementioned OFAT method. QbD proposal is systematic, relating mechanistic understanding of critical material attributes and critical process parameters to drug product critical quality attributes (CQAs). To achieve this goal, multivariate DOE-based experiments are encouraged.
- The manufacturing process is currently fixed, whereas in the QbD format, it is adjustable within the design space. The current process validation is encouraged to follow the lifecycle approach, with continuous process verification (process technology tools (PAT) encouraged), instead of a validation based on a given number of initial full-scale batches.

The QbD framework has been incorporated within the lifecycle concept for product/process development [8]. Briefly, it involves:

- *Phase 1:* Product and process conceptualization, where most of the QbD elements are applied and for which ICH Q8 was specifically addressed.
- *Phase 2:* Technology transfer/scale-up from lab/pilot to commercial scale and process validation. According to ICH Q8, the traditional process validation can be replaced by continuous verification.
- *Phase 3:* Commercial manufacturing and control.
- *Phase 4:* Discontinuation of the product.

Below, the elements of pharmaceutical QbD will be discussed in more detail, and an illustrative example QbD elements will be provided, considering a direct compression process, as a simple introductory example of the concept. It should be highlighted that FDA provided two simulated dossiers as examples to provide guidance to industries on practical aspects that should be reported for submission [9, 10]. These documents provide not only detailed information about Phase 1 to support product development but also clarifications on Phases 2 and 3.

1.2 QbD Elements

An illustrative flowchart of QbD elements is shown in Fig. 1.1, with the elements associated with Phases 1, 2, and 3 of the lifecycle approach.

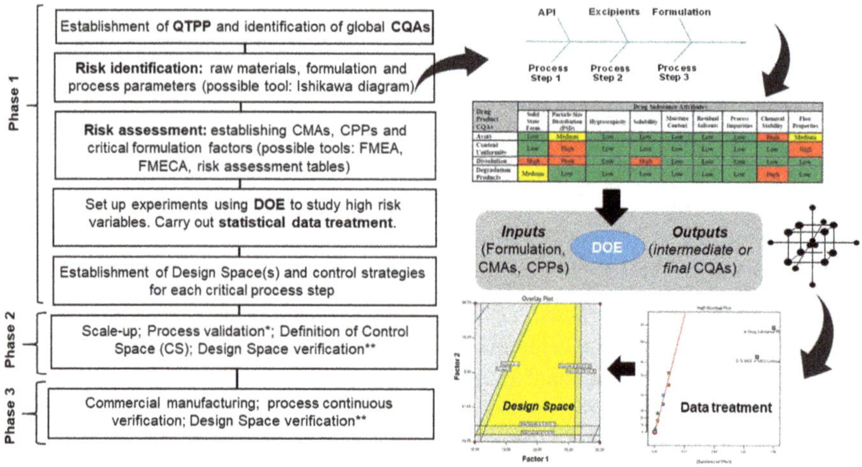

Fig. 1.1 Illustrative flowchart of QbD elements, associated with the product lifecycle concept. **Process validation can be replaced by continuous verification* [2]. ** *Design space verification can be done throughout the lifecycle*

1.2.1 The Quality Target Product Profile (QTPP)

In 2007, the FDA provided a guideline defining the target product profile [11]. The quality target product profile is a subset of TPP and is defined as "A prospective summary of the quality characteristics of a drug product that ideally will be achieved to ensure the desired quality, taking into account safety and efficacy of the drug product" [2]. QTTP should provide an overview of what the product should achieve to perform well, both in vitro and in vivo.

ICH Q8 provides some examples of QTTP elements: intended use in clinical setting; route of administration; dosage form; delivery systems; dosage strength(s); container closure systems; therapeutic moiety release or delivery and attributes affecting pharmacokinetic characteristics (e.g., dissolution, aerodynamic performance) appropriate to the drug product dosage form; and drug product quality criteria (e.g., sterility, purity, stability, and drug release) appropriate for the intended marketed product. In a summary, the QTPP should provide a set of predefined in vitro/in vivo deliverables necessary to achieve a product with the proposed therapeutic effect. Table 1.1 shows a simple illustrative example of a QTTP for a generic product containing a BCS Class I drug.

1.2.2 Global Critical Quality Attributes (CQAs)

The QTTP should contain the "drug product quality attributes," which are attributes related to a specific dosage form and define in vitro measurable quality parameters of the drug product. These quality attributes can be drawn based on pharmacopoeia

Table 1.1 The quality target product profile (QTPP) for a simulated product (oral generic tablet of immediate release)

<table>
<tr><th colspan="2">QTPP elements</th><th>Target</th><th>Justification</th></tr>
<tr><td colspan="2">Dosage form</td><td>Coated tablet</td><td>Pharmaceutical equivalence requirement: same dosage form of reference listed drug product</td></tr>
<tr><td colspan="2">Dosage design</td><td>Immediate-release tablet without a score and with coating</td><td>Pharmaceutical equivalence requirement: same design needed to meet label claims and similar to the reference listed drug product</td></tr>
<tr><td colspan="2">Route of administration</td><td>Oral</td><td>Pharmaceutical equivalence requirement: same route of administration of reference listed drug (product)</td></tr>
<tr><td colspan="2">Dosage strength</td><td>15 mg</td><td>Pharmaceutical equivalence requirement: same strength of reference listed drug (product)</td></tr>
<tr><td colspan="2">Pharmacokinetics</td><td>Absorption is almost complete and independent of food intake (average Tmax 4 h after multiple dosing)
The absolute bioavailability of the drug is expected to be approximately 80%
The elimination half-life (T1/2β) after multiple doses is about 30 h</td><td>Bioequivalence requirement
Needed to ensure efficacy</td></tr>
<tr><td colspan="2">Stability</td><td>At least 24 months at room temperature</td><td>Meet the quality attributes within the established shelf life. Equivalent to or better than RLD shelf life
Internal studies may be conducted for up to 24 or 36 months</td></tr>
<tr><td rowspan="8">Drug product quality attributes</td><td>Physical attributes</td><td rowspan="8" colspan="2">Pharmaceutical equivalence requirement: Must meet the same compendial or other applicable (quality) standards (i.e., identity, assay, purity, and quality)</td></tr>
<tr><td>Identification</td></tr>
<tr><td>Assay</td></tr>
<tr><td>Content uniformity</td></tr>
<tr><td>Dissolution</td></tr>
<tr><td>Degradation compounds</td></tr>
<tr><td>Residual solvents</td></tr>
<tr><td>Microbial limits</td></tr>
</table>

(continued)

Table 1.1 (continued)

QTPP elements	Target	Justification
Container closure system	The packaging material must be qualified as stable for the product The material to be tested for the product will be an aluminum/ aluminum blister	The packaging material needed to achieve the target shelf life and to ensure tablet integrity during shipping
Administration/concurrence with labeling	Similar food effect as reference listed drug (product)	The tablets are taken orally, once a day. The tablets can be taken at any time of the day, with or without food. Preferably always taken at the same time. Swallow the tablets with water without chewing
Alternative methods of administration	As per the instructions in the package insert	None are listed in the reference listed drug product label

monographs, internal specifications and previous results, literature information, etc. Among the quality attributes, those that affect the safety and efficiency of the product and can be affected by raw material attributes and process variables are considered critical. Based on these two criteria, they are classified as "critical quality attributes" (CQAs), a major element in the QbD framework.

According to ICH Q8, CQAs comprise "physical, chemical, biological or microbiological properties or characteristics that should be within an appropriate limit, range, or distribution to ensure the desired product quality"; therefore, they will serve as the basis to guide the rational development seeking for quality. It should be noted that QTPP contains elements related to desired clinical responses and CQAs are those characteristics that reflect the final process and product quality. For example, bioavailability is an element of a product's QTPP, and dissolution may be its associated in vitro CQA [12].

The manufacturing of pharmaceutical products is often carried out in several unit operations (e.g., mixing, granulating, drying, and tableting). CQAs are product- and process-related and can be classified as intermediate (sometimes called IQAs, intermediate quality attributes, or in-process controls) and final (related to final dosage forms). For example, assay, dissolution, and content uniformity are final CQAs for oral dosage forms, whereas granule size, hardness, and friability are typical examples of intermediate CQAs.

As a first step, it is recommended to identify the final CQAs for a given dosage form, based on the QTTP table, as shown in the example (Table 1.2). Later, if the process has more than one unit operation, intermediate CQAs can be established for each of the manufacturing operations. It should be highlighted that each decision of classifying the quality attribute as critical or not should be accompanied by a justification. Moreover, the critical quality attributes that were found upon QTPP can be reassessed later [13]. An example of a CQA list for a solid dosage form can be seen in Table 1.2.

Table 1.2 Identification of critical quality attributes (CQAs) for a simulated product (oral generic tablet of immediate release)

Quality attributes of the drug product		Target	Is this a CQA?	Justification
Physical attributes	Appearance	Color and shape acceptable to the patient. No visual tablet defects observed	No	Color, shape, and appearance are not directly linked to safety and efficacy and are not affected by raw material attributes and process variables. Therefore, they are not critical. The target is set to ensure patient acceptability
	Odor	No unpleasant odor	No	In general, a noticeable odor is not directly linked to safety and efficacy, but odor can affect patient acceptability. For this product, neither the drug substance nor the excipients have an unpleasant odor. No organic solvents will be used in the drug product manufacturing process
	Size	Circular tablets – similar to the reference listed drug (product)	No	For patient acceptability, the tablets must be easy to swallow and comply with the requirements for the treatment; the target for tablet dimensions is set similar to the reference drug product
	Score configuration	Unscored	No	The reference drug product is an unscored tablet; therefore, the generic tablet will be unscored
Identification		HPLC – the main peak of the chromatogram obtained with the sample solution has the same retention time as the main peak obtained with the standard solution, prepared according to the content test	Yes*	Though identification is critical for safety and efficacy, this CQA can be effectively controlled by the quality management system and will be monitored at drug product release. Formulation and process variables do not impact identity. Therefore, this CQA will not be discussed during formulation and process development
Assay		90.0 to 110.0% 100% w/w of label claim	Yes	Assay variability will affect safety and efficacy. Process variables may affect the assay of the drug product. Thus, assay will be evaluated throughout product and process development

(continued)

Table 1.2 (continued)

Quality attributes of the drug product		Target	Is this a CQA?	Justification
Content uniformity		Acceptance value 10 units: VA $\leq$ L1 = 15 30 units: VA $\leq$ L1 = 15 and L2 = 25 No unit is less than $(1 - L2 \times 0.01)$ M or greater than $(1 + L2 \times 0.01)$ M	Yes	Variability in content uniformity will affect safety and efficacy. Both formulation and process variables impact content uniformity, so this CQA will be evaluated throughout product and process development
Dissolution		Not less than 85% in 30 min (Q = 80%)	Yes	Dissolution testing may impact drug efficacy, and failure to meet the dissolution specification can impact bioavailability. Both formulation and process variables affect the dissolution profile. This CQA will be investigated throughout formulation and process development
Degradation compounds		Compound A: Maximum 0.3% Compound B: Maximum 0.5% Compound C: Maximum 0.5% Compound E: Maximum 0.2% Unspecified individual impurities: maximum 0.2% Total impurities: Maximum 2.0%	Yes	Formulation and process variables can impact degradation products. Therefore, degradation products will be assessed during product and process development
Residual solvents		Pharmacopoeial requirements	No	Residual solvents can affect safety. However, no solvents are used in the manufacturing process of this drug. Therefore, the formulation and process variables have no impact on this CQA
Microbial limits	Total aerobic bacteria count	Not more than 1.000 CFU/g	Yes*	Non-compliance with microbial limits will impact patient safety. However, in this case, the risk of microbial growth is very low because direct compression manufacturing process is utilized for this product. Therefore, this CQA will not be discussed in detail during formulation and process development
	Total mold and yeast count	Not more than 100 CFU/g		
	Pathogen research: *Escherichia coli*	Absence of *Escherichia coli* in 1 g		

*Formulation and process variables are unlikely to impact this CQA. Therefore, this CQA does not need to be investigated and discussed in detail in subsequent risk assessment and pharmaceutical development. However, this CQA remains a target element of the drug product profile.

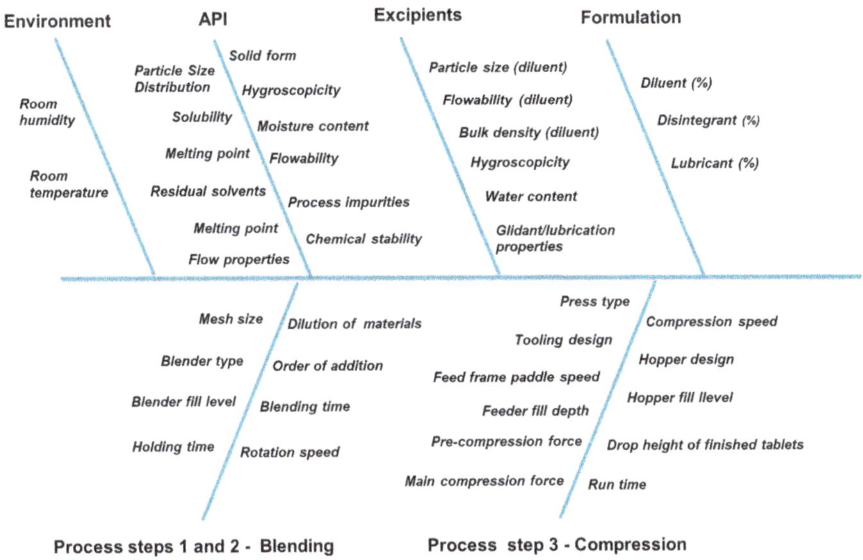

Fig. 1.2 Ishikawa diagram for the simulated example

1.2.3 *Risk Analysis*

The next step on the QbD workflow is to carry out the risk analysis (RA), which is divided into two steps: risk identification and risk assessment. The central goal of this step is to identify and indicate a risk for all possible input variables of the product, including raw material attributes (both active pharmaceutical ingredient (API) and excipients), formulation composition (typically the percentages), and process parameters that might affect the CQAs. The fishbone diagram (or Ishikawa diagram, Fig. 1.2) is a simple though very effective tool to raise all possible variables involved in product development. Further on, formal risk analysis tools such as FMEA (*failure modes and effects analysis*), FMECA (*failure mode, effects, and criticality analysis*), HACCP (*hazard analysis and critical control points*), HAZOP (*hazard and operability analysis*), and PHA (*preliminary hazard analysis*) can be used, or the risk can be already indicated in the Ishikawa diagram.

After risk identification, risk analysis is the next step. Formal risk analysis tools are promptly available nowadays. Probably the failure modes and effects analysis (FMEA) is the most common, which evaluates the severity, probability, and detection using a scale of 1–5 in terms of risk. The risk priority number (RPN) is calculated by multiplying these three parameters. Nevertheless, it should be noted that according to ICH Q9 Quality Risk Management, "it is neither always appropriate nor always necessary to use a formal risk management process. The use of informal risk management tools (using empirical tools and/or internal procedures) can also be considered acceptable." Risk assessment tables, in which a risk is indicated as low, medium, or high based on a color scale, are examples of such informal risk management tools. In order to have a rational basis for the risk presentation, it can be split into raw materials (drug substance and excipients), formulation and process risk analysis. Tables 1.3, 1.4, 1.5, and 1.6 show the risk analysis and the corresponding justification for drug substance and excipients.

Materials RA (Drug Substance and Excipients)

Initial Formulation Development
Before studying the manufacturing process, it is necessary to have a tentative formulation, i.e., to determine the proportions of the excipients to accomplish the major product CQAs (dissolution for solids, density/viscosity for liquids, etc.) and match a reference product, if available. This tentative formulation can be reassessed later on, after process evaluation, if necessary. A "formulation" DOE can be carried out in this step, to be used in the next steps. In this example, the high-risk variables to be studied could be the amount of microcrystalline cellulose (as diluent and major component of the formulation that can significantly affect flow properties and therefore content uniformity), croscarmellose sodium, and magnesium stearate that can affect mainly dissolution. Silicon dioxide and talc are used in very small amounts; therefore, it is assumed their influences on the CQAs are smaller.

Manufacturing Process RA
With a tentative formulation developed in the previous step, it is possible to study the manufacturing process in more details. As a first step, the global CQAs should be evaluated as a function of each operation unit to have an overview of critical steps (Table 1.7). Then each operation unit should be evaluated, considering the process parameters and material attributes involved, as functions of intermediate CQAs (ICQAs) – if they are present – and global CQAs (in this case for the last operation unit of the process). If only one operation unit is present, then there will be no intermediate CQAs to be described. For this purpose, the process map (Fig. 1.4) is a very useful and informative tool.

In Fig. 1.3, the possible critical process parameters (CPPs) of each step and the possible intermediate critical quality attributes (ICQAs) and final CQAs are highlighted in bold. The justification is described in Table 1.8.

1.2.4 Multivariate Analysis: Design of Experiments (DOE)

The steps mentioned above (defining QTTP, CQAs, and risk analysis for input variables) were carried out without experimentation. The QbD mindset is to go to the laboratory or pilot plant having in mind the important variables to be studied after this extensive brainstorming exercise and literature survey. This is an important change of paradigm over the current strategy, in which one goes to the laboratory to generate results that will guide the next steps. In the pharmaceutical QbD scenario, high-risk variables (CMAs, CPPs, and formulation variables) are recommended to be studied by design of experiments (DOE), a multivariate methodology well

Table 1.3 Initial risk assessment of the drug substance attributes

Drug product CQAs	Drug substance attributes									
	Solid form	Particle size distribution (PSD)	Hygroscopicity	Solubility	Moisture content	Melting point	Residual solvents	Process impurities	Chemical stability	Flow properties
Assay	Low	Medium	Low	Low	Low	Low	Low	Low	High	Medium
Content uniformity	Low	High	Low	Low	Low	Low	Low	Low	Low	High
Dissolution	Low	Low	Low	Low	Low	Low	Low	Low	Low	Low
Degradation compounds	Medium	Low	Low	Low	Low	Low	Low	Low	High	Low

Table 1.4 Initial risk assessment of the excipient's attributes

Drug product CQAs	Diluent (MCC)				Disintegrant (croscarmellose sodium)	Glidant/lubricant	
	Particle size	Density	Flow properties	Hygrosc.	Water content	SiO_2/talc glidant/flow properties	Magnesium stearate surface area
Assay	Medium	Medium	Medium	Low	Low	Low	Low
Content uniformity	High	High	High	Low	Low	Medium	Low
Dissolution	Low	Medium	Medium	Low	Medium	Low	High
Degradation compounds	Low	Low	Low	Low	Low	Low	Low

Table 1.5 Justification of the risk indicated in Table 1.3 for drug substance

Drug substance attributes	Drug product CQA	Justification
Solid form	Assay, content uniformity	The solid state does not affect the content and content uniformity as these tests are performed with organic solvents in which the API is highly soluble. The risk is low
	Dissolution	API belongs to BCS Class I, therefore it has high solubility in water. The different polymorphic forms are all very soluble in water – the risk is low
	Degradation compounds	Differences in polymorph synthesis routes could influence the generation of degradation products. The risk is medium
Particle size distribution	Assay	Small particle sizes can impact the homogeneity and flowability of the mixture. In extreme cases, poor flowability and lack of homogeneity may cause an assay failure. The risk is medium
	Content uniformity	Particle size has a direct impact on drug substance flowability and mixture homogeneity and consequently on product content uniformity. Since this drug substance has a small particle size and the process is direct compaction, the risk is high
	Dissolution	The drug substance is a BCS Class I compound. Due to its high solubility, the particle size should not impact the dissolution. The risk is low
	Degradation compounds	The effect of particle size reduction on drug substance stability has been evaluated by the DMF holder. The milled drug substance did not exhibit increased amounts of impurities. Furthermore, the degradation products were studied in the preformulation stage and will be monitored in the release of raw materials for the production of batches. The risk is low
Hygroscopicity	Assay, content uniformity, dissolution, degradation compounds	This drug substance is not hygroscopic. The risk is low
Solubility	Assay, content uniformity, degradation compounds	Solubility does not affect tablet assay, uniformity content, and degradation compounds. The risk is low
	Dissolution	Solubility can affect dissolution, however, this drug is BCS Class I (high solubility/high permeability); therefore, the risk is low

(continued)

Table 1.5 (continued)

Drug substance attributes	Drug product CQA	Justification
Moisture content	Assay, content uniformity, dissolution	The water content of drug substance is low (around 0.3–0.4%); therefore, it is unlikely to affect uniformity and dissolution content. The risk is low
	Degradation compounds	The drug substance is not sensitive to moisture and its water content is low. The risk is low
Melting point	Assay, content uniformity, dissolution, degradation compounds	The drug substance has a melting point between 146 and 154 °C. The risk is low
Residual solvents	Assay, content uniformity, dissolution	Residual solvents at ppm level do not impact these CQAs; the risk is low
	Degradation compounds	The presence of residual solvents does not cause degradation of the API; therefore the risk is low
Process impurities	Assay, content uniformity, dissolution	Synthesis impurities are in low concentrations (NMT 1.0%) and therefore do not impact these CQAs. The risk is low
	Degradation compounds	Synthesis impurities were not related to degradation products in the preformulation step; the risk is low
Chemical stability	Assay	API is susceptible to dry heating and basic and oxidative degradation. Under these conditions, degradation can affect the content; the risk is high and must be controlled throughout the manufacturing process
	Content uniformity	Content uniformity, as measured by % RSD, is not impacted by degradation
	Dissolution	If the dissolution is corrected for the assay, there is no impact of degradation on dissolution
	Degradation compounds	The risk is high – see justification for assay
Flow properties	Assay	The drug substance presents poor flow properties; in extreme cases, poor flowability can impact the assay. The risk is medium
	Content uniformity	Poor flow properties can impact tablet content uniformity. The risk is high
	Dissolution	The flowability of the drug substance is not related to its degradation pathway or solubility, therefore, the risk is low
	Degradation compounds	

Table 1.6 Justification of the risk indicated in Table 1.4 for excipients

Excipient attribute	Drug product CQA	Justification
Microcrystalline cellulose particle size	Assay	The particle size of the diluent (which is the major component of this formulation) can impact the homogeneity and fluidity of the mixture. In some cases, poor flowability and inhomogeneity may cause an assay failure. The risk is medium
	Content uniformity	The particle size of the diluent in direct compression process can impact the homogeneity and fluidity of the mixture and, consequently, the content uniformity of the product. Poor flowability and inhomogeneity may affect average tablet weight variation and consequently content uniformity; therefore, the risk is high
	Dissolution	The particle size of the diluent can influence the compressibility and hardness of the tablets. However, in a formulation in which the active ingredient belongs to Class I in the Biopharmaceutical Classification System, that is, it has high solubility, the diluent particle size will not impact the API solubility, and the hardness can be controlled during the in-process control. The risk is low
	Degradation compounds	The particle size of microcrystalline cellulose is not able to affect the active ingredient degradation compounds. Since microcrystalline cellulose and active substance are compatible, there will be no impact on drug degradation; the risk is low
Microcrystalline cellulose density	Assay	The density of microcrystalline cellulose is related to the porosity and particle size of the material. The inadequate density of the diluent used in formulations for direct compression can impact the homogeneity and flowability of the mixture, which can cause an assay failure. The risk is medium
	Content uniformity	A very low density of microcrystalline cellulose used in direct compression processes can affect the flow of the mixture and lead to a high variation in tablet weight in the compression step. This factor can directly impact the content uniformity of active substance in the tablets, so the risk is high
	Dissolution	The diluent density alone is not capable of affecting the dissolution of the product. However, since the density of microcrystalline cellulose can impact the assay and content uniformity of the tablets, if there is a problem with the assay and lack of content uniformity in the tablets, dissolution may be affected; therefore the risk is medium
	Degradation compounds	The density of microcrystalline cellulose is not related to the degradation compounds of the product. The risk is low

(continued)

Table 1.6 (continued)

Excipient attribute	Drug product CQA	Justification
Microcrystalline cellulose flow properties	Assay	The diluent flow properties are closely linked to the size and morphology of the particles that can impact the homogeneity and flowability of the mixture. In some cases, poor flowability and lack of homogeneity may cause an assay failure. The risk is medium
	Content uniformity	Microcrystalline cellulose is the diluent and the excipient in the highest concentration in this formulation. Poor flow property can affect the homogeneity of the mixture and lead to a high variation in tablet weight in the compression step. These two factors can directly impact the content uniformity of active substance in tablets, so the risk is high
	Dissolution	The flow property of the diluent alone is not capable of affecting the dissolution of the product. However, since the flow property of microcrystalline cellulose can impact the assay and content uniformity of the tablets, if there are a failure with the assay and lack of uniformity in the tablets, the dissolution may be affected; due to this, the risk is medium
	Degradation compounds	The flow property of microcrystalline cellulose is not able to affect the degrading compounds of the product. The risk is low
Microcrystalline cellulose Hygroscopicity	Assay	Different grades of microcrystalline cellulose can contain varying amounts of water. In case of a moisture-sensitive active substance, the hygroscopicity of the cellulose can impact the formulation leading to difficulties in the manufacturing process and/or degradation of the active substance. No drug compatibility problem with cellulose was found, and there was no drop in assay during the preliminary stability studies, so the risk is low
	Content uniformity	Microcrystalline cellulose hygroscopicity does not affect neither the chemical stability of the API nor the flow properties of the powder. The risk is low
	Degradation compounds	Since there is no incompatibility between the active substance and microcrystalline cellulose, the hygroscopicity does not affect the degradation of the product. The risk is low
	Dissolution	The hygroscopicity of microcrystalline cellulose does not affect the dissolution of the product. The risk is low
Croscarmellose sodium water content	Assay Content uniformity	Croscarmellose sodium is used in the formulation as a disintegrant and in low concentration. The water content present in croscarmellose is not capable of affecting the assay and content uniformity of the product. The risk is low
	Dissolution	Croscarmellose sodium promotes its disintegrating action through its water absorption and swelling capacity. The high water content in the raw material reduces its water absorption capacity, which can impact the disintegration of the tablet and, consequently, the dissolution of the drug. The risk is medium
	Degradation compounds	Croscarmellose sodium is compatible with the active substance, and the water content present in the raw material is not capable of impacting the degradation of the drug. So the risk is low

SiO_2 glidant/flow properties	Assay	Silicon dioxide is used in very low amounts in the formula. The glidant property improves the flowability of the powder mix, however, it is unlikely to impact product assay. The risk is low
	Content uniformity	The glidant properties of silicon dioxide and talc provide desirable flowability for the dry powders. Poor flowability could cause CU issues. The risk is medium
	Dissolution	The glidant properties of silicon dioxide/talc are not capable of affecting the dissolution of the product. The risk is low
	Degradation compounds	The glidant properties of silicon dioxide/talc do not affect product degradation since no incompatibilities were found in preformulation studies. The risk is low
Magnesium stearate surface area	Assay	Magnesium stearate is used in low amounts in the formula, and its surface area is not capable of affecting assay and content uniformity. The risk is low
	Content uniformity	
	Dissolution	Magnesium stearate is a very fine powder with hydrophobic characteristics. The surface area of magnesium stearate may result in a greater or lesser contact of this excipient with the active substance, affecting its wettability and, therefore, its dissolution. Therefore the risk is high
	Degradation compounds	Since there was no incompatibility between magnesium stearate and the active substance in the drug/excipient compatibility study, the surface area of this raw material does not impact the degradation of the product. The risk is low

Table 1.7 Initial risk assessment of the global manufacturing process

Drug product CQAs	Process steps		
	Calibration and blending	Blending – lubrication	Compression
Assay	High	Low	Medium
Content uniformity	High	Low	High
Dissolution	Medium	Medium	High
Degradation compounds	Low	Low	Medium

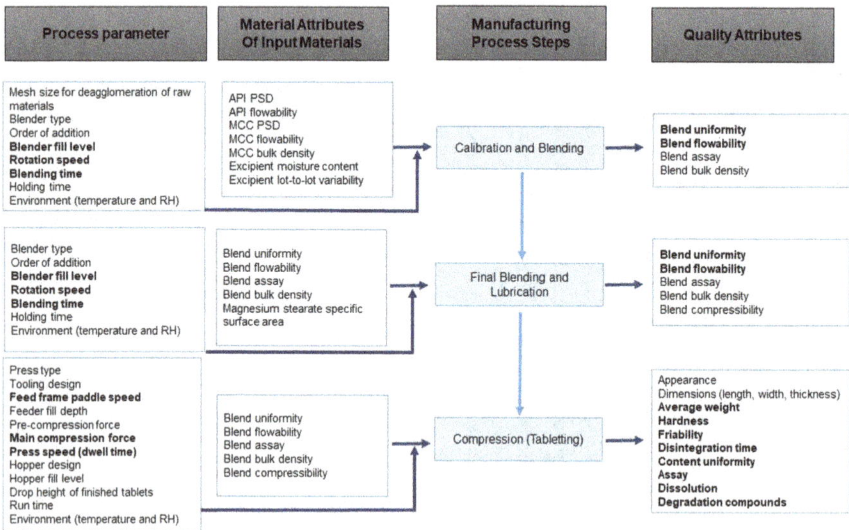

Fig. 1.3 Process map for the simulated example

established and broadly used nowadays in many different areas of knowledge. Statistical thinking and principles of DOE, including factorial designs and regression analysis, were introduced by Ronald A. Fisher in the 1920–1930s, followed by the work of George E. P. Box and K. B. Wilson in the 1950s, especially in the field of response surface methodology (RSM). DOE found its place in industrial applications by Taguchi's work in the 1970s. The readers can find detailed explanation of DOE fundamentals in many reference books and papers [14–17] as well as other chapters of this book.

The recommendation of multivariate DOE-based methods within the QbD framework is due to the several advantages over the traditional one-factor-at-time (OFAT) or one-variable-at-a-time (OVAT) approach, such as:

- DOE are based on combination of levels of factors. Therefore, this setup allows estimating interactions among the variables. Interaction is a major concept often dismissed by the researchers and not possible to be detected by the OFAT procedure. Variable A has an interaction with the variable B if the influence of

Table 1.8 Description of possible CPPs, ICQAs, and CQAs for each manufacturing step and the corresponding justification

Process steps	Possible CPPs	Possible ICQAs	Drug product CQA	Justification
Calibration and blending	Blender fill level; rotation speed; blending time	Blend uniformity, blend flowability	Assay	Insufficient blending time, inadequate rotation speed, and blender fill level can cause lack of homogeneity of the mixture, affecting the average weight of the tablets, therefore causing uniformity content problems – in extreme cases, it can affect assay. Moreover, material attributes can cause flowability problems, therefore affecting content uniformity and ultimately assay. The risk is high
			Content uniformity	
			Dissolution	As indicated above, lack of homogeneity could impact assay and content uniformity, which could impact dissolution, so the risk is medium
			Degradation compounds	Process and material attributes are unlikely to generate degradation products in this unit operation. The degradation routes were previously studied in the preformulation step. The risk is low
Blending – lubrication	Blender fill level; rotation speed; blending time	Blend uniformity, blend flowability	Assay	The uniformity of the mixture is controlled in the previous step; the small percentage of lubricant is not likely to affect the assay and content uniformity. The risk is low
			Content uniformity	
			Dissolution	Excessive blending time of the magnesium stearate lubricant can affect dissolution of the tablets. The risk is medium
			Degradation compounds	Process and material attributes are unlikely to generate degradation products in this unit operation. The degradation routes were previously studied in the preformulation step. The risk is low

(continued)

Table 1.8 (continued)

Process steps	Possible CPPs	Possible ICQAs	Drug product CQA	Justification
Compression	Feed frame paddle speed; main compression force; press speed (dwell time)	Average weight, hardness, friability, disintegration time	Assay	Blend flow issues, blend uniformity, and compression process variables might affect average weight and therefore assay. Since assay is measured by a bulk of several dosage units, it is less affected than content uniformity. The risk is medium
			Content uniformity	Blend flow issues, blend uniformity, and compression process variables can affect average tablet weight and, in turn, impact content uniformity. The risk is high
			Dissolution	Compression process variables can impact intermediate CQA hardness, friability, and disintegration time and therefore impact dissolution. The risk is high
			Degradation compounds	In the compression process, the heating of the cores can cause the tablets to adhere to the punches. Due to the heating, degradation compounds could be generated. The risk is medium

A on the response (CQA) depends on the level of B. The pharmaceutical development is complex, and interactions are normally expected in many steps. One should anticipate interactions between material attributes and formulation variables, interactions between material attributes and process variables, and interaction between formulation variables and process parameters.

- Allows the construction of models relating the input variables to the CQAs (intermediate or final). These models can be used to establish the so needed cause-and effect relationships required to build quality into the product during its development.
- The models generated can be used to describe the behavior of the CQAs all over the studied region and not only at the points where experiments were actually made, as in the OFAT approach. This allows building response surfaces to have a "whole picture" of the CQA behavior as a function of factor levels and to establish safe regions for future work ("sweet spots," design spaces), as well as the edges indicating failure, as will be discussed in the following topic.

"Design of experiments" is a very broad term that encompasses different methods that can be chosen according to the number of variables and the goal of the study. Figure 1.4 summarizes the most important classes of DOE methods and their characteristics. Chapter 2 of this book will cover methods for screening whereas Chap. 3 methods for optimization. A suggestion for the rational use of DOE within the QbD framework will be described in Chap. 5. Often the studies are split into two major parts: formulation development and process development. In both cases, high-risk material attributes should be included if they represent high-risk factors that could interact with the formulation/process factors.

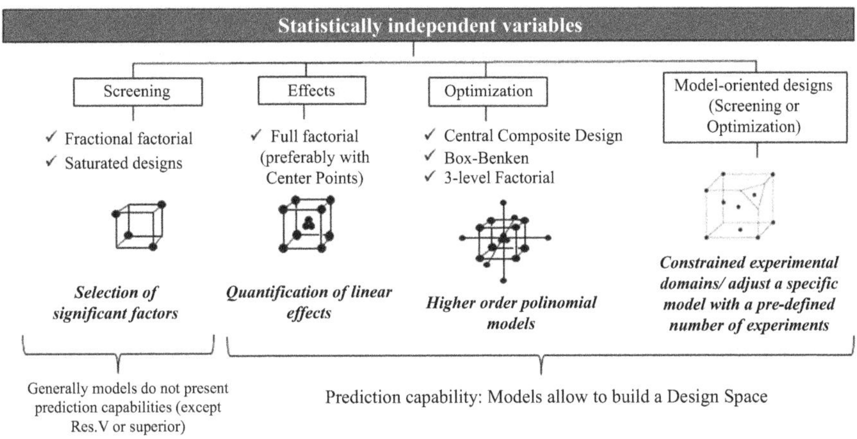

Fig. 1.4 Illustrative examples of DOE methods according to the goal of the study

1.2.5 Design Space(s)

Depending on the DOE method used, the generated models will take into account linear and higher other effects of the factors, as well as interactions among them. These models should be validated, using, for example, ANOVA (analysis of variance), analysis of residuals, and prediction capability to establish a region where the CQA specifications are met. ICH Q8 defines the design space as "The multidimensional combination and interaction of input variables (e.g., material attributes) and process parameters that have been demonstrated to provide assurance of quality." One should note that there will be many "design spaces" over the pharmaceutical development, considering the different variables involved. The level of knowledge gained by the DS establishment allows the regulatory agencies to guarantee flexibility, as also described in the same guideline: "Working within the Design Space is not considered as a change. Movement out of the design space is considered to be a change and would normally initiate a regulatory post approval change process." This regulatory flexibility is very welcome by companies and also by regulators, and it is only possible due to the level of knowledge gained by the DS establishment.

Design spaces can be built by either the overlay or desirability procedures, both "multicriteria decision-making" tools. ICH Q8 states that "Design space is proposed by the applicant and is subject to regulatory assessment and approval." The original design space determined by the abovementioned tools is normally an irregular region, and according to the ICH guidance, a smaller but regular region – therefore easier to be described – can be proposed to the regulatory agency. For manufacturing processes, an even smaller region, often called normal operating conditions (NOR), can be set within the proposed design space around the operating conditions. NOR limits are based on the normal experimental variability of the manufacturing process. This is indicated in Fig. 1.5.

The design space(s) are proposed at the lab/pilot scale in Phase 1; however, they will be used in the subsequent steps of the product lifecycle (scale-up and manufacturing) in commercial scale. Sometimes, the scale-up failure is due to a lack of understanding of the surrounding conditions around the set point. According to FDA and EMA, it is not necessary to verify the design space at the commercial scale during submission as this can be done during the product lifecycle instead [18].

1.3 Control Strategy

Control strategy is defined in the ICH Q8 (R2) document as follows: "A planned set of controls derived from current product and process understanding that ensures process performance and product quality. The controls can include parameters and attributes related to drug substance and drug product materials and components, facility and equipment operating conditions, in-process controls, finished product

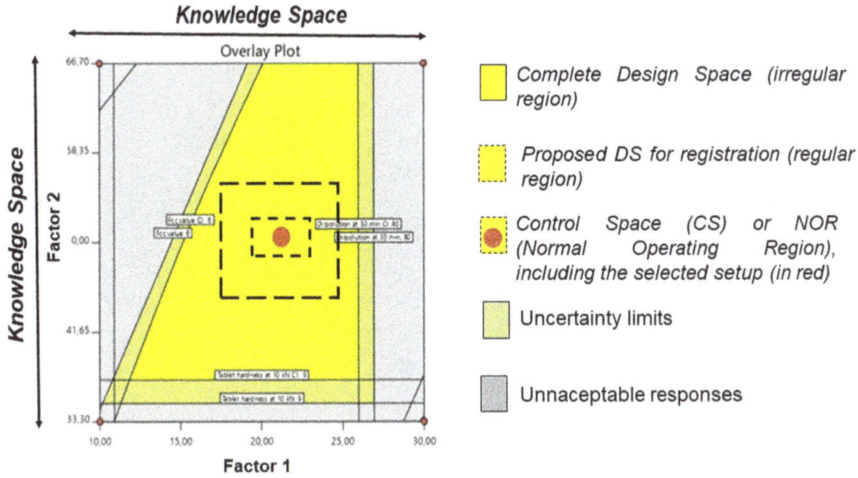

Fig. 1.5 Illustration of knowledge space (KS), design space (irregular), proposed design space (regular), control space (CS or normal operating conditions, NOR), and the set point

specifications, and the associated methods and frequency of monitoring and control. (ICH Q10)." The control strategy is derived from DOE and, when applied, from the design space(s). It represents in-process controls for each step of the manufacturing process, finished product testing, and any other specifications that control final product quality and consistency.

1.4 Regulatory Aspects

The enhanced approach of QbD for product development generates a lot of data and information, and therefore, it is natural to wonder what would be the level of detail the company should describe in the development report. FDA provided examples of development reports for Abbreviated New Drug Applications (ANDAs) [9, 10]. Several additional documents, such as "Points to Consider" [19], "Questions and Answers (Q & A)" [18, 20, 21], and "Lessons Learnt from training sessions" [22, 23], were provided by FDA, EMA (European Medicines Agency), and ICH. These documents contain important clarification on how to proceed in many practical aspects.

The document "Questions and answers on level of detail in the regulatory submissions" [20] brings information on risk analysis and DOE outcomes that are expected to be present in the development report. Regarding the latter, it indicates that the level of detail should be commensurate with the significance of the outcome of the DOE to the selection of the product design, commercial manufacturing process, and control strategy. For example, a DOE to define operating ranges for an important unit operation would normally be considered of high significance. The information provided in such cases should include detailed statistics on the models. On the other hand, this level of detail is not expected for screening design in an initial step of development, for example.

The difference between normal operating region (NOR), proven acceptable ranges (PAR), and design space is described in the document "Improving the understanding of NORs, PARs, DS and normal variability of process parameters" [21]. Clarification on scale-up and design space verification at commercial scale can be found in "Questions and Answers on Design Space Verification" [18]. In this document, both agencies agree that the design space verification at commercial scale is not necessarily complete at the time of submission of the application but should occur over the lifecycle of the product and process. Initial design space verification at commercial scale often occurs solely at or near the target operating ranges [18].

References

1. Department of Health and Human Services, Food and Drug Administration (FDA) (2004) Pharmaceutical cGMPs for the 21st century – a risk-based approach. Available in https://www.fda.gov/media/77391/download. Accessed 5 Aug 2022
2. The International Council for Harmonisation of Technical Requirements for Pharmaceuticals for Human Use (ICH) (2009) ICH harmonised tripartite guideline: Q8(R2) pharmaceutical development. Available in https://www.ich.org/page/quality-guidelines. Accessed 5 Aug 2022
3. The International Council for Harmonisation of Technical Requirements for Pharmaceuticals for Human Use (ICH) (2009) ICH harmonised tripartite guideline: Q9 quality risk management. Available via https://www.ich.org/page/quality-guidelines. Accessed 5 Aug 2022
4. The International Council for Harmonisation of Technical Requirements for Pharmaceuticals for Human Use (ICH) (2009) ICH harmonised tripartite guideline: Q10 pharmaceutical quality system. Available via https://www.ich.org/page/quality-guidelines. Accessed 5 Aug 2022
5. Juran JM (1986) The quality trilogy: a universal approach to managing for quality. Paper presented at the ASQC 40th annual quality congress in Anaheim, California, 20 May 1986
6. Juran JM (1992) Juran on quality by design: the new steps for planning quality into goods and services. Free Press, New York
7. Juran JM, De Feo JA (2010) Juran's quality handbook: the complete guide to performance excellence. McGraw Hill, New York
8. Parenteral Drug Association (PDA) (2013) Technical report 60 – process validation: A lifecycle approach
9. Food and Drug Administration (FDA) (2012) Quality by design for ANDAs: an example for immediate-release dosage forms. Available in https://wwwpharmaexcipientscom/quality-by-design-for-andas-an-example-for-immediate-release-dosage-forms/. Accessed 5 Jan 2023
10. Food and Drug Administration (FDA) (2012) Quality by design for ANDAs: an example for modified release dosage forms
11. U.S. Department of Health and Human Services Food and Drug Administration Center for Drug Evaluation and Research (CDER) (2007) Guidance for Industry And Review Staff Target Product Profile – a strategic development process tool
12. Van Buskirk GA, Asotra S, Balducci C et al (2014) Best practices for the development, scale-up, and post-approval change control of IR and MR dosage forms in the current quality-by-design paradigm. AAPS PharmSciTech 15:665–693
13. Mesut B, Özsoy Y, Aksu B (2015) The place of drug product critical quality parameters in quality by design (QBD). Turk J Pharm Sci 12:75–92

14. Montgomery DC (2001) Design and analysis of experiments. Wiley, Hoboken
15. Box GEP, Hunter WG, Hunter JS (2005) Statistics for experimenters: design, innovation and discovery. Wiley, Hoboken
16. Neto BB, Scarminio IS, Bruns RE (2005) Statistical design – chemometrics, Data handling in science and technology. Elsevier, Amsterdam
17. Dejaegher B, Vander Heyden Y (2011) Experimental designs and their recent advances in set-up, data interpretation, and analytical applications. J Pharm Biomed Anal 56:141–158
18. European Medicines Agency (EMA) – Food and Drug Administration (FDA) (2013) Question and answer on design space verification. https://www.ema.europa.eu/en/documents/other/questions-answers-design-space-verification_en.pdf. Accessed 5 Jan 2023
19. The International Council for Harmonisation of Technical Requirements for Pharmaceuticals for Human Use (ICH) (2012) ICH quality-IWG – points to consider for ICH Q8/Q9/Q10 guidelines. Available in https://www.ema.europa.eu/en/documents/scientific-guideline/international-conference-harmonisation-technical-requirements-registration-pharmaceuticals-human-use/q9/q10-guidelines_en.pdf. Accessed 5 Jan 2023
20. European Medicines Agency (EMA) – Food and Drug Administration (FDA) (2014) Questions and answers on level of detail in the regulatory submissions. Available in: https://www.ema.europa.eu/en/documents/other/questions-answers-level-detail-regulatory-submissions_en.pdf. Accessed 5 Jan 2023
21. European Medicines Agency (EMA) (2017) Questions and answers: improving the understanding of NORs, PARs, DSp and normal variability of process parameters. Available in: https://www.ema.europa.eu/en/documents/scientific-guideline/questions-answers-improving-understanding-normal-operating-range-nor-proven-acceptable-range-par_en.pdf. Accessed 5 Jan 2023
22. European Medicines Agency (EMA) – Food and Drug Administration (FDA) (2013) Human medicines development and evaluation. EMA-FDA pilot program for parallel assessment of quality-by-design applications: lessons learnt and Q & A resulting from the first parallel assessment. Available in: https://www.ema.europa.eu/en/documents/other/european-medicines-agency-food-drug-administration-pilot-programme-parallel-assessment-quality_en.pdf. Accessed 5 Jan 2023
23. Department of Health and Human Services Food and Drug Administration Center for Drug Evaluation and Research (CDER), Food and Drug Administration (FDA) (2012) Guidance for industry: Q8, Q9, & Q10 questions and answers – appendix Q & As from training sessions. https://www.fda.gov/media/83904/download. Accessed 5 Jan 2023

Chapter 2
Fundamentals of Design of Experiments and Optimization: Designs for Factor Screening and Data Analysis in Pre-Response Surface Methodology

Fabricio A. Chiappini, Carla M. Teglia, Silvana M. Azcarate, and Hector C. Goicoechea

Abstract Chapter 2 presents an overview of the most important designs implemented for factor screening, including data analysis in pre-response surface methodology (pre-RSM). This chapter mainly focuses on practical aspects considering a group of experimental factors at different levels at the same time, with a minimum number of experiments. Then it is discussed how to carry out studies in order to build an empirical model that correlates the information about factors and responses. Data analysis in pre-RSM is mostly represented by the well-known statistical method of multiway analysis of variance (ANOVA), whose basics are presented, together with the analysis of the validity of the results. In addition, two graphical strategies, the Pareto chart and the diagram of normal probability, are discussed. An example is used to illustrate the concepts. Suitable references are provided.

Keywords Factor screening · Pre-response surface methodology · Design of experiments · Multivariate optimization

---

F. A. Chiappini (✉) · C. M. Teglia (✉) · H. C. Goicoechea
Laboratorio de Desarrollo Analítico y Quimiometría (LADAQ), Cátedra de Química Analítica I, Facultad de Bioquímica y Ciencias Biológicas, Universidad Nacional del Litoral, Ciudad Universitaria, Santa Fe, Santa Fe, Argentina

Consejo Nacional de Investigaciones Científicas y Técnicas (CONICET), Ciudad Autónoma de Buenos Aires, Buenos Aires, Argentina
e-mail: fchiappini@fbcb.unl.edu.ar

S. M. Azcarate
Instituto de Ciencias de la Tierra y Ambientales de La Pampa (INCITAP), and Facultad de Ciencias Exactas y Naturales, Universidad Nacional de La Pampa, Santa Rosa, La Pampa, Argentina

Consejo Nacional de Investigaciones Científicas y Técnicas (CONICET), Ciudad Autónoma de Buenos Aires, Buenos Aires, Argentina

M. C. Breitkreitz, H. C. Goicoechea (eds.), *Introduction to Quality by Design in Pharmaceutical Manufacturing and Analytical Development*, AAPS Introductions in the Pharmaceutical Sciences 10, https://doi.org/10.1007/978-3-031-31505-3_2

2.1 Introduction

Design of experiments (DOE) and optimization together constitute a series of mathematical, statistical, and computational techniques that aim to establish optimum experimental conditions that enable obtaining the best possible outcomes, minimizing the number of experimental runs. Generally, the outcome is represented by one or more experimental responses of interest that depend on a group of experimental factors. Therefore, the philosophy of DOE and optimization consists in generating experimental data according to a proper statistical design and then building an empirical model that correlates the information about factors and responses [1].

The area of DOE and optimization has also received the name response surface methodology (RSM), and this will be the preferred nomenclature throughout the following three chapters. RSM constitutes a robust and straightforward line of action that can be, in principle, applied for the optimization of any multivariate experimental system or process. For this reason, it has gained great relevance in the most diverse areas, among both researchers and industry [2–5].

The multivariate approach of RSM enables the researcher to simultaneously study a group of experimental factors at different levels at the same time, with a minimum number of experiments. This also results in the possibility of uncovering factor interactions. Conversely, when one factor at a time is varied, keeping the others constant, a large number of experimental runs are generally required, generating only local information in a narrower experimental domain and without taking into account any factor interactions. This univariate approach is often referred to as one-variable-at-a-time (OVAT) methodology and, surprisingly, is still frequently used by the scientific community. Evidently, the RSM approach should be the strategy of choice because it generates more profitable experimental information with minimum effort, time, and resources.

2.2 General Overview of RSM and Basic Terminology

The main objective of RSM is to optimize a given experimental system or process, with minimum effort. In this sense, the experimenter generally identifies one or more experimental responses of interest to be optimized, i.e., a dependent variable y, that depends on k experimental factors, i.e., independent variables x_1, x_2, …, x_k. Evidently, the decision about responses and factors is up to the analyst and will closely depend on the purpose of the study and the amount of previous knowledge about the system. Moreover, as usual in any experiment, the independent variables can be controllable or uncontrollable. Therefore, RSM aims to generate proper experimental data to build an empirical model, according to the general expression [2]:

$$y = f(x_1, x_2, \ldots, x_k) + \varepsilon \tag{2.1}$$

where x_1, x_2, ..., x_k represent the controllable factors that are included in the experimental design by the researcher and ε represents response variations owing to the uncontrollable variables.

From the mathematical point of view, Eq. (2.1) can be conceived as a stochastic model, where the second term collects the information due to experimental random error. Subsequently, the generated model is used to evaluate the relationship between the factors and the response. At this step, different diagnostic tools can be implemented to decide which factors (or combination of factors) exert a significant effect on the response. Non-significant factors are then ignored during optimization. Finally, different mathematical strategies are applied to find the optimal combination of factors that generate the best possible yield for the response, according to predefined criteria (generally, maximization, minimization, or keeping the response at a target value). When several responses are to be optimized, this problem is usually referred to as multiple response optimization, where the aim is to find a compromise solution for all the responses. Specific mathematical strategies have been developed for this purpose. In addition, individual factors can also be included during the optimization procedure.

As can be noticed, the implementation of an RSM-based approach for the optimization of a given experimental system or process can be conceived as a systematic sequence of steps, which can be summarized as follows:

(i) Step 1: definition of the experimental system, which involves defining factors and their experimental domain, and the responses of interest.
(ii) Step 2: generation of experimental data according to a statistical design.
(iii) Step 3: generation of an empirical model.
(iv) Step 4: optimization of an individual or multiple responses.

When the number of selected factors is very large at the beginning of an RSM study, the number of experimental runs is greatly increased. Fortunately, on many occasions, not all factors exert a significant impact on the response of interest, i.e., the response depends on a fewer number of factors than originally thought. However, this information might not be a priori known to the experimenter. In these scenarios, RSM is preceded by a screening step, usually referred to as factorial screening or pre-RSM.

The main objective of factor screening is to identify which of the considered experimental variables at the beginning are influential on the response. From the statistical point of view, a factor is considered to exert a significant effect on the response when the latter undergoes a statistically significant change with a variation in the level of the factor. Generally, the statistical designs that are used in pre-RSM generate information for each combination of factors only at two levels, i.e., low and high. As the number of experiments increases exponentially with the number of factors, this property of being simple permits the analyst to handle an important number of factors, with a minimum number of runs. Moreover, if properly selected, screening designs enable scrutinization of possible factor interactions and/or factor curvatures, i.e., a non-linear dependency between y and x_k. The experimental data that are generated from screening designs constitute the input for the data analysis step. Data analysis in pre-RSM is mostly represented by the well-known statistical method of multiway analysis of variance (ANOVA), whose basics will be presented in the following sections.

Screening designs are meant to generate proper data to perform factor screening by implementing data analysis tools. In most cases, after factor screening, the number of factors is reduced. Therefore, with fewer factors, it is possible to proceed to the RSM step. In RSM, more sophisticated statistical designs are implemented to deeply investigate the relation between responses and factors. These designs generate enough experimental information to build more complex models that include factor interactions and curvatures.

Before selecting a specific screening design, the first step consists in defining the factors to be studied and their experimental domain. At this point, the previous information about the system under study plays a critical role. The factors considered in an RSM problem can be qualitative, i.e., non-measurable variables that represent a feature or a property of the experiment, or quantitative, i.e., factors that can be described by a numeric quantity.

Screening designs are generally used before response surface designs to explore the effect of both quantitative and qualitative variables, investigated at two levels. The levels define the extremes of the experimental domain and can be coded as -1 (lower level) and $+1$ (upper level). Coding is a useful strategy to standardize all factors to an adimensional scale avoiding effect values that depend on different dimensional units that could hamper the interpretation of factorial results.

On the other hand, RSM designs include three or more levels, often, intermediate points between -1 and $+1$. The different combinations of factors established by the design that characterize each of the experimental runs are arranged in an n (number of runs or experiments) $\times k$ (number of factors) matrix, known as the design matrix.

Factors are commonly represented in many software by capital letters, e.g., A, B, C, etc. For instance, following this nomenclature and considering an experiment with two factors A and B, A refers to the main effect of factor A, B refers to the main effect of factor B, AB refers to the interaction between both factors, and A^2 and B^2 represent the curvatures of A and B, respectively.

Another important aspect is related to the order of the experiments. Once the design matrix is obtained, it is desirable to perform the experiments in random order. Randomization ensures the independence of the results, avoiding the impact of systematic errors or drifts on effect values and minimizing the effects of uncontrolled factors, which could lead to wrong conclusions. In general, designs that generate experiments to be randomly performed are known as completely randomized designs. However, in certain scenarios, it is impossible to perform the complete set of experiments from a random design as suggested due to practical concerns. In those cases, non-randomized designs can be implemented.

The present chapter is dedicated to screening designs and data analysis tools for the pre-RSM step. In the following section, the most popular completely randomized designs at two levels for factor screening are introduced. Later, basic data analysis tools for pre-RSM are presented in Sect. 2.4. Finally, in the last part of this chapter, non-randomized designs are briefly described.

2.3 Completely Randomized Designs for Factor Screening

2.3.1 *Two-Level Full and Fractional Factorial Designs*

One of the simplest experimental designs for screening is the well-known full factorial. Particularly, a two-level full factorial design contains all possible combinations between the k factors and their two levels, leading to a total number of runs or experiments equal to 2^k. This design allows estimating which factors have a significant effect on the response (factor main effect) and if the variation of a factor influences the effect of other factors (factor interactions). The design matrix, made by 2^k rows and k columns, is easily built by alternating the -1 and $+1$ levels of each factor in an appropriate fashion until all possible combinations are generated.

The points investigated by a full factorial can be geometrically represented in a k-dimensional space defined by the factors and occupy the vertices of a square, for $k = 2$; a cube, for $k = 3$; or a hypercube, for $k > 3$. Generic examples for 2^2 and 2^3 are depicted in Fig. 2.1a and b, respectively.

Although full factorial designs are simple and useful, they can become impractical since the number of experiments grows exponentially with the number of factors. In these circumstances, an alternative design that derives from the full factorial consists in implementing a fractional factorial. A two-level fractional factorial design contains a fraction of the full factorial design and allows examining k factors at two levels in 2^{k-p} experiments, where 0.5^p represents the fraction of the full factorial and p is a small natural number. For example, if $p = 1$, the number of runs results in $0.5\ 2^k = 2^{k-1}$. These types of fractional designs are also known as half-fraction factorial designs. Figure 2.1c shows the geometric representation of experimental points from a half-fraction factorial design with three factors.

Evidently, because a fractional factorial design is obtained by partitioning a full factorial, some information is lost, i.e., not all main and interaction effects can be properly estimated. Therefore, the value of p should be carefully selected to obtain a compromise design that enables the analyst to screen a large number of factors with a minimum number of runs while avoiding generating poor-quality information. In

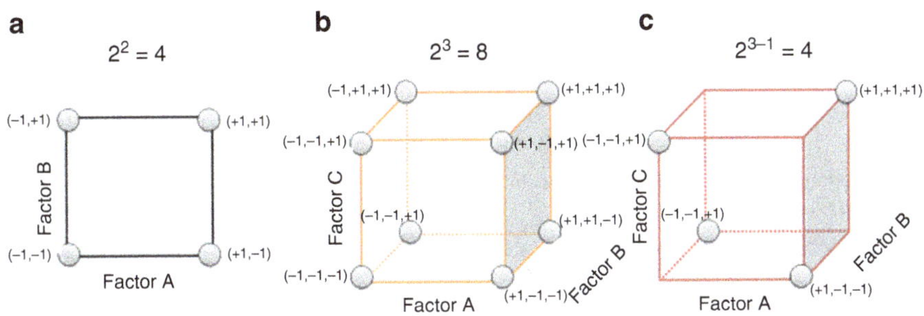

Fig. 2.1 Geometric representation of factorial designs: (**a**) 2^2 full factorial; (**b**) 2^3 full factorial; and (**c**) 2^{3-1} fractional factorial

this sense, it is common in fractional designs that some effects are confounded, which means that they are estimated together. This situation is commonly referred to as a factor alias and should be thoroughly handled to avoid wrong conclusions. For example, if the effect of a factor is aliased with the effect of a two-factor interaction, the experimenter cannot decide whether the considered factor is truly significant or not.

Fractional designs are based on two principles: (i) the Pareto principle which establishes that, although a large number of factors can be included at the beginning of the study, only a few of them are significant for the variability of the response and (ii) hierarchy of effects, which states that there exists a hierarchy in the significance of the effects, i.e., when significant, the effect of an individual factor is more important than first-order (or binary) interactions, followed by ternary interactions and so on.

Another term that characterizes a fractional factorial design is "resolution." The resolution indicates the level of confoundment (aliases) that can be present in the estimation of the main and interaction effects. The resolution is commonly expressed in Roman numbers, e.g., I, II, III, etc. The smaller the number, the greater the loss of information and, therefore, the lower the resolution. The resolution also determines the number of experiments, i.e., low resolution leads to a smaller number of runs. For example, a 2^{k-1} design represents a resolution V, in which principal effects and binary interactions can be confounded with interactions of third or higher order. On the other hand, a 2^{k-2} design has a resolution IV, where principal effects are not confounded between each other nor with binary interactions, but these can be confounded with each other. A 2^{k-3} design represents a resolution III, where the principal effects are not confounded among themselves, but principal effects can be confounded with binary interactions. Resolutions III and IV in the fractional factorial designs imply highly fractionated designs and have been widely used for early-stage screening experimentation.

2.3.2 *Plackett-Burman Designs*

Another alternative design that can be implemented for a large number of factors is the so-called Plackett-Burman design (PBD). PBD constitutes an alternative to the full factorial design and even to a fractional design with low resolution. This design has resolution III and exhibits a high degree of aliases. However, it is a useful strategy when the analyst is only interested in the main effect of the factors. Consequently, this design should not be used to evaluate the effects of factor interactions.

As for the case of factorial designs, PBD also has two levels for each factor. In this design, the number of experiments n is equal to the first multiple of 4 greater than k. Then, n experiments can be used for a different number of factors, up to $n - 1$. For instance, up to 7 factors can be studied with 8 experiments, up to 11 factors with 12 experiments, and so on. The construction of the experimental matrix for a PBD is based on the cyclic permutation of the first row. The next row will be formed by the cyclical shifting of the previous row to the right. The process will continue until several rows equal to the number of experiments to be performed minus 1 are generated. Finally, the last step is to add a row of low level (-1) for all factors to the design. In this way, each factor is equally investigated at the low and high levels.

Other designs for implementing the screening step can be found in the literature. Among these designs, Taguchi's experimental design (TED) and the definitive screening design (DSD) can be mentioned. The description of these designs is beyond the scope of this chapter, but the reader is recommended to review Refs. [6] and [7].

2.4 Data Analysis in Pre-RSM

Once the data have been generated, the next step is to proceed to their analysis. As aforementioned, the main objective is to discriminate between significant and non-significant factors to simplify the number of variables for the RSM step.

Mathematically, the effect of the i-th factor E_i on the response y for n experimental runs can be estimated as follows:

$$E_i = \frac{\sum_{j=1}^{n/2} y_j^+ - \sum_{j=1}^{n/2} y_j^-}{n} \tag{2.2}$$

where the superscripts + and – indicate, respectively, the measured response at the high and low levels of the factor. It is important to remark that, during screening, the effects of all the factors are computed assuming a linear relation with the response. This assumption simplifies the required number of experimental data. Occasionally, a central point can be added to the experimental design to study curvature, which, if significant, is an indicator that there is a second-order relationship between the factors and the response.

Despite being simple, Eq. (2.2) is very useful to obtain preliminary information about the influence of the factors. To illustrate this idea, consider the following example. Suppose that a given response was measured from a 2^3 full factorial design. As previously shown, with only eight runs, the effect of main effects and interactions can be estimated without aliases. The design matrix augmented by columns showing factor interactions (binary and ternary) and the corresponding measured response for each run is shown in Table 2.1. In the last row, the effects of all factors and interactions have been computed according to Eq. (2.2).

For example, the obtained value for factor A means that increasing this factor from the low to the high level causes an average increase in the response of 61.3 units. As can be observed, Eq. (2.2) also considers the sign of each effect. Therefore, factors A and C and the interactions between AC and BC have positive

Table 2.1 Design matrix from a generic example of a 2^3 full factorial design, augmented by columns showing factor interactions and the corresponding measured responses (in an arbitrary unit) for each run

Run	Factor A	Factor B	Factor C	AB	AC	BC	ABC	Response
1	−1	−1	−1	+1	+1	+1	−1	30
2	+1	−1	−1	−1	−1	+1	+1	20
3	−1	+1	−1	−1	+1	−1	+1	0
4	+1	+1	−1	+1	−1	−1	−1	0
5	−1	−1	+1	+1	−1	−1	+1	40
6	+1	−1	+1	−1	+1	−1	−1	300
7	−1	+1	+1	−1	−1	+1	−1	60
8	+1	+1	+1	+1	+1	+1	+1	300
Effect	61.3	−3.8	81.3	−1.3	63.8	8.9	−3.4	–

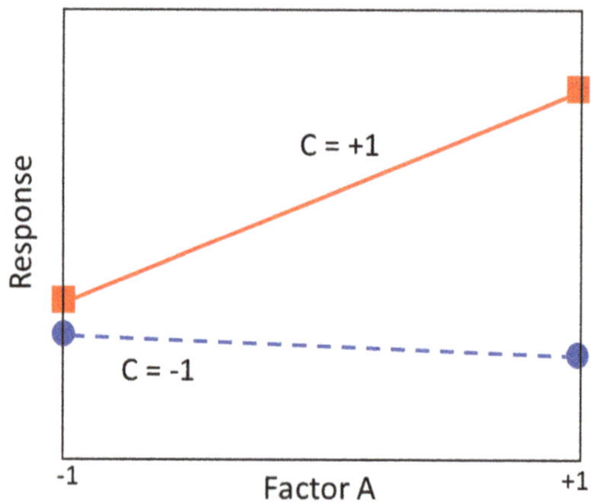

Fig. 2.2 Interaction graph showing the effect of factor C on the level of A (example of Table 2.1). The solid line represents the change in the response when A changes from −1 to +1 at level + 1 of factor C. The dashed line shows the change in the response when A changes from −1 to +1 at level − 1 of factor C

effects on the response. On the other hand, B and the interactions AB and ABC have negative effects. It can be noted that factor C exerts a higher effect on the response (in a positive way), while interaction ABC exerts a lower effect on the response, in a negative fashion.

As was mentioned before, an interaction occurs when the response is different depending on the settings of two variables. This effect can be appreciated by the number computed with Eq. (2.2) and also, in a graphic manner, the interaction plot. If the interaction is important, the plot will show two non-parallel lines, indicating that the effect of one factor depends on the level of the other. If not, the plot will show two parallel lines. As an example, Fig. 2.2 shows the interaction plot corresponding to the variation of the response when changing the level of factor A from −1 to +1 at two different levels of factor C (data corresponding to Table 2.1). In solid line, it is observed the change on the response when C is at level + 1. Interestingly, the change on the response is more marked than the one produced when C is at level − 1 (dashed line).

Once the effects of all factors (and interactions, if appropriate) have been computed, the next step is to decide whether these effects are statistically significant or not. The quintessential method to carry out this task is the well-known multiway ANOVA, whose basics are briefly introduced in the next sections.

Performing a multiway ANOVA implies fitting a special kind of statistical linear model, where the number of terms is closely related to the number of effects and interactions that are to be studied. However, as it has been seen before, the more variables the analyst wants to include, the greater the number of experimental runs is needed. For this reason, ANOVA models need to be carefully refined and validated to generate reliable conclusions. To facilitate this task, many software packages for RSM come bundled with a series of exploratory and graphical tools that assist in the choice of significant terms for the construction of the ANOVA model. The two most relevant strategies based on graphical analysis are presented in the next section. Afterward, key concepts regarding ANOVA modeling are introduced in Sect. 2.4.2.

2.4.1 Graphical Analysis

Two graphical strategies are mostly used, the Pareto chart and the diagram of normal probability, which generic examples are depicted in Fig. 2.3a and b, respectively.

In the first place, the Pareto chart consists in a graphic of horizontal or vertical bars. The bars show a standardized value of the computed effect for all the possible factors and interactions included in the design. The standardization procedure can

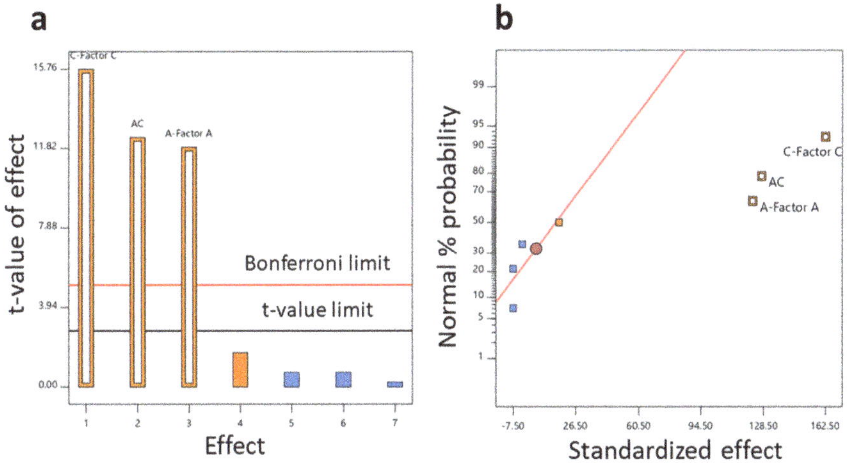

Fig. 2.3 Graphical tools for the evaluation of the significance of experimental factors during pre-RSM data analysis. (**a**) Pareto chart, horizontal lines represent statistical limits calculated according to different criteria; (**b**) diagram of normal probability, squares that are not close to the straight line are considered to have a significant effect on the response. In both sub-figures, orange and blue squares represent factors or interactions having a positive and a negative effect, respectively

vary from one software to another. For example, in the Pareto chart shown in Fig. 2.3a (corresponding to data presented in Table 2.1), each value obtained through Eq. (2.2) was divided by the corresponding standard deviation. Factor/interaction effects are standardized in order to minimize the possibility of under- or overestimation due to differences in the experimental error among the runs.

The Pareto chart is also supplied with a threshold value, represented as a straight line. According to the software, different statistical criteria can be implemented to calculate this boundary. For example, in the considered example, two bounds are calculated. The highest limit corresponds to the Bonferroni *t*-critical value, whereas the lower limit shows the critical value of hypothesis tests for individual factors, based on a Student's distribution. Those factors exhibiting values that are greater than these limits can be considered significant.

Secondly, another popular graphical tool for factor selection is the diagram of normal probability. This strategy is commonly used as a method to check the normal distribution of a given variable in a dataset. To build this diagram, standardized factor effects are plotted against the expected values that would be obtained if the data came from a normal distribution. Therefore, those factors that follow a normal distribution are displayed around the identity line. Conversely, those factors whose values are distant do not come from a normal distribution and are considered to have a significant effect on the response (Fig. 2.3b).

Finally, in the considered examples, both graphics take also into account the sign of each effect, i.e., positive and negative effect values are depicted in orange and blue in this particular case, respectively.

2.4.2 Multiway ANOVA

As previously mentioned, the graphical tools are useful to have a preliminary idea of which factors are considered in the experiment and/or their interactions are significant to the response. However, the selected variables need to be submitted to a rigorous test to decide which of them are statistically significant. In this way, multiway ANOVA is generally the selected method.

ANOVA is a collection of statistical models and associated methods that can be conceived as a hypothesis test for multiple comparisons. In particular, multiway ANOVA is an extension of the well-known one-way ANOVA test. The general philosophy of this methodology consists in partitioning the total variance from a set of observations in the contribution of each source of variation. In the context of pre-RSM, the idea is to decompose the variability due to a given factor effect from the variability due to the random error, to test the hypothesis that the factor significantly alters the mean value of the response variable [8].

If the response would depend on only one factor (univariate experiment) at k levels, then the j-th observation of the response y corresponding to the i-th level of the factor (also referred to as treatment in the statistic literature) can be expressed according to the following model equation:

$$y_{ij} = \mu + \alpha_i + \varepsilon_{ij} \tag{2.3}$$

where μ is the overall mean value of y for all the treatments of factor combinations, α_i is known as the effect of the i-th level of the factor, and $\varepsilon_{\mathrm{ij}}$ represents the random error, which is assumed to follow a normal distribution with mean zero and constant variance σ^2. Therefore, the associated hypothesis test of interest is:

$$\begin{aligned} H_0 &: \alpha_1 = \alpha_2 = \ldots = \alpha_k = 0, \\ H_1 &: \text{at least one } \alpha_i \neq 0 \end{aligned} \tag{2.4}$$

The ANOVA test is based on the following decomposition of the total variability (also known as sum of squares identity):

$$\sum_{i=1}^{k} \sum_{j=1}^{n} \left(y_{ij} - \overline{y}_{..}\right)^2 = n \sum_{i=1}^{k} \left(\overline{y}_{i.} - \overline{y}_{..}\right)^2 + \sum_{i=1}^{k} \sum_{j=1}^{n} \left(y_{ij} - \overline{y}_{i.}\right)^2 \tag{2.5}$$

where $\overline{y}_{i.}$ denotes the mean of all observations under the i-th level of the factor and $\overline{y}_{..}$ denotes the mean of all the observations. In Eq. (2.5), the left term is also known as the total sum of squares (SS_T). On the right side, the first term is referred to as the sum of squares due to the factor (SS_F), and the second term is the sum of squares due to the error (SS_E).

Now, consider a factorial design with only two factors, A and B, with a levels of factor A and b levels of factor B and n runs. The k-th observation of the response y corresponding to the i-th level of factor A and j-th level of factor B can be written as:

$$y_{ijk} = \mu + \tau_i + \beta_j + (\tau\beta)_{ij} + \varepsilon_{ijk} \tag{2.6}$$

where μ is the overall mean value of y, τ_i is the effect of factor A, β_j is the effect of factor B, $(\tau\beta)_{ij}$ is the effect of the interaction between both factors, and ε_{ijk} represents the random error, which is assumed to follow a normal distribution with mean zero and constant variance σ^2. Therefore, three associated hypothesis tests can be formulated as:

1.
$$\begin{aligned} H_0 &: \tau_1 = \tau_2 = \ldots = \tau_a = 0, \\ H_1 &: \text{at least one } \tau_i \neq 0 \end{aligned} \tag{2.7}$$

2.
$$H_0 : \beta_1 = \beta_2 = \ldots = \beta_b = 0,$$

$$H_1 : \text{at least one } \beta_j \neq 0 \tag{2.8}$$

3.
$$H_0 : (\tau\beta)_{ij} = 0 \text{ for all } i, j,$$

$$H_1 : \text{at least one } (\tau\beta)_{ij} \neq 0 \tag{2.9}$$

Analogous to the previous definitions, let $y_{i..}$ denote the sum of all observations under the i-th level of factor A, $y_{.j.}$ denote the sum of all observations under the j-th level of factor B, $y_{ij.}$ denote the sum of all observations in the ij-th levels of factors A and B, and $y_{...}$ denote the grand total of all the observations. These quantities can be defined by the following equations:

$$y_{i..} = \sum_{j=1}^{b}\sum_{k=1}^{n} y_{ijk} \qquad \bar{y}_{i..} = \frac{y_{i..}}{bn} \qquad i = 1, 2, \ldots, a \tag{2.10}$$

$$y_{.j.} = \sum_{i=1}^{a}\sum_{k=1}^{n} y_{ijk} \qquad \bar{y}_{.j.} = \frac{y_{.j.}}{an} \qquad j = 1, 2, \ldots, b \tag{2.11}$$

$$y_{ij.} = \sum_{k=1}^{n} y_{ijk} \qquad \bar{y}_{ij.} = \frac{y_{ij.}}{n} \qquad i = 1, 2, \ldots, a; \quad j = 1, 2, \ldots, b \tag{2.12}$$

$$y_{...} = \sum_{i=1}^{a}\sum_{j=1}^{b}\sum_{k=1}^{n} y_{ijk} \quad \bar{y}_{...} = \frac{y_{...}}{abn} \tag{2.13}$$

where $\bar{y}_{i..}$, $\bar{y}_{.j.}$, $\bar{y}_{ij.}$, and $\bar{y}_{...}$ represent the corresponding mean values. Hence, in the multivariate version of ANOVA, the SS_T can be expressed as:

$$\begin{aligned}
\sum_{i=1}^{a}\sum_{j=1}^{b}\sum_{k=1}^{n} (y_{ijk} - \bar{y}_{...})^2 &= \sum_{i=1}^{a}\sum_{j=1}^{b} \\
&\times \sum_{k=1}^{n} [(\bar{y}_{i..} - \bar{y}_{...}) + (\bar{y}_{.j.} - \bar{y}_{...}) + (\bar{y}_{ij.} - \bar{y}_{i..} - \bar{y}_{.j.} + \bar{y}_{...}) + (y_{ijk} - \bar{y}_{ij.})]^2 \\
&= bn\sum_{i=1}^{a} (\bar{y}_{i..} - \bar{y}_{...})^2 + an\sum_{j=1}^{b} (\bar{y}_{.j.} - \bar{y}_{...})^2 + n\sum_{i=1}^{a} \\
&\times \sum_{j=1}^{b} (\bar{y}_{ij.} - \bar{y}_{i..} - \bar{y}_{.j.} + \bar{y}_{...})^2 + \sum_{i=1}^{a}\sum_{j=1}^{b}\sum_{k=1}^{n} (y_{ijk} - \bar{y}_{ij.})^2
\end{aligned} \tag{2.14}$$

As it can be observed, SS_T is partitioned into a sum of squares due to factor A (SS_A), a sum of squares due to factor B (SS_B), a sum of squares due to the interaction between A and B (SS_{AB}), and a sum of squares due to the error (SS_E), i.e., $SS_T = SS_A + SS_B + SS_{AB} + SS_E$. If each sum of squares is divided by the corresponding degrees of freedom, four variance values or mean squares (MS) are obtained, i.e., $MS_A = S_A^2$, $MS_B = S_B^2$, $MS_{AB} = S_{AB}^2$, and $MS_E = S_E^2$, whose expected values obey the following expressions:

$$E\left(S_A^2\right) = E\left(\frac{SS_A}{a-1}\right) = \sigma^2 + \frac{bn\sum_{i=1}^{a}\tau_i^2}{a-1} \tag{2.15}$$

$$E\left(S_B^2\right) = E\left(\frac{SS_B}{b-1}\right) = \sigma^2 + \frac{an\sum_{j=1}^{b}\beta_i^2}{b-1} \tag{2.16}$$

$$E\left(S_{AB}^2\right) = E\left(\frac{SS_{AB}}{(a-1)(b-1)}\right) = \sigma^2 + \frac{n\sum_{i=1}^{a}\sum_{j=1}^{b}(\tau\beta)_{ij}^2}{(a-1)(b-1)} \tag{2.17}$$

$$E\left(S_E^2\right) = E\left(\frac{SS_E}{ab(n-1)}\right) = \sigma^2 \tag{2.18}$$

It should be noted that if the null hypotheses stated in Eqs. (2.7, 2.8, and 2.9) are true, then S_A^2, S_B^2, S_{AB}^2, and S_E^2 are all independent and unbiased estimators of σ^2. To test the null hypothesis for factor A, the statistic $F_{0,\,A}$ is computed as:

$$F_{0,A} = \frac{S_A^2}{S_E^2} \tag{2.19}$$

Under the assumption that the model is adequate, and the errors are normally and independently distributed with constant variance σ^2, then $F_{0,\,A}$ is an F random variable with $a - 1$ and $ab(n - 1)$ degrees of freedom, when H_0 is true. Therefore, H_0 from Eq. (2.8) is rejected if $F_{0,\,A} > F_{a-1,\,ab(n-1)}$ for a given level of significance α. In terms of associated probability (p-value approach), H_0 is rejected when p-value is less than α. Analogously, to test the null hypothesis for factor B and the interaction term AB, equivalent statistics are calculated:

$$F_{0,B} = \frac{S_B^2}{S_E^2} \tag{2.20}$$

for factor B, and

$$F_{0,AB} = \frac{S_{AB}^2}{S_E^2} \tag{2.21}$$

for the interaction term AB. Then, H_0 from Eq. (2.8) is rejected if $F_{0,\ B} > F_{b\ -\ 1,\ ab(n\ -\ 1)}$, and H_0 from Eq. (2.9) is rejected when $F_{0,\ AB} > F_{(a\ -\ 1)(b\ -\ 1),\ ab(n\ -\ 1)}$.

The validity of the results from ANOVA is subjected to the fulfillment of three key assumptions: the error terms belong to a normal distribution, the variance among the treatments is constant (hypothesis of homoscedasticity), and the observations are independent of each other. During the implementation of an ANOVA, a model of the form of Eq. 2.6 is fitted, and the corresponding hypothesis tests over the terms of interest are carried out. Then, for each observation, it is possible to calculate a residual, as the difference between the real and the predicted value of y. Considering an experiment of two factors and the model expressed by Eq. 2.6, the error residual e corresponding to the ijk-th observation is:

$$e_{ijk} = y_{ijk} - y_{ijk} \tag{2.22}$$

where y_{ijk} represents the predicted value of the ijk-th observation.

The set of residuals for all observations constitute the so-called model residuals, and they are useful to check the assumptions of the ANOVA test. In this regard, the most common and straightforward way to check the model assumptions is using a graphical analysis of the residuals. Firstly, the normality can be verified through a diagram of normal probability (Fig. 2.4a). If most of the residual values tend to lay on the identity line in a normal probability plot, then it can be expected that they come from a normal distribution. Secondly, homoscedasticity can be investigated by analyzing a graphic of the residuals versus predicted values. To verify that the variance is constant among the treatments, the pairs (y_{ijk}, e_{ijk}) should be randomly distributed along the graphic, i.e., they must not show any trend (Fig. 2.4b). Finally, the statistical independence can be scrutinized by plotting the residuals as a function

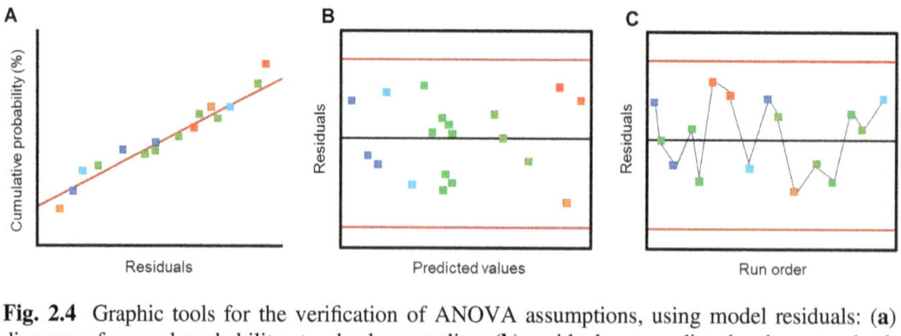

Fig. 2.4 Graphic tools for the verification of ANOVA assumptions, using model residuals: (**a**) diagram of normal probability, to check normality; (**b**) residuals vs predicted values, to check homoscedasticity; and (**c**) residuals vs run order, to check independence

of the experiment order number (run order). Once more, the independence of the observations is verified when no trend along the graphic is observed, i.e., absence of correlation. Violation of this assumption generally indicates deficiencies in experimental planning or execution (Fig. 2.4c).

2.5 Non-randomized Designs

2.5.1 Blocks

During factor screening, it is expected that the observed changes in the response are due to the factors that the analyst wants to study and not due to other variables or factors not taken into account in the design. However, when the experimenter is interested in eliminating systematic effects due to factors not included in the study, a technique called blocking can be implemented. A block represents a spurious factor that is artificially explicit and, therefore, is then included in the ANOVA model. This information is necessary to be included in the model because the spurious factor can affect the response and, consequently, mask the effect of the factor or factors of interest. Blocking enables the analyst to consider the effect of an undesirable factor, with a minimal increase in the number of experiments concerning a design with the same number of factors without blocks.

The most common block factor is time, e.g., when sets of experiments are carried out on different days. Blocking this factor enables the analyst to evaluate whether there is a systematic effect on the response over time and, hence, to decide if all experiments need to be done on the same day or not. This aspect is especially interesting when the number of experiments that can be carried out per day is limited by material or equipment restrictions.

In principle, any kind of pre-RSM design (and even RSM designs) can be blocked. There are different types of blocking strategies. For instance, in a randomized complete block design (RCBD), each block contains all the treatments, and experiments are performed randomly within the block. In this sense, blocks form homogenous experimental sub-units on which treatments or factors are compared, whereas the variability among the blocks is eliminated.

To model the effect of a block, a block term can be incorporated in a model of the type of Eq. (2.5), as follows:

$$y_{ijkl} = \mu + \tau_i + \beta_j + (\tau\beta)_{ij} + \gamma_\kappa + \varepsilon_{ij\kappa l} \tag{2.23}$$

where γ_κ represents the effect of the κ-th block. In this model, it is assumed that the interaction between blocks and treatments is negligible. Equivalently, a term representing the sum of squares due to the block effect is included in Eq. (2.5), and the corresponding *MS* (variance) value for the block is computed. To analyze the significance of the factors and the interaction term, the hypothesis formulation, and ANOVA test are exactly the same as those presented above. However, to decide about the effect of the block, a statistic F_0 for the block cannot be built because it does not represent a F random variable. Therefore, the significance of the block effect is analyzed through the direct inspection of the variance value *MS*. As a rough rule of thumb, if the block variance value is three or four times the residual variance, then it can be considered that there exists a block effect in the experiment.

2.5.2 Split-Plot Designs

Split-plot is another non-randomized design useful when one or more factors are difficult to manipulate or they tend to be kept at fixed values, which could hinder the possibility of randomizing the order of the experiments. A common case is represented by temperature. For instance, when the temperature cannot be easily changed in a given experiment, the analyst can run multiple combinations of the other factors, keeping the hard-to-change variable at a fixed value. This particular type of factorial design is called split-plot design.

The terminology associated with split-plot designs is inspired by agricultural sciences. In a split-plot design, the hard-to-change value is associated with a so-called whole plot (parcel). This factor is referred to as the "whole plot factor" (WP). Then, the easy-to-change factors are applied at different levels to smaller sections of the whole plot, constituting the so-called subplot factors (SP). This scheme is reproduced for the two levels of the WP factor and the corresponding number of replicates, as appropriate. It can be noticed that the levels of the WP and the SP factors are applied at different times. Hence, this type of experimental design can be conceived as two experiments superimposed on each other.

The statistical model for a split-plot design is:

$$y_{ijk} = \mu + \tau_i + \varepsilon_{WP} + \beta_j + (\tau\beta)_{ij} + \varepsilon_{SP} \tag{2.24}$$

where τ_i refers to the whole plot factor, ε_{WP} is the whole plot error, β_j refers to the subplot factor, ε_{SP} is the whole subplot error, and $(\tau\beta)_{ij}$ represents an interaction between a whole plot factor and a subplot factor. The model has two errors terms since two randomizations are involved. Therefore, the error structure of this design is represented by two variances, i.e., σ^2_{WP} and σ^2_{SP}. Under this scenario, whole plot factors are tested against σ^2_{WP}, whereas subplot factors and their interactions with whole plot factors are tested against σ^2_{SP}. In general, the error associated with whole plots is greater than the error within subplots, since the latter tend to be more homogeneous.

References

1. Montgomery DC (2012) Design and analysis of experiments, 8th edn. John Wiley & Sons, New York
2. Myers RH, Montgomery DC (2009) Response surface methodology: process and product optimization using designed experiments. Wiley, New York
3. Vera Candioti L, De Zan MM, Cámara MS, Goicoechea HC (2014) Experimental design and multiple response optimization. Using the desirability function in analytical methods development. Talanta 124:123–138
4. Dejaegher B, Vander Heyden Y (2011) Experimental designs and their recent advances in set-up, data interpretation, and analytical applications. J Pharm Biomed Anal 56:141–158
5. Ferreira SL, Bruns RE, da Silva EG, Dos Santos WN, Quintella CM, David JM, de Andrade JB, Breitkreitz MC, Jardim IC, Neto BB (2007) Statistical designs and response surface techniques for the optimization of chromatographic systems. J Chromatogr A 1158:2–14
6. Taguchi G (1990) Introduction to quality engineering. McGraw-Hill, New York
7. Jones B, Nachtsheim CJ (2011) A class of three-level designs for definitive screening in the presence of second-order effects. J Qual Technol 43:1–15
8. Walpole RE, Myers RH, Myers SL, Ye K (2012) Probability and statistics for engineers and scientists, 9th edn. Pearson, Boston

Chapter 3
Fundamentals of Design of Experiments and Optimization: Experimental Designs in Response Surface Methodology

Silvana M. Azcarate, Carla M. Teglia, Fabricio A. Chiappini, and Hector C. Goicoechea

Abstract Designing an experiment to fit a response surface model typically involves selecting a design among several candidates. There are often many competing criteria that could be considered to properly choose the final design. This chapter presents a description of the most reported designs involved in the data collection process. The basic principles and characteristics of the designs, highlighting their strengths and weaknesses, are described in detail and compared in terms of their characteristics and efficiency. The information is divided into process designs, mixture designs, and designs with the combination of process variables with mixture variables. Additionally, the use of constraints in the delimitation of experimental spaces in both process and mixture designs is also introduced and discussed. Moreover, a brief description of non-randomized designs is presented.

Keywords Response surface methodology · Process designs · Mixture designs · Multivariate optimization

S. M. Azcarate (✉)
Instituto de Ciencias de la Tierra y Ambientales de La Pampa (INCITAP), and Facultad de Ciencias Exactas y Naturales, Universidad Nacional de La Pampa, Santa Rosa, La Pampa, Argentina

lConsejo Nacional de Investigaciones Científicas y Técnicas (CONICET), Ciudad Autónoma de Buenos Aires, Buenos Aires, Argentina
e-mail: sazcarate@exactas.unlpam.edu.ar

C. M. Teglia · F. A. Chiappini · H. C. Goicoechea
Laboratorio de Desarrollo Analítico y Quimiometría (LADAQ), Cátedra de Química Analítica I, Facultad de Bioquímica y Ciencias Biológicas, Universidad Nacional del Litoral, Ciudad Universitaria, Santa Fe, Santa Fe, Argentina

lConsejo Nacional de Investigaciones Científicas y Técnicas (CONICET), Ciudad Autónoma de Buenos Aires, Buenos Aires, Argentina

M. C. Breitkreitz, H. C. Goicoechea (eds.), *Introduction to Quality by Design in Pharmaceutical Manufacturing and Analytical Development*, AAPS Introductions in the Pharmaceutical Sciences 10, https://doi.org/10.1007/978-3-031-31505-3_3

3.1 Introduction

Multivariate design of experiments (DOE) and optimization of processes through the well-known response surface methodology (RSM) are issues of paramount importance in real-world applications since, as is described in Chap. 2, its implementation consumes less time and requires fewer efforts and resources than the use of univariate procedures for the same purpose. DOE is a specific set of experiments defined by a matrix composed of the different level combinations of the studied factors. DOE is a powerful tool able to simultaneously investigate the effect of several factors on a response or several responses, detect interactions among them, and predict the response value(s) under experimental conditions that were not tested. In fact, from the DOE and RSM combination, it is possible to generate empirical models that express the response as a function of the variables. They are characterized by different complexities regarding the selected design. Then, to attain a reliable optimization, it is relevant to select the proper factors and experimental domain and to choose a design that satisfies the expected needs to find a good model that properly fits the data and provides the simplest mathematical approximation of the process or system under study. Choosing the relevant factors and the appropriate domain are not always straightforward tasks, so relevant knowledge from theoretical principles, preliminary tests, or previous studies from the literature is always needed. Besides, it is necessary to look at the available times and costs in order to choose the most suitable design and the number of experiments to be performed. Then, the choice of the design will depend on the scope of the study (screening, investigation of the effects, optimization) or, in other words, on the level of depth needed to understand the problem and, if required, optimize the response.

The first step to implementing the RSM is the selection of an experimental design that will define the experiments to be performed in the studied experimental domain. There are different types of factors that can be involved in a DOE, which can be classified into process variables and mixture variables. Process variables are independent factors of each other that usually have their levels combined without mathematical constraints. On the other hand, mixture designs differ from designs for optimization of process variables since the response generated is a function of the proportions of each component being each variable dependent on another.

The real relationship between the factors and the response is normally unknown. However, within an appropriate experimental domain, the mathematical relationship can be approximated by simple polynomial functions. In this context, linear models could not be ideal to express the response as a function of the variables when dealing with complex processes or systems. In fact, if a curvature in the response surface is present, linear models are not valid. In such cases, designs that permit the building of more complex models are necessary. The simplest non-linear models are the quadratic ones, which are particularly recommended in the case of optimizations since they may allow identifying the optimum. Designs for quadratic models are based on investigating at least three levels of each variable to detect curvature in the response surface. Only quantitative variables can be studied with this kind of design since the response surfaces are continuous functions.

According to the latter concerns and considering the different kinds of factors that can be investigated (process or mixture), a wide variety of designs are available in order to assess the interaction among factors and to find the best conditions for conducting the experiments.

Choosing a design that is suitable for the aim of the experimenter is important when the RSM is implemented. Each type of response surface design is characterized by its own features, which help in the selection. First, the number of experiments required by each specific design must be considered; indeed, in the case of expensive procedures, even a reduction of one observation can make a difference. The same consideration applies to the selection of the replicates (usually in the central point): triplicate experiments are considered a minimum, but a higher number should be chosen, if possible, to have a better estimation of the experimental variability. Then, the selection of an appropriate design needs to take into consideration some issues, such as the number of variables and interactions to be studied, the complexity of each design, the effectiveness of data modeling, the feasibility, and the costs involved in its execution or, in other words, on the level of depth needed to understand the problem and to optimize the response. However, its application will depend on the analysis aims, for instance, the need to carry out a reduced number of experiments, prohibited zones in the experimental region, confidence in the parameters to be adjusted, and so on. Thus, the specific design will depend on the experimenter's criteria, who can select one design over another according to its own advantages and disadvantages and mainly the analysis objectives.

In addition, for a comprehensive and comparative study on the RSM, some relevant properties of each design should be considered: orthogonality, rotatability, and efficiency. Orthogonality can be considered the most important parameter from a mathematical point of view: to pursue the objective of minimizing the variance of the regression coefficients. Orthogonal columns of the design matrix are interpreted as independent vectors, and, consequently, the estimated coefficients in the fitted model are uncorrelated with each other (the mathematical reason will be developed in the next chapter). The rotatability concept is an interesting property that ensures a reasonably stable distribution of the prediction variance in the entire design space; in other words, a design is rotatable if the leverage is symmetrical with respect to the center of the domain. The efficiency parameter was proposed for establishing a comparison among the RSM designs. It is defined as the number of coefficients in the estimated model divided by the number of experiments required by the design [1].

This chapter presents a description of the most reported designs involved in the data collection process used to fit response surfaces. The basic principles of the designs are explained by giving concise theoretical information as well as practical hints. The most popular response surface designs are described in detail.

3.2 Response Surface Methodology Designs with Independent Factors

The real relationship between the variables and the response(s) acquired from a DOE is normally unknown. However, within an appropriate experimental domain, the mathematical relationship can be approximated by simple polynomial functions. The experimental domain of a real problem often contains maxima, minima, or a saddle point. Therefore, it is necessary to add some quadratic terms to have an adequate description. When the DOE involves independent factors, the RSM allows the determination of the nominal values of the factors (maximum or minimum), also referred to as critical conditions [1]. For two factors, the quadratic model generated is described by Eq. (3.1), from which curvature and interactions can be evaluated.

$$y=\beta_0+\beta_1x_1+\beta_2x_2+\beta_3x_1^2+\beta_4x_2^2+\beta_5x_1x_2+\varepsilon \tag{3.1}$$

where y is an experimental response; x_1 and x_2 are the experimental factors; β_0, β_1, β_2, β_3, β_4, and β_5 are the populational regression coefficients; and ε is the experimental random error. The general theory for fitting models will be deepened in the next chapter.

The experimental designs used to fit a response surface can be divided into symmetrical and asymmetrical designs, depending on their appropriateness to be used in an asymmetrical domain [2].

3.2.1 *Symmetrical Designs*

To estimate the parameters in Eq. (3.1), the experimental design must assure that all studied variables are assayed in at least three-factor levels. Symmetrical designs cover a symmetrical experimental domain, and, among the best-known designs that can fulfill these rules, the following can be mentioned: (a) full factorial design (FFD), (b) central composite design (CCD), (c) Box-Behnken design (BBD), and (d) Doehlert design (DD). These symmetrical designs differ from one another regarding the selection of experimental points, number of levels for variables, and number of runs.

3.2.1.1 Full Factorial Design

FFD is an extension of the full factorial 2^k presented in Chap. 2 for screening designs. For this design, the factors are investigated at the coded levels −1, 0, and +1. Thus, the number of experiments is $n = 3^k$, where 3 indicates the number of levels and k the number of factors [3]. The experimental matrix, made by 3^k rows and k columns, is easily built by alternating in an appropriate way the levels of each

Table 3.1 Experimental matrix (in coded values) corresponding to full factorial design (FFD) and central composite design (CCD) with two factors (A and B) and Box-Behnken design (BBD) and Doehlert design (DD) with three factors (A, B, and C)

Experiment	FFD		CCD		BBD			DD		
	A	B	A	B	A	B	C	A	B	C
1	−1	−1	−1	−1	−1	−1	0	1	0	0
2	−1	0	+1	−1	+1	−1	0	+0.5	+0.866	0
3	−1	+1	−1	+1	−1	+1	0	+0.5	+0.289	+0.816
4	0	−1	+1	+1	+1	+1	0	−1	0	0
5	0	0	α	0	−1	0	−1	−0.5	−0.866	0
6	0	+1	α	0	+1	0	−1	−0.5	−0.289	−0.816
7	+1	−1	0	α	−1	0	+1	+0.5	−0.866	0
8	+1	0	0	α	+1	0	+1	+0.5	−0.289	−0.816
9	+1	+1	0	0	0	−1	−1	0	+0.577	−0.816
10					0	+1	−1	−0.5	+0.866	0
11					0	−1	+1	−0.5	+0.289	+0.816
12					0	+1	+1	0	+0.577	+0.816
13					0	0	0	0	0	0

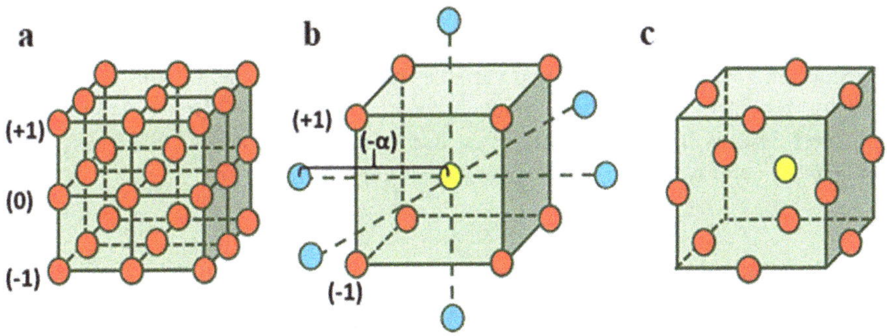

Fig. 3.1 Schematic representation of (**a**) a full factorial design, (**b**) a central composite design, and (**c**) a Box-Behnken design. All of them are for three factors. (Extracted from Ref. [3])

factor, so as to obtain all the possible combinations; as an example, Table 3.1 displays a matrix for a two-factor design ($n = 9$). Moreover, the points investigated by an FFD can be geometrically represented in a k-dimensional space defined by the factors and occupy the vertices of a square ($k = 2$), cube ($k = 3$), or hypercube ($k > 3$). Figure 3.1a shows the representation of a three-level FFD for the optimization of 3 variables, for which at least 27 experiments are required.

Despite this design being completely orthogonal and exploring all the possible combinations among the factor levels, it comprises a high number of experiments (for studying three or more than three factors). For this reason, FFD has had limited application in RSM when the number of factors is higher than 2 owing to the very large number of experiments required, thereby losing its efficiency in the modeling

Table 3.2 Summary of the main characteristics of the most common response surface designs: full factorial design (FFD), central composite design (CCD), Box-Behnken design (BBD), and Doehlert design (DD)

Design	Factor levels	Numbers of experiments (n)	Orthogonality	Rotatability
FFD	3	3^k	Completely orthogonal	No
CCD	5	$2^k + 2k + Cp$	Yes-No[a]	Yes-No[b]
BBD	3	$2k(k - 1) + Cp$	Yes	Yes
DD	Different for each factor	$k^2 + k + Cp$	No	No
D-optimal	Different for each model irregular experimental domain	Selected subset of all possible combinations	No	Yes

[a]May be orthogonal if $Cp = \left(\alpha\sqrt{2^k}\right) + 4 + 2^k$, being Cp the number of center points and k number of factors
[b]May be rotatable if $\alpha = 2^{(k/4)}$, being α the star point distance and k number of factors

of quadratic functions. Nevertheless, for two variables, the efficiency is comparable with designs such as central composite (see below). Table 3.2 provides comparative information about the most relevant features of all designs and their main properties, including the calculation of the number of experiments.

Since an FFD for more than two variables requires experimental runs that cannot usually be accommodated in practice, central composite, Box-Behnken, and Doehlert designs have been more often used since they present higher efficiency and can be employed with a lower number of experiments.

3.2.1.2 Central Composite Design

CCD presented by Box and Wilson is one of the most commonly used designs [2–4]. The classical CCD explores the factors at five levels coded as $-\alpha$, -1, 0, +1, and $+\alpha$. It can be considered as an extension of a two-level full factorial design adding experiments that include each one of the factors at $-\alpha$ and $+\alpha$ levels, namely, axial or star points. Thus, this design is built with a two-level (-1 and +1) factorial design points (*Fp*), star (axial) points (*Sp*), and center points (*Cp*). Then, it requires a number of experiments according to $n = 2^k + 2k + Cp$, where k is the number of factors and *Cp* represents the number of replicates at the central point. The experimental matrix for a CCD with two factors is provided in Table 3.1. As can be observed, the experimental runs required for this design are exactly the same as for the FFD. However, with the incorporation of one more factor, the difference of n changes drastically and, consequently, their efficiency, being from $k = 3$ higher for the CCD. Table 3.3 shows the comparison of the efficiency values of the different symmetrical designs as the value of k increases.

The α-value determines the location of the *Sp* in the design and usually varies from 1 to $\sqrt{k}$. The former allocates the axial points on the face of the cube or

Table 3.3 Comparison of the efficiency (computed for a second-order model) of full factorial design (FFD), central composite design (CCD), Box-Behnken design (BBD), and Doehlert design (DD)

		Number of experiments (n)				Efficiency			
Factors (k)	Number of coefficients	FFD	CCD	BBD	DD	FFD	CCD	BBD	DD
2	6	9	9	–	7	0.67	0.67	–	0.86
3	10	27	15	13	13	0.37	0.67	0.77	0.77
4	15	81	25	25	21	0.18	0.60	0.60	0.71
5	21	243	43	41	31	0.09	0.49	0.61	0.68
6	28	729	77	61	43	0.04	0.36	0.46	0.65
7	36	2187	143	85	57	0.02	0.25	0.42	0.63

hypercube and is known as face-centered CCD ($\alpha = 1$; each factor is thus investigated at three levels). $\sqrt{k}$ results in experimental points placed in the domain of a cube or hypercube. In Fig. 3.1b, the experimental points of a CCD with three factors are geometrically represented. For $k = 3$ and 1 Cp, at least 15 experiments are required, although replications of the central points are recommended in order to evaluate the lack of fit of the models (see Chap. 4). On the other hand, α values depend on the number of variables and can be calculated by $\alpha = 2^{(k/4)}$, and, in this condition, the design will be rotatable. This is an interesting property, since the experimenter does not know at the outset where the desirable response may lie within the design space. Rotatability ensures a reasonably stable distribution of the prediction variance over all the design space. However, for $k > 3$, rotatable designs require a large α-value, not always being practical from an operational point of view. In such cases, a spherical design, which results in near-rotatability, is preferable to stabilize the variance in a narrower and acceptable experimental region. In the case of the CCD, rotatability is achieved by making a proper choice of α, the axial distance. Furthermore, it is important to note that rotatability is achieved regardless of the number of center-point runs. For rotatable CCD, α achieves values of 1.41, 1.68, and 2.00 for two, three, and four variables, respectively. On the other hand, orthogonality depends on the number of replications at the center points; then, CCD can be orthogonal if $Cp = \left(\alpha\sqrt{2^k}\right) + 4 + 2^k$.

3.2.1.3 Box-Behnken Design

Box and Behnken suggested how to select points from the three-level factorial arrangement, which allows the efficient estimation of the linear and quadratic coefficients of the mathematical model [1, 2]. BBD is characterized by appreciable statistical properties. In fact, it is rotatable and defined as quasi-orthogonal: the covariance among the coefficients is zero for most terms and very close to zero for the others. This design is more efficient than its corresponding 3^k designs, and it also is slightly more efficient than CCD (see Table 3.3).

In BBD, a minimum of three factors are explored at three levels (−1, 0, +1), and the experimental matrix is easily built. For instance, the matrix for a three-factor design (12 points plus the central point) is composed of three blocks of four experiments, in which one variable is fixed at the 0 level and the others have the sign alternation characteristic of a two-level and two-factor full factorial design (Table 3.1). The experimental points are located on a hypersphere equidistant from the central point, as exemplified by a three-factor design, in Fig. 3.1c. The number of experiments for this design is $n = 2k(k - 1) + Cp$, i.e., for 3 factors ($k = 3$) and 1 central point, at least 13 experiments are required, and all factor levels are adjusted only to 3 levels (−1, 0, +1) with equally spaced intervals between these levels.

As it can be observed, this design requires a smaller (or equal) number of experiments than the two designs previously presented. Thus, for six factors ($k = 6$), the minimal number of required experiments for the mentioned experimental designs will be the following: FFD (729), CCD (77), and BBD (49). BBD has been widely applied; however, its application is still much smaller in comparison with CCD.

Another advantage of BBD is that it does not contain any run for which all factors are simultaneously at their highest or lowest levels. Therefore, it is a suitable design when unsatisfactory results might occur at the extreme points of the experimental domain. However, caution should be used if the theoretical optimum falls in those regions because the obtained model may not properly represent them.

3.2.1.4 Doehlert Design

Doehlert design (DD) is a practical alternative to other experimental designs. It is a polyhedron based on hypertriangles with a hexagonal structure in the simplest case. This means they have uniform space-filling properties with an equally spaced distribution of points lying on concentric spherical shells. The geometrical representation of a DD for two and three factors is shown in Fig. 3.2a, b. Thus, this design describes a circular domain for two variables, spherical for three variables, and hyperspherical for more than three variables, which accents the uniformity of the studied factors in the experimental domain. Although the design matrices are not rotatable as previous designs, it presents some advantages, such as requiring few experimental points for its application and high efficiency. The required experiment number is calculated by $n = k^2 + k + Cp$. A particularly important characteristic is that each variable is studied at a different number of levels, which is relevant when some of them are subjected to restrictions such as cost and/or instrumental constraints or when it is interesting to study a variable at a major or minor number of levels or when the intervals among its levels present a uniform distribution. Moreover, the displacement of the experimental matrix to another experimental region can be achieved using previous adjacent points.

For two variables, the DD is represented by a central point surrounded by six points from a regular hexagon (Fig. 3.2a). The latter figure also shows some possibilities of displacement from the original experimental conditions. For three factors, this design is represented by a geometrical solid called a cuboctahedron (Fig. 3.2b) [5].

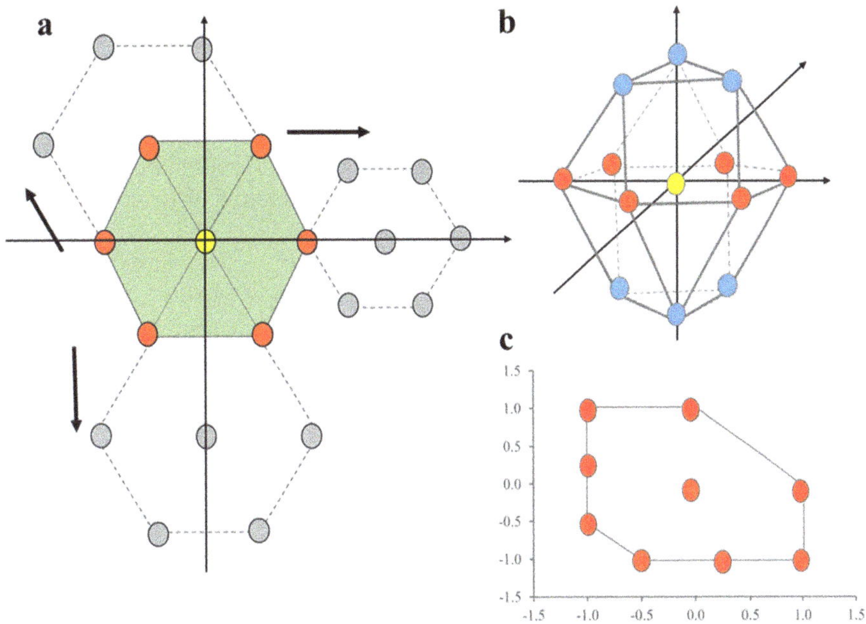

Fig. 3.2 Schematic representation of Doehlert designs for the optimization of (**a**) two variables and some possibilities for the displacement from the initial design using previous points; (**b**) three variables; and (**c**) a constrained region in two variables constructing a *D*-optimal design (constraints removing two corners of the square, producing an irregular experimental region)

3.2.2 Asymmetrical Designs: D-*Optimal Design*

Sometimes symmetrical domain of a design can be lost for several reasons, and, consequently, an asymmetrical domain should be examined. Therefore, the classical designs above described are not able to work under strict constraints of the experimental domain since symmetric designs in an asymmetric domain are problematic. Moreover, in extreme conditions, they become too large and require experiments in an unfeasible area or they are too small and then a considerable part of the domain is not covered. Then, another sort of design must be implemented to cope with multifactor constraints. These designs are called asymmetrical because when an asymmetrical domain is inspected, the plotted experiments generate an asymmetrical shape.

The optimality concept can be applied to select a design when the classical symmetrical designs cannot be used, such as when the experimental region is irregular in shape, the number of experiments chosen by a classical design is too large, or it is required to apply models that deviate from the usual linear or quadratic ones [1, 6].

Design optimality criteria are characterized by letters of the alphabet, and some of them focus on good estimation of model parameters, while others focus on good prediction in the design region. The best-known and most often used criterion to choose a particular design is *D*-optimality. The latter generates a particular kind of design that must be performed in complex situations as those previously mentioned.

The *D*-criterion states that among all matrices that can be chosen from the matrix of candidate points, the one that leads to a model matrix **X** that minimizes the determinant of the so-called dispersion matrix $(\mathbf{X}^T\mathbf{X})^{-1}$ is optimal (herein, the superscript "T" indicates the mathematical operation of matrix transposition). To construct a *D*-optimal design with the objective of studying *k* factors, firstly, the kind of model to be built that requires minimal n_{min} experiments (to enable estimating the model coefficients) is defined. Secondly, the number of experiments, *n*, to be performed is defined ($n \geq n_{min}$). The experimental domain is represented by several candidate experiments (n_{grid}) forming a grid over the domain. The number of experimental runs of the *D*-optimal design is selected as the combination with the maximal determinant for $(\mathbf{X}^T\mathbf{X})$ (equal to *D*-optimality). As will be shown in Chap. 4, the latter operation is performed to obtain the regression coefficients. The selection is achieved by applying an algorithm that calculates the determinant of all the possible information matrices and finds the one having the highest determinant (hence the *D* of *D*-optimal). Each computed information matrix refers to a different *n*; therefore, to compare the matrices and decide on the best set of experiments, the normalized determinant must be employed. Thus, the matrix of moment $\mathbf{M} = \mathbf{X}^T\mathbf{X}/n_{min}$ is used to compute its determinant as $|\mathbf{M}| = |\mathbf{X}^T\mathbf{X}|/n_{min}p$ [6], where | | indicates the determinant operation and *p* the number of regression coefficients. Finally, once the appropriate design and domain have been chosen and the experimental matrix has been built, the experiments must be carried out randomly. For the reasons already explained, it is also fundamental to perform at least three replicates at the central point randomly distributed in time, to have an actual estimation of the experimental variance over the entire time-lapse of the experiments.

As an example of an irregular experimental region, Fig. 3.2c shows the experimental domain that results from applying the $-1.5 \leq A + B \leq 1$ constraint to design a matrix built by two factors. Notice that this constraint effectively removes two corners of the square, producing an irregular shape. There is no standard response surface design that fits exactly into this region. Table 3.4 displays the experimental matrix acquired by constraint application resulting in a matrix with 12 experimental runs.

3.2.3 Split-Plot Designs (SPD)

In Chap. 2, the SPD was introduced as a technique for running experiments in a more economical manner if some of the factors are hard to change, referred to as the whole plot factor (WP). Time or cost limitations lead to considering a split-plot structure

Table 3.4 *D*-optimal design matrix corresponding to the constrained region represented in Fig. 3.2c

Experiments	A	B
1	−0.50	−1.00
2	+1.00	0.00
3	−0.08	−0.08
4	−1.00	+1.00
5	+1.00	−1.00
6	0.00	+1.00
7	−1.00	+0.25
8	+0.25	−1.00
9	−1.00	−0.50
10	+1.00	0.00
11	0.00	+1.00
12	−0.08	−0.08

rather than a completely randomized design. For example, heating an industrial oven from 300 to 500 °C can be very time-consuming. In these situations, it will be much easier to group the experimental runs for each temperature to be investigated instead of waiting for the oven to heat up or cool down each time as in a completely randomized design. A design called "split-plot" allows taking into account this type of constraint. It should be noted that to get the appropriate ANOVA *p*-values from a split-plot design, specialized statistical tools such as restricted maximum likelihood must be executed [7].

Interestingly, the SPD building involves two randomization processes. The WP combinations are randomly assigned to whole plot units based on the whole plot design. Within each whole plot, the easy-to-change subplot factors (SP) are assigned to the subplot units with separate randomization for each whole plot. Consequently, two error terms will be obtained by applying these two levels of randomization, a whole plot error, and a subplot error. Care should be taken to ensure that both types of error can be estimated and that the terms of the model are assessed relative to the appropriate error term. Typically, the main, interaction, and quadratic effects involving whole plot factors are whole plot effects. Terms of the model involving just subplot factors or combinations of whole and subplot factors are generally subplot effects.

The above considerations lead to designs with a more complex structure; therefore, a completely randomized experiment may involve a large number of runs to be executed. Thus, the optimal design criteria can be used to build more economical designs [1, 8]. As an example, Table 3.5 shows an SPD built for one WP, oven temperature (300–500 °C), and two SP, time (5.0–50.0 min) and reagent concentration (2.0–10.0 g L^{-1}). As can be appreciated in the table, applying the *D*-optimal criterion, 15 experimental runs are necessary, divided into 3 groups, each of them corresponding to 3 oven temperatures (300, 400, and 500 °C).

Table 3.5 *D*-optimal design matrix corresponding to an SPD built for one hard-to-change factor (oven temperatures) and two easy-to-change factors (time and reagent concentration)

Group	Runs	WP: oven temperature (°C)	SP: time (min)	SP: concentration (gL^{-1})
1	1	500	5.0	2.0
1	2	500	27.5	2.0
1	3	500	50.0	6.0
1	4	500	5.0	10.0
1	5	500	50.0	10.0
2	6	400	5.0	2.0
2	7	400	50.0	2.0
2	8	400	50.0	10.0
2	9	400	27.5	6.0
2	10	400	50.0	2.0
3	11	300	27.5	2.0
3	12	300	5	10.0
3	13	300	5	2.0
3	14	300	50	10.0
3	15	300	50	6.0

3.3 Response Surface Methodology Designs with Dependent Factors: Mixture Designs

In the experimental designs discussed previously, the levels chosen for any factor in the experimental domain are independent of the levels chosen for the other factors. In contrast, a mixture experiment is a special type of design in which the factors are components of a mixture, and the response is a function of the proportions of each component. RSM applied to mixture experiments allows researching the proportion influence of each component that constitutes a mixture, and that is part of some stage of a method, helping to improve its performance. Mixture designs shape a sort of designs with dependent factors that differ from designs for optimization of process variables (designs with independent factors) since the response generated is a function of the proportions of each component and their levels are not independent of one another.

Mixture designs allow studying the effect of the variation of the ratios among the variables. The sum of the proportions of all mixture components must always be constant, and, due to this constraint, the levels of the factors cannot be chosen independently. Thus, the proportional amounts of each component are typically measured by weight, volume, mole ratio, and so on. The domain of a mixture designs for non-constrained regions is a regular figure having as many vertices as components, in a space with dimensionality equal to the number of components minus one [1]. The experimental domains for mixture designs for three and four components are shown in Fig. 3.3a, b. In this kind of design, the experimental points

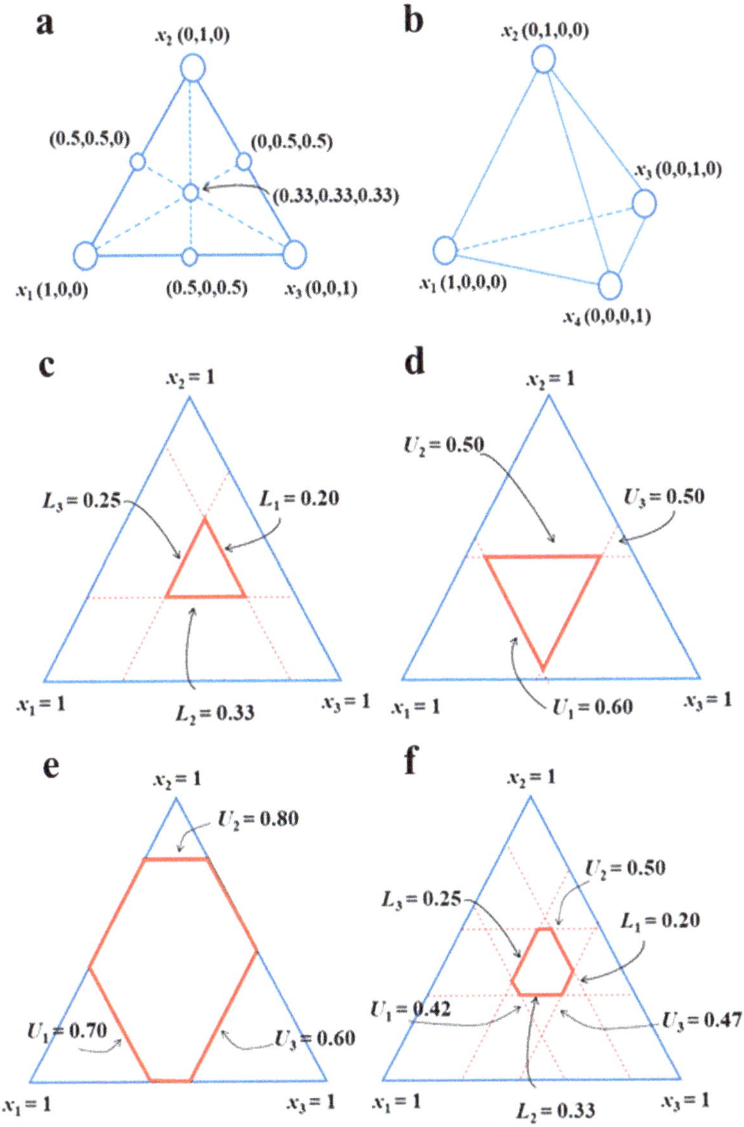

Fig. 3.3 Graphical representations of mixture designs with (**a**) three and (**b**) four components, respectively. Graphical representation of constrained designs: (**c**) with lower limit imposition, $L_1 = 0.20$, $L_2 = 0.33$, and $L_3 = 0.30$; (**d**) with upper limit imposition, $U_1 = 0.60$, $U_2 = 0.50$, and $U_3 = 0.50$; (**e**) upper limits imposed for the three components: $U_1 = 0.70$, $U_2 = 0.80$, and $U_3 = 0.60$; and (**f**) upper and lower impositions, $L_1 = 0.20$, $L_2 = 0.33$, and $L_3 = 0.25$ and $U_1 = 0.42$, $U_2 = 0.55$, and $U_3 = 0.47$. (Extracted from Ref. [10])

can be described by a straight line, an equilateral triangle, or a regular tetrahedron form, for two, three, and four components, respectively. Above four components, these designs are represented in abstract n-dimensional spaces with more than three dimensions. Algebraically, it is possible to apply mixture design for optimization of a mixture with many components. Nevertheless, its practical application is scarce, and the interpretation of the results becomes quite complicated since the number of components (q) increases. For the triangle used for modeling ternary mixtures, any point on one of the vertices represents the pure component (100% of that component), the points located on the sides, binary mixtures, and any point on the internal region, ternary mixtures. The same reasoning can be applied to the tetrahedron used in the study of quaternary mixtures, and its interior represents mixtures of four components.

It should be noted that the models used in mixture designs differ from the polynomials used in response surfaces for independent variables. Another important difference regarding a standard response surface experiment is that the form of the mixture polynomial is slightly different from the standard polynomials used in RSM for DOE with independent factors. Much of the work in this area was originated by Scheffé [9]. The mathematical models are the well-known Scheffé polynomials, which can be linear, quadratic, full cubic, and special cubic. They will be developed in Chap. 4.

In this section, the designs that allow an appropriate response surface model to be fitted over the entire mixture space are introduced. Moreover, the basic principles and characteristics of these designs, and the use of constraints in the delimitation of experimental spaces associated with the use of pseudo-components to simplify the mathematical treatment, are presented and commented on. The combination of process variables with mixture variables and the combination between two mixture designs for simultaneous optimization are also approached.

3.3.1 Simplex Designs

Lattice simplex design for a mixture experiment can be considered the equivalent of factorial design for process variables in the sense that experimental points are taken at the extremes of the experimental domain and, for more than two levels, they are evenly spaced along with the coordinates that represent the variable. A simplex lattice is a uniformly spaced set of points on a simplex. $A(q, m)$ simplex lattice design for q components contains points defined by the following coordinate settings (Eq. 3.2), equally spaced between levels 0 and 1, and all possible combinations (mixtures) of the proportions from this equation are used:

$$x_i = 0, \frac{1}{m}, \frac{2}{m}, \ldots, 1 \qquad i = 1, 2, \ldots, q \tag{3.2}$$

As it will be seen in the subsequent chapter, the notation (q, m) implies a simplex lattice design with q components that will support a mixture polynomial of degree m.

Simplex lattice designs present some disadvantages since they require a large number of experimental points as the degree of polynomial increases with a consequent laborious application. On the other hand, when the mathematical model is fitted, only points corresponding to pure components and binary mixtures acquired from experimental data are used to calculate the regression coefficients, which could cause the contributions, for instance, of ternary mixtures to be underestimated.

Later, to overcome these limitations, the centroid-simplex design was introduced. In this design, one central point is included to allow modeling and evaluation of the influence of possible synergistic or antagonistic effects that could appear in the presence of all components of the mixture. For instance, for the study of ternary mixtures, a special cubic model has been proposed (see Chap. 4). This polynomial function was formulated to include only terms that represent experimental points related to the presence of components in equal proportions within the experimental domain. In this kind of design, the number of experimental points and their configuration allow that the special cubic results in the most efficient fitted model.

In Fig. 3.3a, b, graphical representations of mixture designs with three and four components, respectively, can be seen. In the first case, the design is an equilateral triangle whose vertices correspond to the pure components. Each of the three sides represents a binary mixture, and the internal points correspond to ternary mixtures. Usually, these designs are augmented with extra points in the interior of the experimental region in order to evaluate the model and even to fit the complete cubic model. Unfortunately, the centroid-simplex design still has some disadvantages presented by the simplex lattice design. Although the number of terms in the particular model equation is smaller than for the full cubic model, the number of points also increases disproportionately with increasing of q. Most of the observations are collected at design points located at the extremes of the experimental domain affecting the representativeness of the studied domain unless more points are incorporated within the domain.

3.3.2 Simplex Designs with Constraints in the Component Proportions: Pseudo-Components

Sometimes, some mixture experiments also need constraints, in this case, on the constituent proportions. Accordingly, upper and/or lower limits can be established for some or all components of a mixture regarding the following form:

$$Li \leq xi \leq Ui \qquad i = 1, 2, \ldots, q \tag{3.3}$$

where Li and Ui are the lower and the upper bound for the i-th component, respectively, and $x_1 + x_2 + x_3 = 1$. The lower bound represents a minimum proportion of the i-th component, while the upper bound is a maximum proportion of the i-th component that must be present in the mixture. Figures 3.3c, d represent two practical examples of constrained mixture problems for three components. The constraints have been set by practical bounds on the relative proportion of components as $Li \geq 0$ and $Ui \leq 1$, for $i = 1, 2, 3$. In the first case, shown in Fig. 3.3c, $L_1 = 0.20$, $L_2 = 0.33$, and $L_1 = 0.25$. On the other hand, Fig. 3.3d shows the mixture components with upper limit impositions: $U_1 = 0.60$, $U_2 = 0.50$, and $U_3 = 0.50$. As can be observed, these constraints do not modify the original simplex shape. Then, by way of simplification of the design building in addition to its posterior modeling, a new set of components, usually named pseudo-components, considering values from 0 to 1 over the feasible region can be defined. Therefore, pseudo-components are the new coordinates obtained by the transformation of the original proportions and that greatly simplify the mathematical treatment of mixtures that present this type of constraint on their components.

3.3.3 Designs with Constraints in the Component Proportions When the Simplex Shape Has Been Lost: D-Optimal Criterion

Sometimes only upper bound constraints of the form $xi \leq Ui$ are placed on the component proportions, which can lead to an experimental region where the simplex shape is lost. In such cases, asymmetrical designs based on the D-optimal criterion presented before are logical design alternatives. Regarding this, other examples are displayed in Fig. 3.3e, f. The first of them, represented in Fig. 3.3e, is a design in which upper limits have been imposed for the three components: $U_1 = 0.70$, $U_2 = 0.80$, and $U_3 = 0.60$. Similar behavior will be present in those cases in which both lower and upper bounds are imposed. Figure 3.3f shows the latter case, in which the following constraints were imposed: $L_1 = 0.20$, $L_2 = 0.33$, and $L_1 = 0.25$ and $U_1 = 0.42$, $U_2 = 0.50$, and $U_3 = 0.47$.

D-optimal designs can be a useful strategy to select points for a mixture design in a constrained domain in those cases in which the simplex shape has been lost, as in those cases shown previously in Fig. 3.3e, f. In the same way, as for irregular designs with independent factors, design points are selected from a list of candidate points so that the variances of the model regression coefficients are minimized through the point exchange procedure.

For instance, in an example described by Myers and Montgomery [1], a shampoo with lower and upper bounds as $0.20 \leq x_1 \geq 0.30$, $0.07 \leq x_2 \geq 0.10$, and $0.13 \leq x_3 \geq 0.20$, where $x1 + x2 + x3 = 0.5$, is formulated. In this mixture design, a total of 6 of the 13 candidate points must be chosen to fit a quadratic model and some additional replicated runs to estimate the error and adequacy of the model (in this case, 4 replicated and 3 distinct points), giving in this case 13 experimental combinations. For a quadratic model, 4 vertices of the region, 4 edge centers, overall centroid, and 4 axial points that lie midway between the centroid and the vertices of

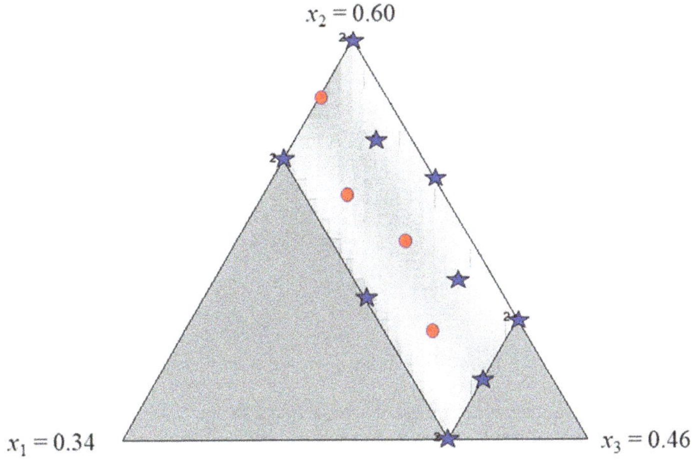

Fig. 3.4 Representation of the *D*-optimal design using the pseudo-components built for a shampoo formulation with lower and upper bounds given by $0.20 \leq x_1 \geq 0.30$, $0.07 \leq x_2 \geq 0.10$, and $0.13 \leq x_3 \geq 0.20$ (where $x_1 + x_2 + x_3 = 0.5$). Candidate points (stars and dots) and those selected by the *D*-optimal criterion (stars). (Extracted from Ref. [10])

the constrained region are included as candidate points (13 points). Figure 3.4 exhibits the design built in terms of pseudo-components and shows the candidate points (stars and dots) and those selected by the *D*-optimal criterion (stars). It is important to note that other optimality criteria can also be used, such as A, I, G, etc.

3.3.4 Mixture-Process and Mixture-Mixture Designs

In an optimization stage, where the experimental conditions generate the best response(s) for a given process or system, sometimes, it can be necessary to include a combination of process and mixture variables. Thus, it is possible to study them simultaneously using combined experimental designs that include process variables in mixture experiments or vice versa [10]. The process variables can be optimized by full or fractional two-level factorial designs, by designs suitable for quadratic surfaces (CCD, BBD, DD, or *D*-optimal), while the mixture variables can be optimized by lattice simplex designs, centroid simplex designs, and constrained designs, among others. Then, all these designs can be optimally combined to acquire a single experimental matrix that allows the simultaneous modeling of the data.

In an example reported by Alcaraz et al. [11], it is demonstrated how performing an experiment with mixtures, the process variables can affect the blending properties. Here, optimum electrophoretic conditions were set by experimental design and

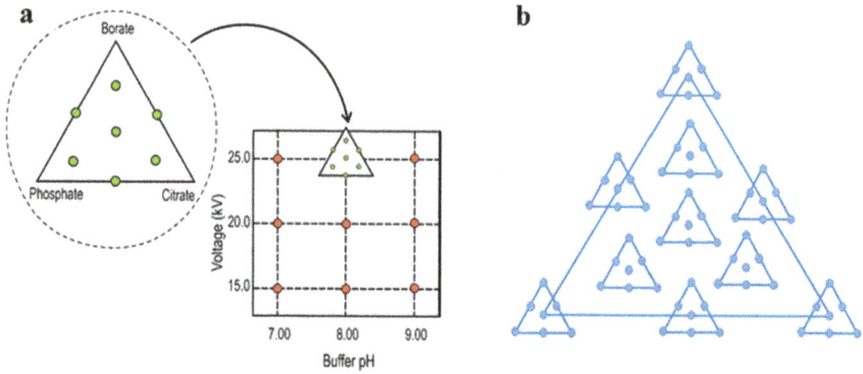

Fig. 3.5 (**a**) Experimental mixture-process design scheme of the optimization presented in Ref. [11]. The red dots represent the factorial design points, and the green dots represent the mixture design points. (**b**) Graphical representation of a combined design built for a combination of two mixture designs. (Extracted from Ref. [10])

optimization combining process variables (voltage and pH) and a mixture of background electrolyte (BGE) salts. The design was built by combining simplex lattice and factorial designs. The entire design was performed with 63 experimental runs. For the mixtures, the amounts of three different salts used to prepare the BGE were varied using the simplex lattice design, in the range of 0 to 15 mmol L^{-1}, with the restriction that no pure salt should be used and with a maximum concentration of 10 mmol L^{-1}. This design consists of seven combinations: one central point, three points in the interior of the simplex, and the last three points in the center of the edges. As can be appreciated in Fig. 3.5a, the factorial process design consisted of nine combinations of pH (between 7.00 and 9.00) and voltage (between 15 and 25 kV). From a combined design, the experimental responses were obtained and, then, fitted to a combined model. The mathematical equations necessary to simultaneously fit mixture and process variables are described in Chap. 4.

Another sort of combined design can be built for a combination of two mixture designs. As an example, a crossed mixture design was built for the optimization of the mobile phase and solvent extraction media [12]. First, a simplex-centroid design with three axial mixture points was selected to investigate the influence of the solvent proportions on extract preparation; based on that, solvents from different groups in the triangle present different selectivity characteristics. Then, a mixture of water and two organic solvents was carried out using a simplex-centroid design considering that the type and concentration of the organic modifiers are of prime importance. The crossed design is illustrated in Fig. 3.5b leading to 70 different experimental combinations.

3.4 Desirable Properties of Response Surface Designs

There are several classes of DOE and many criteria on which DOE are based. Nowadays, many computer packages give optimal designs based on special criteria and inputs from the user. Nevertheless, it is important to review a set of properties that should be considered when the selection of a response surface design is performed. Myers and Montgomery [1] suggest that a good design should:

1. Result in a good fit of the model to the data
2. Provide sufficient information to allow a test for lack of fit
3. Allow models of increasing order to be constructed sequentially
4. Provide an estimate of "pure" experimental error
5. Be insensitive (robust) to the presence of outliers in the data
6. Be robust to errors in the control of design levels
7. Be cost-effective, that is, not require too many runs
8. Allow for experiments to be done in blocks
9. Provide a check on the homogeneous variance assumption
10. Provide a good distribution of the variance of the predicted response throughout the design region

Certainly, not all the above properties are required in every RSM application. Anyhow, most of them should be given serious consideration on each occasion in which a design of experiment is performed. Most of the properties are self-explanatory. Finally, it is of paramount importance to consider that designing an experiment is not necessarily easy and should involve balancing multiple aims, not just focusing on a single characteristic.

References

1. Myers RH, Montgomery DC, Anderson-Cook CM (2016) Response surface methodology: process and product optimization using designed experiments, 4th edn. Wiley, Hoboken
2. Dejaegher B, Vander Heyden Y (2011) Experimental designs and their recent advances in set-up, data interpretation, and analytical applications. J Pharm Biomed Anal 56:141–158
3. Goicoechea HC (2018) Experimental design. In: Fingerprinting techniques in food authentication and traceability, 1st edn. CRC Press, Boca Raton
4. Vera Candioti L, De Zan MM, Cámara MS, Goicoechea HC (2014) Experimental design and multiple response optimization. Using the desirability function in analytical methods development. Talanta 124:123–138
5. Sarabia LA, Ortiz MC (2009) Response surface methodology. In: Comprehensive chemometrics. Elsevier, Amsterdam
6. de Aguiar PF, Bourguignon B, Khots MS, Massart DL, Phan-Than-Luu R (1995) D-optimal designs. Chemom Intell Lab Syst 30:199–210
7. Wolfinger R, Tobias R, Sall J (1994) Computing Gaussian likelihood and their derivatives for general linear mixed models. SIAM J Sci Comput 15:1294–1310
8. Cortes LA, Simpson JR, Parker PA (2018) Response surface split-plot designs: a literature review. Qual Reliab Eng Int 34:1374–1389

9. Scheffé H (1958) Experiments with mixtures. J R Stat Soc B 20:344–366
10. Azcarate SM, Pinto L, Goicoechea HC (2020) Applications of mixture experiments for response surface methodology implementation in analytical methods development. J Chemom 34:e3246
11. Alcaráz MR, Vera-Candioti L, Culzoni MJ, Goicoechea HC (2014) Ultrafast quantitation of six quinolones in water samples by second-order capillary electrophoresis data modeling with multivariate curve resolution-alternating least squares. Anal Bioanal Chem 406:2571–2580
12. de Almeida AA, Scarminio IS (2007) Statistical mixture design optimization of extraction media and mobile phase compositions for the characterization of green tea. J Sep Sci 30:414–420

Chapter 4
Fundamentals of Design of Experiments and Optimization: Data Modeling in Response Surface Methodology

Fabricio A. Chiappini, Silvana M. Azcarate, Carla M. Teglia, and Hector C. Goicoechea

Abstract The experimental data collected according to a proper statistical design constitute the input for data modeling, which represents the last step in response surface methodology (RSM). Data modeling consists in applying a set of statistical methods that enable the analyst to thoroughly study the relation between experimental factors and responses. Essentially, this task is carried out by building empirical models, which are then used to make predictions and investigate possible optimal experimental regions. In this chapter, important concepts of multivariate statistics related to data modeling in RSM are introduced. In particular, models based on multiple linear regression, MLR (parametric), and artificial neural networks, ANN (non-parametric), are presented, which are the two most important modeling approaches in RSM. Additionally, relevant issues regarding model validation, outlier diagnosis, prediction, and interpretation are discussed, and mathematical methods for single and multiple response optimization are briefly described. Finally, some of the most popular software for RSM implementation are summarized.

Keywords Response surface methodology · Multivariate optimization · Empirical models · Least-squares regression · Artificial neural networks

F. A. Chiappini (✉) · C. M. Teglia · H. C. Goicoechea
Laboratorio de Desarrollo Analítico y Quimiometría (LADAQ), Cátedra de Química Analítica I, Facultad de Bioquímica y Ciencias Biológicas, Universidad Nacional del Litoral, Ciudad Universitaria, Santa Fe, Santa Fe, Argentina

Consejo Nacional de Investigaciones Científicas y Técnicas (CONICET), Ciudad Autónoma de Buenos Aires, Buenos Aires, Argentina
e-mail: fchiappini@fbcb.unl.edu.ar

S. M. Azcarate
Instituto de Ciencias de la Tierra y Ambientales de La Pampa (INCITAP), and Facultad de Ciencias Exactas y Naturales, Universidad Nacional de La Pampa, Santa Rosa, La Pampa, Argentina

Consejo Nacional de Investigaciones Científicas y Técnicas (CONICET), Ciudad Autónoma de Buenos Aires, Buenos Aires, Argentina

M. C. Breitkreitz, H. C. Goicoechea (eds.), *Introduction to Quality by Design in Pharmaceutical Manufacturing and Analytical Development*, AAPS Introductions in the Pharmaceutical Sciences 10, https://doi.org/10.1007/978-3-031-31505-3_4

4.1 Introduction

From a practical point of view, the main purposes of response surface methodology (RSM) are achieved by means of statistical data modeling. This step is as crucial as data generation in the laboratory. Good data analysis criteria should be orientated not only by the scientific knowledge about the system under study but also by good statistical methodology. In Chap. 2, general concepts about RSM have been introduced, and main topics regarding factor screening (pre-RSM) have been discussed. Besides, the most commonly used designs for RSM were introduced in Chap. 3. The experimental data resulting from an RSM design are the input for data modeling. In this chapter, basic concepts of multivariate statistics related to experiment optimization in the context of RSM for individual and multiple experimental responses are introduced.

In general lines, data modeling for experiment optimization consists in finding a mathematical rule (model) that relates each response of interest y (dependent variable) with several experimental factors x (independent variables or predictors). In this sense, the inputs for RSM are a matrix of experimental data $\mathbf{M}$ of size n observations $\times k$ factors (usually referred to as design matrix) and an n-vector $\mathbf{y}$ that contains the experimental measurements of y for each observation (run or experiment). Under this notation, the i-th observation is characterized by a k-vector $\mathbf{x}_i$ containing the levels of the experimental factors and the corresponding value of the response y_i. Usually, the procedure of building a model is referred to as model fitting or training. After validation, the generated model is essentially used to make predictions and investigate possible optimal experimental regions. For these purposes, two modeling approaches are most frequently implemented: (i) multiple linear regression (MLR), as a parametric methodology, and (ii) artificial neural networks (ANNs), as a non-parametric strategy. As usual in statistics, parametric methods are normally the first choice because they are robust, simple to interpret, and very well characterized in terms of statistical theory. However, they rely on relatively hard assumptions about the variables, e.g., normal distribution, which can be a limitation for certain databases. In these scenarios, non-parametric methods, such as ANNs, represent more flexible alternatives. Although they are less characterized than parametric methods, non-parametric strategies have a strong predictive capacity. For this reason, they still can be very convenient for RSM objectives.

From the data modeling perspective, experiment optimization is classified as a problem of supervised learning. This implies that during model training, the information of both response (target) and predictor variables are involved in the fitting or learning algorithm.

The following two sections are dedicated to providing a general overview of parametric and non-parametric modeling approaches, where the main purpose is focused on building an empirical model for a given response of interest. Then, the most relevant issues about the optimization of individual and multiple responses are presented. Finally, some of the most popular software for RSM implementation is mentioned.

4.2 Multiple Linear Regression (MLR)

4.2.1 General Theory

The most important parametric statistical approach for experiment optimization consists in fitting a MLR model. Mathematically, for k experimental factors, a first-order MLR model obeys the following expression [1]:

$$y = \beta_0 + \beta_1 x_1 + \beta_2 x_2 + \ldots + \beta_k x_k + \varepsilon = \beta_0 + \sum_{j=1}^{k} \beta_j x_j \tag{4.1}$$

where y is a given experimental response; $x_1, x_2, \ldots, x_k$ represent the experimental factors; $\beta_0, \beta_1, \beta_2, \ldots, \beta_k$ are the so-called regression coefficients; and ε represents the experimental random error. As mentioned previously, ε is supposed to follow a normal distribution with mean μ equals to zero and variance σ^2. More complex models than Eq. (4.1) can also be formulated if interactions between factors or curvature effects are presumed. For instance, for a two-factor model, a factor interaction term can be included as:

$$y = \beta_0 + \beta_1 x_1 + \beta_2 x_2 + \beta_3 x_1 x_2 + \varepsilon \tag{4.2}$$

where $x_1 x_2$ represents a binary interaction. For more factors, higher-order interaction terms might also be considered. However, significant interactions of more than three factors have been found to be unusual [2].

On the other hand, polynomial terms of order higher than one are possible. For example, a model with interactions and pure quadratic terms can be written for two factors as:

$$y = \beta_0 + \beta_1 x_1 + \beta_2 x_2 + \beta_3 x_1^2 + \beta_4 x_2^2 + \beta_5 x_1 x_2 + \varepsilon \tag{4.3}$$

In order to avoid name conflicts between statistical and RSM fields, it is important to make the following clarifications. On the one hand, from the statistical point of view, any regression model of the type of Eqs. (4.1, 4.2, and 4.3) and equivalents are considered linear models when the response variable y linearly depends on the model parameters, i.e., the regression coefficients. Indeed, if interaction and power terms in Eqs. (4.2) and (4.3) are rewritten by defining $x_1 x_2$, x_1^2, and x_2^2 as x_3, x_4, and x_5, respectively, both equations take the form of Eq. (4.1).

On the other hand, in the context of RSM, models of the form of Eqs. (4.1) and (4.2), where only k independent factors and low-order interactions are included, are generally known as first-order models. Conversely, when pure quadratic terms are also included, a model of the form of Eq. (4.3) is referred to as a second-order model. Under this nomenclature, it can be stated that first-order models are suitable for fitting experimental data that arise from two-level designs, whereas data from RSM designs provide enough experimental points to adequately estimate the higher number of coefficients required by second-order models. Optimization problems are most commonly tackled through second-order models. However, this does not necessarily imply that first-order or even models of orders higher than two cannot be used for optimization too.

Naturally, the above equations represent theoretical (population) models. Sample data and statistical estimation methodologies lead to the estimation of population parameters, i.e., a vector $\widehat{\boldsymbol{\beta}}$ of estimated parameters, which enables calculating the estimated response for the i-th observation y_i as:

$$y_i = \beta_0 + \sum_{j=1}^{k} \beta_j x_{ji} \tag{4.4}$$

The results of Eq. (4.4) are also used to calculate the so-called model residuals $\mathbf{e}$. The residual for the i-th observation is computed as $e_i = y - y_i$. Model residuals are usually used as an estimator of ε. The sum of squared residuals for all observations is usually referred to as total sum of squares SS_T.

Fitting an MLR model consists in calculating the regression coefficients. This task is normally carried out through a classic and well-established statistical estimation technique, known as the least-squares method (LS). Briefly, the LS procedure chooses the parameters β_j so that the sum of the squares of the errors for each observation i is minimized. Mathematically, this is equivalent to minimizing the following least-squares function L for n observations:

$$L = \sum_{i=1}^{n} \left(y_i - \beta_0 - \sum_{j=1}^{k} \beta_j x_{ji} \right)^2 \tag{4.5}$$

In matrix notation, this procedure leads to the following solution for the calculation of $\widehat{\boldsymbol{\beta}}$:

$$\widehat{\boldsymbol{\beta}} = \left(\mathbf{X}^{\mathrm{T}}\mathbf{X}\right)^{-1}\mathbf{X}^{\mathrm{T}}\mathbf{y} \tag{4.6}$$

where $\mathbf{X}$ is a matrix of size $n \times (k + 1)$, i.e., the so-called augmented design matrix, consisting of a column of ones and the levels of the independent variables according to the model form, and T and -1 superscripts indicate, respectively, transpose and inverse matrix. Formal mathematics regarding LS methodology are out of the scope of this text and can be found in the specific literature [1].

As it can be seen in Eqs. (4.1, 4.2, and 4.3), the more complex the model, the greater the number of parameters that must be estimated. In this sense, the total number of model parameters, according to the selected model complexity, should be balanced in relation to the amount of the experimental data available to ensure adequate statistical properties of the LS estimators. As previously discussed, the generation of balanced matrix designs is one of the main characteristics of the specific designs presented in the former chapters.

4.2.2 Model Inference, Validation, and Diagnosis

After model fitting, as usual in many statistical practices, model inference and validation are important tasks that should be done before using the model for interpretation and/or prediction steps. Regarding model inference, different hypothesis tests can be performed to measure the usefulness of the fitted model and the significance of the estimated parameters [2]. Firstly, the significance of regression can be evaluated by applying the following hypothesis test:

$$H_0 : \beta_1 = \beta_2 = \cdots = \beta_k = 0$$
$$H_1 : \beta_j \neq 0, \text{ for at least one } j \tag{4.7}$$

where H_0 and H_1 are the mean null and alternative hypotheses, respectively. Rejection of H_0 implies that at least one of the regressor variables contributes significantly to the model. Usually, a p-value approach through the ANOVA method is implemented for this test, because it involves partitioning SS_T into two contributions:

$$\mathrm{SS}_T = \mathrm{SS}_R + \mathrm{SS}_E \tag{4.8}$$

where the first term is the error due to the model (sum of squares explained by model) and the second, due to experimental error (residual sum of squares related to modeling error). It is possible to show that both $\frac{\mathrm{SS}_R}{\sigma^2}$ and $\frac{\mathrm{SS}_E}{\sigma^2}$ are distributed as χ^2 with k and $n - k - 1$ degrees of freedom, respectively. Thus, the statistic F_0 can be computed as:

$$F_0 = \frac{\mathrm{SS}_R/k}{\mathrm{SS}_E/(n-k-1)} \tag{4.9}$$

and H_0 is rejected if F_0 exceeds $F_{\alpha,\, k,\, n\, -\, k\, -\, 1}$ for a given level of significance α. In terms of p-values, H_0 is rejected when the associated p-value is less than α.

On the other hand, individual estimated coefficients can also be subjected to hypothesis tests, in order to analyze their marginal significance. In the context of RSM, this procedure is essential to decide which experimental factors and/or their interactions or polynomial terms are statistically significant for the response. In order to test any coefficient β_j, a generic hypothesis test is formulated as:

$$H_0 : \beta_j = 0$$
$$H_1 : \beta_j \neq 0, \text{for at least one } j \tag{4.10}$$

For this test, a t-Student statistic is computed as:

$$t_0 = \frac{\beta_j}{\sqrt{\sigma^2 C_{jj}}} \tag{4.11}$$

where C_{jj} is the diagonal element of $\mathbf{X}^T\mathbf{X}$ corresponding to β_j and σ^2 is the estimator of σ^2, calculated as:

$$\sigma^2 = \frac{SS_E}{n - p} \tag{4.12}$$

where p is the total number of model coefficients, i.e., $p = k + 1$. Thus, H_0 is rejected if $|t_0|$ is greater than $t_{\alpha/2,\ n-k-1}$. It is interesting to mention that, in RSM, a more general approach to test the significance of individual coefficients consists in building a so-called partial F-test, since it is useful to test subsets of r number of predictors. For this purpose, an F_0 statistic equivalent to Eq. (4.9) can be built. For instance, considering a single predictor x_j, if the value of the increase in SS_R due to adding x_j in a model that already includes $x_1, \ldots, x_{j-1}, x_{j+1}, \ldots, x_k$, i.e., $SS_R(\beta_j | \beta_0, \beta_1, \ldots, \beta_{j-1}, \beta_{j+1}, \ldots \beta_k)$, divided by r degrees of freedom, is included in the numerator of Eq. (4.9), then H_0 is rejected if $F_0 > F_{\alpha,\ r,\ n-p}$. For one predictor, the partial F-test is equivalent to the t-test represented by Eqs. (4.10, 4.11, and 4.12), as the square of a t random variable with ν degrees of freedom is an F random variable with 1 and ν degrees of freedom. Moreover, in both approaches, an associated p-value is often used to accept or reject H_0.

It is important to remark that these types of hypotheses tests indicate the relative or marginal significance of the factors, e.g., a given factor could be non-significant only when another is present. Moreover, to maintain the model consistency, if an interaction or polynomial term is significant, the main effect should also be included, even if they are non-significant. This rule is known as the hierarchy principle in MLR models [3].

The computed values of SS_R and SS_E are also useful to build another indicator related to the model usefulness, i.e., the coefficient of multiple determination R^2, which is defined as:

$$R^2 = \frac{SS_R}{SS_T} \tag{4.13}$$

This indicator reflects the amount of variability in the response that is explained by the independent variables kept in the model and varies between 0 and 1. It should be pointed out that adding new terms to the model always increases the value of R^2. However, a high level of this parameter does not necessarily mean that the model is satisfactory. If many non-significant variables are added to the model without any criteria, poor predictions are generally obtained. This situation is often referred to as model overfitting [3].

In order to generate a more realistic indicator, a modified version of R^2 has been introduced, which is called adjusted R^2, R^2_{adj}. This parameter penalizes the value of R^2 by the total number of variables p included in the model and is calculated as:

$$R^2_{\text{adj}} = 1 - \frac{n-1}{n-p}\left(1 - R^2\right) \tag{4.14}$$

The second important topic in this section corresponds to model validation. From the statistical point of view, this task implies examining the model adequacy and verifying that the three LS assumptions are fulfilled, i.e., independence of the observations, normality, and homoscedasticity of the residuals. Otherwise, it is very likely that the conclusions that arise from model analysis and predictions are biased or incorrect [2]. There exists a relatively extensive list of diagnostic strategies that can be performed for this purpose. Most important and useful graphical strategies for checking independence, normality, and homoscedasticity have been presented in Chap. 2, i.e., residual versus run, normal plot, and residual versus predicted values, respectively. These analysis tools are equivalent for both ANOVA and MLR models. Another useful graphical strategy to evaluate the model adequacy is to plot fitted vs nominal values. The better the predictions, the more the fitted vs nominal pairs will be aligned in a straight line of slope 1 and intercept 0 (identity line). In many cases, graphical methods can be accompanied by statistical ad hoc tests. For instance, the normality of the residuals can be checked by the Shapiro-Wilk test, whereas variance homogeneity might be analyzed through Bartlett's or Levene's test. All these methods are based on hypothesis tests, which can also be associated with a p-value approach and can be easily implemented in many statistical software.

Another important issue that needs to be thoroughly explored at this point is the presence of atypical values, i.e., potential outliers. On many occasions, outliers (which normally represent a very small subset of observations) can have a large influence on the estimation of the model parameters, and, in consequence, they can significantly affect the regression. This scenario is not desirable because it affects the model reliability, especially, when the model is intended to be used for predictions. Hence, the influence of such atypical observations on the model fitting should be addressed. For this purpose, several diagnostic tools have been developed. The most important strategies are briefly presented here. For further details, please refer to the specialized literature [4].

In the first place, a useful and straightforward way to identify potential outliers is by inspecting the model residuals. Usually, the residuals are first submitted to a scaling operation to standardize the values that arise from the LS procedure and facilitate the interpretation. For instance, the standardized residual for the i-th observation r_i is defined as:

$$r_i = \frac{e_i}{\widehat{\sigma}} \tag{4.15}$$

The complete set of standardized residuals has a mean of zero and approximately unit variance. This facilitates the visual inspection of potential outliers when standardized residuals are plotted against observation numbers. Other kinds of residual normalizations are also possible. In addition, the Bonferroni method constitutes a very well-known hypothesis test that can be carried out to statistically check the model residuals and detect atypical observations.

After the identification of potential outliers, it is very important to address the influence of these values on the model quality, i.e., not all outliers may have the same degree of influence on the regression. For this purpose, there are two statistical parameters of particular importance: leverage and Cook's distance. On the one hand, the leverage of a given observation h_i measures its distance to the mean value in the x-space, i.e., the space spanned by the regressor variables. Mathematically, taking into account Eqs. (4.4) and (4.6), it can be derived for the complete set of observations that the fitted values of a given model can be expressed as [5]:

$$\widehat{\mathbf{y}} = \mathbf{X}\widehat{\boldsymbol{\beta}} = \mathbf{X}\left(\mathbf{X}^{\mathrm{T}}\mathbf{X}\right)^{-1}\mathbf{X}^{\mathrm{T}}\mathbf{y} = \mathbf{H}\mathbf{y} \tag{4.16}$$

where $\mathbf{H}$ is known as a hat matrix, of size $n \times n$. The leverage value of the i-th observation is given by the diagonal elements of $\mathbf{H}$, i.e., $h_i = [\mathbf{H}]_{ii}$. For a one-predictor model, the leverage of the i-th observation can be written as:

$$h_i = \frac{1}{n} + \frac{(x_i - \overline{x})^2}{\sum_{k=1}^{n} (x_k - \overline{x})^2} \tag{4.17}$$

Since the denominator of the second term in Eq. (4.17) resembles the calculation of a variance, intuitively, the leverage can also be conceived as the contribution of a given observation to the variance in the x-space. Moreover, it can be directly deduced from Eq. (4.17) that the bigger the difference (distance) between the atypical observation and $\overline{x}$, the larger the leverage of x_i. In general, observations with high leverages need to be individually analyzed, since they could dramatically affect the estimated parameters of the regression. However, a sample with a high leverage might not be necessarily considered as an influential point. For this purpose, an ad hoc statistic to measure the influence of a given observation on the estimation of model parameters has been proposed, known as Cook's distance. Cook's distance for the i-th observation $D_{C,\,i}$ is a measure of the squared distance between the LS solution for $\widehat{\boldsymbol{\beta}}$ calculated with the complete set of observations and the solution obtained by eliminating i, $\widehat{\boldsymbol{\beta}}(i)$ [4]. One way of calculating this quantity is as:

$$D_{C,i} = \frac{(\beta(\mathrm{i}) - \beta)^{\mathrm{T}}\mathbf{X}^{\mathrm{T}}\mathbf{X}(\beta(\mathrm{i}) - \beta)}{p\sigma\ 2} \tag{4.18}$$

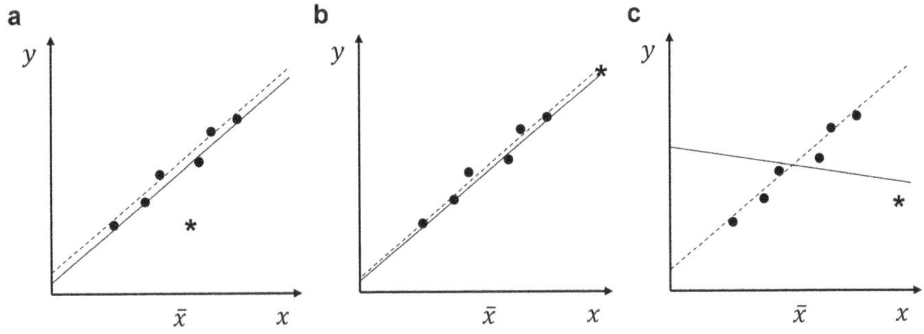

Fig. 4.1 Outlier diagnosis. Differences between leverage and influence in regression. Dots represent (x, y) observations in a one-predictor regression problem. Asterisks represent atypical runs. The lines represent a fitted model $y = \beta_0 + \beta_1 x_1 + \varepsilon$ built with the complete set of observations (solid) and after removing the atypical value (dashed). (**a**) The atypical observation does not significantly influence the regression model (low $D_{C,\,i}$ because it has a high residual but a low leverage, since it is close to $\bar{x}$). (**b**) The atypical observation does not significantly influence the regression model (low $D_{C,\,i}$ because it presents a high leverage, since it is not close to $\bar{x}$, but a low residual). (**c**) The atypical observation does significantly influence the regression model (high $D_{C,\,i}$, because it has a high residual and a high leverage, since it is not close to $\bar{x}$)

Observations with large $D_{C,\,i}$ have a high influence on the estimation of $\widehat{\boldsymbol{\beta}}$. The rule of thumb indicates that points of $D_{C,\,i}$ greater than 1 are considered influential.

The interpretation of leverage and influence can become abstract in a multivariate context. For this reason, a generic univariate regression example has been considered to illustrate the difference between these two concepts for outlier diagnosis (Fig. 4.1). Fig. 4.1.A to C displays cases of dispersion plots built with different (x, y) pairs, for the same one-predictor regression problem, i.e., $y = \beta_0 + \beta_1 x_1 + \varepsilon$. In each sub-figure, the asterisk is an atypical observation. Besides, the solid line represents the fitted model with all observations, while the dashed line is the fitted model without the atypical point. As it can be seen, the atypical value in Fig. 4.1a presents a high residual value (considering the vertical distance to the fitted lines). However, its leverage is not high because it is near $\bar{x}$ in the horizontal dimension. Low leverage observations are not considered influential for the regression model (low $D_{C,\,i}$). Conversely, in Fig. 4.1b, the atypical point presents a high leverage, because the distance to $\bar{x}$ in the horizontal axis is high. However, as this point is very close to the fitted model, the observation is also not influential (low $D_{C,\,i}$). Finally, Fig. 4.1c presents an example where an atypical observation has high leverage and a high difference between the nominal and predicted value (high $D_{C,\,i}$). When both things occur, the observation exerts a significant influence on the regression line, as can be evidenced due to the great disparity between the two fitted lines.

Although there is no strict relation, the elimination of outlier in some cases may improve the quality of the model, e.g., true outliers may affect the normality and/or the variance homogeneity of the residuals. However, the analyst should investigate outliers very carefully before removing them to avoid biasing the model.

After outlier exclusion, deviations from MLR assumptions can still be detected through graphical and statistical diagnostic tools. On some occasions, considering more complex models can improve the quality of model assumptions. For this purpose, when true observation replicates are processed during the generation of experimental data, there is another useful hypothesis test to check model adequacy, known as the lack of fit test. The lack of fit (LOF) gives a statistical criterion to decide whether a model contains the correct regression parameters, i.e., usually when interaction and/or polynomial terms need to be included. Experimental designs with replicates permit calculating a model-independent estimation of error variance σ^2. The statistical basis of this test consists in partitioning the term SS_E from Eq. (4.8) into two contributions, as follows:

$$SS_E = SS_{PE} + SS_{LOF} = \sum_{i=1}^{m} \sum_{j=1}^{n_i} (y_{ij} - y_i)^2 + \sum_{i=1}^{m} n_i (\bar{y}_i - y_i)^2 \tag{4.19}$$

where SS_{PE} is the pure error sum of squares and SS_{LOF} is the sum of squares due to lack of fit; n_i represents the number of replicates at the i-th level of predictors $\mathbf{x}_i$, with $i = 1, \ldots, m$; and y_{ij} is the j-th measured response at $\mathbf{x}_i$. Under the assumption of constant variance, the first term represents the pure error since it only depends on the values of y at each level of $\mathbf{x}_i$. The second term is a weighted sum of squared deviations between the mean response and the corresponding fitted value, at each $\mathbf{x}_i$. Usually, the term for SS_{LOF} is calculated as the difference between SS_E and SS_{PE}. The formulation for this hypothesis test is:

H_0: a simple (reduced) model fits the data,
H_1: a simple model does not fit the data and a more complex model is needed, and the test statistic is: (4.20)

$$F_0 = \frac{SS_{LOF}/(m-p)}{SS_{PE}/n-m} \tag{4.21}$$

The null hypothesis will be rejected when F_0 is greater than $F_{\alpha,\, m-p,\, n-m}$. Normally, the lack of fit test is coupled to regression ANOVA.

Furthermore, another useful strategy to improve model adequacy is to apply a transformation of the response variable. The main purpose of this procedure is to stabilize the response variance, so that this variable is closer to a normal distribution. In principle, many mathematical operations could be considered. However, power transformation of the form $y^* = y^\lambda$, where y^* denotes the transformed response variable and λ is a non-zero transformation parameter to be determined, has shown to be an adequate option. In this context, Box and Cox proposed a methodology for the systematic selection of λ parameter in a given dataset, based on a maximum likelihood approach [6]. Briefly, it consists of computing the SS_E of the regression on y^* for different values of λ. Most software packages offer a straightforward implementation of this technique. Usually, the values of SS_E (or log SSE) are plotted

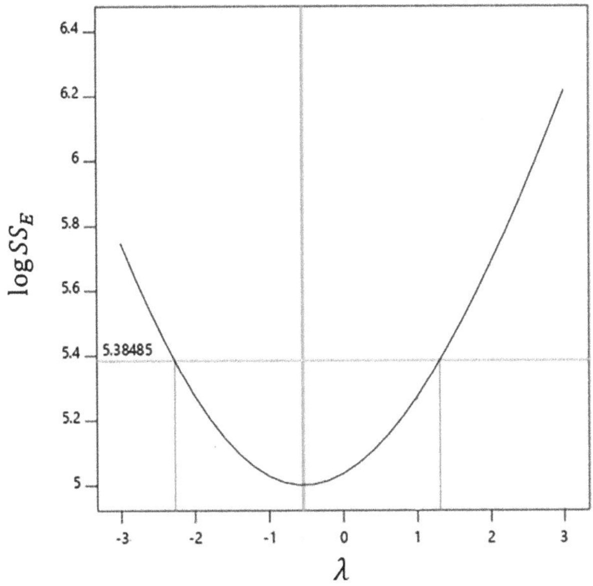

Fig. 4.2 Graphical output from Box-Cox methodology for response power transformation $y^* = y^\lambda$. $\log SS_E$ as a function of the transformation parameter λ. The green line indicates the best λ value, while the red lines show the boundaries of a 95% confidence interval for λ

against λ to graphically identify the value of λ that minimizes SS_E. In addition, a 100 $(1-\alpha)\%$ confidence interval for λ can be calculated as $SS_E\left(1+\frac{t^2_{\alpha/2,\upsilon}}{\upsilon}\right)$, for υ degrees of freedom. If the interval includes the value of 1, this implies that no specific transformation is necessary. A generic example for the graphic inspection of λ values through the Box-Cox method is depicted in Fig. 4.2. The green line indicates the best λ value, while the red lines show the boundaries of a 95% confidence interval for λ.

It is also interesting to point out that, in general, simple values of λ are chosen as candidates in order to obtain a more interpretable transformation, e.g., 0.5 corresponds to the root square of y. Besides, the Box-Cox method also considers logarithmic transformation that corresponds to the special case of $\lambda = 0$.

4.2.3 *Model Interpretation and Prediction*

After model validation, the last step before proceeding to find the optimum experimental conditions consists in interpreting the model, i.e., exploring the relations between the variables, and making statistical predictions. As it can be evidenced, the more effort the analyst makes during the validation step, the more robust and reliable will be the resulting model.

One of the greatest advantages of MLR models is that they are easily interpretable. In this sense, the final model formulation can be very beneficial to unveil non-evident associations between the experimental factors with each other and with the response of interest. It is important to remark that statistical models are not intended to provide causal or mechanistic equations that explain these associations. However, they can be very useful in interpreting how y varies at different combinations of the regressors. The relative magnitude for which a given factor influences the response variable is directly associated with the value of the corresponding regression coefficient. In this context, the interpretation of the model can be facilitated through graphical tools.

A graphical representation of a model of the type of Eqs. (4.1, 4.2, and 4.3) is obtained when the response variable is plotted against the regressor variables. For a two-factor model, a three-dimensional surface is generated, i.e., y vs x_1 and x_2. Generic examples for different model complexities are shown in Fig. 4.3. When the response variable depends linearly on the experimental factors (first-order model), the 3D representation corresponds to a plane (Fig. 4.3a1). Conversely, if first-order interactions are significant, a more complex landscape is generated, as can be appreciated in Fig. 4.3b1 (first-order with low-order interactions). In addition, if second-order polynomial terms are included (second-order model), the representation looks like a paraboloid surface (Fig. 4.3c1). Another useful graphic is the

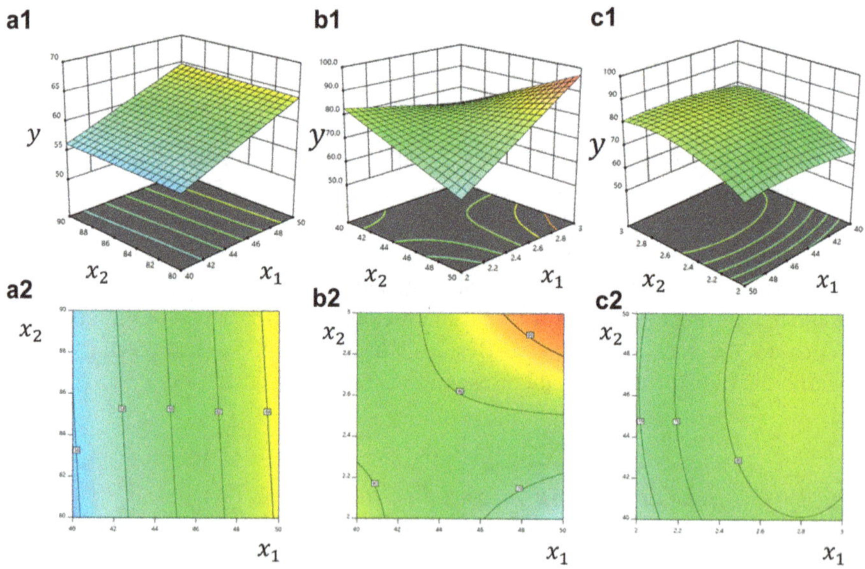

Fig. 4.3 Graphical representations of fitted two-factor models of the type of Eq. (4.1), corresponding to different complexity: (**a**) first-order model (only main effects x_1 and x_2 are significant for y); (**b**) first-order model with low-level interactions (first-order interaction x_1x_2 is significant for y); and (**c**) second-order model (second-order polynomial terms x_1^2 and x_2^2 are also significant for y)

so-called contour plot, i.e., constant y slices are plotted in a 2D format. Figure 4.3a2–c2 shows the contour plots corresponding to a linear, linear with first-order interactions, and second-order model, respectively.

On the other hand, when more than two factors are optimized, it is not possible to generate a complete representation of the model. Usually, in those cases, the response variable is plotted against a given pair of factors, keeping the others at constant values. Hence, several marginal representations can be generated through this strategy.

Another important issue in search of an experimental optimum is related to model predictions. A suitable statistical model should be capable of estimating the value of the response variable at specified values of the regressors. Mathematically, in the case of MLR models, a point prediction y_0 is obtained as:

$$y_0 = \widehat{y}(\mathbf{x}^*) = \mathbf{x}^* \widehat{\boldsymbol{\beta}} \tag{4.22}$$

where $\mathbf{x}^*$ represents a $k + 1$ vector containing particular levels of the experimental factors. Ideally, point predictions should be accompanied by the corresponding confidence interval. Therefore, a $100(1 - \alpha)\%$ prediction interval for y_0 is given by:

$$\begin{aligned} \widehat{y}(\mathbf{x}^*) - t_{\alpha/2,n-p}\sqrt{\widehat{\sigma}^2\left(1 + \mathbf{x}^*\left(\mathbf{X}^{\mathrm{T}}\mathbf{X}\right)^{-1}\mathbf{x}^*\right)} \leq y_0 \leq \widehat{y}(\mathbf{x}^*) \\ + t_{\alpha/2,n-p}\sqrt{\widehat{\sigma}^2\left(1 + \mathbf{x}^*\left(\mathbf{X}^{\mathrm{T}}\mathbf{X}\right)^{-1}\mathbf{x}^*\right)} \end{aligned} \tag{4.23}$$

It is of the utmost importance that the values included in $\mathbf{x}^*$ are within the experimental domain. Statistical models in general are not intended to make predictions by extrapolating beyond the region containing the observations that were used to fit the model.

4.2.4 *Mixture Experiments: The Scheffé Model*

As described in the previous chapter, in mixture experiments, the factor levels are not independent of each other. Mathematically, this implies the constraint $x_1 + x_2 + \ldots + x_q = 1$, where q is the number of components in the mixture. For this reason, the data that arise from mixture designs cannot be modelled through classical MLR. Instead, Scheffé has proposed modified versions of the polynomial expressions presented above [7]. For instance, considering Eq. (4.1), if the term β_0 is multiplied by $x_1 + x_2 + \ldots + x_q$, the following expression is obtained:

$$y = \beta_0\left(x_1 + x_2 + \ldots + x_q\right) + \sum_{j=1}^{q} \beta_j x_j = \sum_{j=1}^{q} \beta_j^* x_j \tag{4.24}$$

where $\beta_j^* = \beta_0 + \beta_j$. Equation (4.24) is known as canonical or Scheffé polynomial for a first-order mixture model. Higher-order canonical model has also been proposed, i.e., quadratic,

$$y = \sum_{i=1}^{q} \beta_i x_i + \sum \sum_{i<j=2}^{q} \beta_{ij} x_i x_j \tag{4.25}$$

full cubic,

$$y = \sum_{i=1}^{q} \beta_i x_i + \sum \sum_{i<j=2}^{q} \beta_{ij} x_i x_j + \sum \sum_{i<j=2}^{q} \delta_{ij} x_i x_j (x_i - x_j) + \sum \sum \sum_{i<j<k=3}^{q} \beta_{ijk} x_i x_j x_k \tag{4.26}$$

and special cubic,

$$y = \sum_{i=1}^{q} \beta_i x_i + \sum \sum_{i<j=2}^{q} \beta_{ij} x_i x_j + \sum \sum \sum_{i<j<k=3}^{q} \beta_{ijk} x_i x_j x_k \tag{4.27}$$

In Eq. (4.24), the coefficient β_j^* and, in Eqs. (4.25, 4.26, and 4.27), β_i represent the expected response for the pure mixture, i.e., $x_i = 1$ and $x_j = 0$, $i \neq j$. Besides, the first term is called the linear blending portion. When blending is additive, this implies that there are no interactions between the components and Eq. (4.24) is an appropriate model. The quadratic term $\sum \sum_{i<j=2}^{q} \beta_{ij} x_i x_j$ represents the excess response due to interaction between two components, often referred to as synergism or antagonism. Figure 4.4 shows a typical response surface for a quadratic mixture model with three factors.

On the other hand, in combined problems like process-mixture and mixture-mixture, two models are taken into account to generate a unique matrix design. Then, a combined model can be fitted from the experimental data. Mathematically, for two groups of components x and z, this can be expressed as:

$$y = f(x) \times g(z) + \varepsilon \tag{4.28}$$

where f and g represent two polynomial models. For instance, in a process-mixture problem, considering a quadratic model in the mixture components and a model with both main and 2-factor interaction in the process components, the 24-parameter combined model results in:

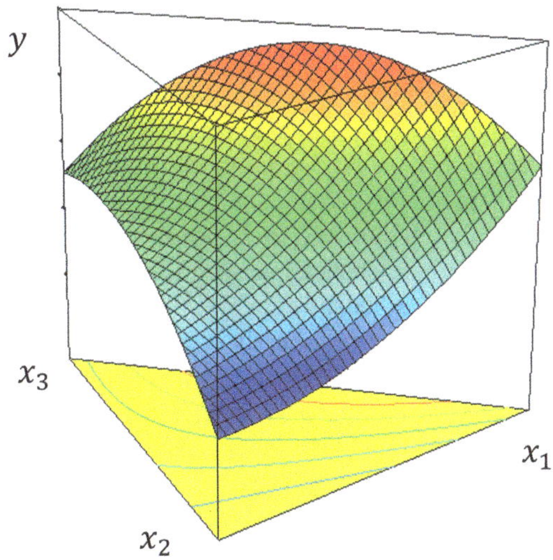

Fig. 4.4 Graphical representation of a response surface obtained with a quadratic three-factor model (Eq. 4.25)

$$\begin{aligned} y= & \sum_{i=1}^{3} \beta_i x_i + \sum\nolimits_{i<j=2}^{2} \sum_{i<j=2}^{3} \beta_{ij} x_i x_j \\ & + \sum\nolimits_{k=1}^{2} \left[\sum_{i=1}^{3} \alpha_{ik} x_i + \sum \sum_{i<j=2}^{3} \alpha_{ijk} x_i x_j \right] z_k \\ & + \left[\sum_{i=1}^{3} \delta_{i12} x_i + \sum \sum_{i<j=2}^{3} \delta_{ij12} x_i x_j \right] z_1 z_2 \end{aligned} \tag{4.29}$$

In a mixture-mixture design, if two quadratic Scheffé models are considered, the combined model is:

$$\begin{aligned} y= & \sum_{i} \sum_{j} \beta_i^j x_i z_j + \sum_{i} \sum_{j<l} \beta_i^{jl} x_i z_j z_l + \sum_{i<k} \sum_{j} \beta_{ik}^{j} x_i z_k z_j + \sum_{i<k} \\ & \times \sum_{j<l} \beta_{ik}^{jl} x_i x_k z_j z_l \end{aligned} \tag{4.30}$$

where $i, k = 1, \ldots I, i < k$ and $j, l = 1, \ldots, J, j < l$. The parameters β are the result of the product of individual quadratic models. The combined model has a total number of 36 parameters.

After model fitting, all the previously described steps for model inference, validation, and prediction are equivalent.

4.3 Artificial Neural Networks (ANNs)

From the computational point of view, ANNs are a group of algorithms of artificial intelligence that belong to the family of machine learning methods. The main characteristic of ANNs is that they are capable of learning a rule that correlates a set of predictor variables with a desired response from a dataset, without being expressly programmed for that purpose.

From the statistical perspective, ANNs can be conceived as non-parametric regression methods. Therefore, they do not assume any a priori model form, like Eq. (4.1). For this reason, these algorithms have been demonstrated to be suitable to handle non-linear problems or with datasets that lack certain properties or structures. As usual in non-parametric approaches, normality and homoscedasticity are not required assumptions, making ANNs much more flexible than MLR models. In contrast to parametric methods, ANNs have been less characterized and still maintain a certain "black box" nature. For this reason, they are less interpretable and tend to be more sensitive to overfitting. However, when properly trained, ANNs present very good prediction capacity, making them attractive for RSM purposes, in particular, for those optimization problems where the dataset does not conform to the assumptions required by MLR models.

From the beginning of the 1990s, a vast diversity of ANN algorithms has been developed. In particular, two types of feed-forward multilayer networks that have been extensively reported in the literature are the so-called multilayer perceptron (MLP) and radial basis function multilayer networks (RBF). Both types of ANNs have proven to be suitable for solving non-linear regression problems [8].

Although each type of network has its own particularities, they share the same philosophy. They consist in projecting a set of input variables (predictors) onto a non-linear function and adjusting their free parameters automatically through an iterative procedure of error minimization. The operational unit of these types of ANNs is the so-called node or neuron. The sets of neurons are generally disposed of in three layers, i.e., input, hidden, and output layers. The number of nodes per layer constitutes the so-called network architecture or topology, and it is in direct relation to the total number of adjustable parameters of the model.

Input and output layers handle predictor and response variables, respectively, whereas the hidden layer represents the core of the algorithm. The training proceeds iteratively, one array $(\mathbf{x}_i, y_i)$ at a time in a hierarchical manner, from input to output layers. In general, two mathematical operations can take place in a given neuron. First, the input variables are weighted by a set of free parameters, known as synaptic weights. Second, the resulting linear combination is projected onto a function called transfer or activation function. Mathematically, the output O of the j-th neuron that projects m input variables s from the i-th neuron in the layer above can be expressed as:

$$O_j = f\left[\sum_{i=1}^{m} s_i w_{ij} + w_{bj}\right] \tag{4.31}$$

where w_{ij} represents the connection or synaptic weight between the neurons i and j and w_{bj} represents the bias parameter added to the j-th neuron. The latter parameter is included to increase the model flexibility. It is important to remark that there might exist some divergence among the specific literature, regarding the nomenclature of ANN layers. Generally, the activation function in the output layer is linear, whereas neurons in the hidden layer project the weighted sum onto non-linear functions. In particular, MLP uses sigmoidal functions, e.g., logistic or hyperbolic tangent, while RBF, as their name suggests, are characterized by radial basis functions, e.g., Gaussian-type functions. In the output layer, the re-scaled image of the transfer function yields the network prediction $\widehat{y}$.

The complete dataset corresponding to experimental factors and measured responses constitute the so-called training set. During training, all observations are forward propagated through the network, and each estimate y_i is compared with its nominal value y_i to obtain a measurement of the global prediction error. This quantity is then backward propagated to automatically adjust the synaptic weights by following an error gradient-based procedure, i.e., a learning algorithm. The forward and back propagation of all samples constitutes an iteration of the algorithm, also referred to as epoch. The whole procedure is then repeated iteratively until a pre-established convergence error criterion is achieved.

As aforementioned, in contrast to parametric models, ANNs are more sensitive to overfitting. To avoid this, a common strategy is to split the training dataset into two subsets. The first subset (training set) is used to train the network. The remaining samples (also known as the monitoring set) are used to make predictions at the end of each epoch in order to calculate the statistic root mean square error of predictions in the monitoring set (RMSEM), according to:

$$\text{RMSEM} = \sqrt{\frac{1}{n_{\text{m}}}\sum_{i=1}^{n_{\text{m}}} (y_i - y_i)^2} \tag{4.32}$$

where n_{m} is the number of observations in the monitoring set. Thus, reaching a minimum RMSEM constitutes a common strategy to stop the training algorithm.

Another distinct feature of ANNs is that the architecture must be optimized to obtain a model with a satisfactory prediction ability. In general, the optimum network topology, related to the total number of free parameters, is not known a priori. For this purpose, diverse strategies have been proposed, and, in this sense, many software packages present diverse automatic optimization strategies [9]. After architecture optimization and training, the network is ready to be used to make predictions of the response and, therefore, to proceed to the step of experiment optimization.

4.4 Optimization

The last step of RSM consists of experiment optimization. During experiment optimization, the experimental factors are further examined to determine optimal conditions for the response. The conception of optimum will vary according to the problem and the system under study. In most cases, the purpose is to maximize or minimize the response of interest. In addition, the necessity of an optimum response can be associated with a specific target value or, in a more flexible context, with a region of interest. Likewise, the factors can also be included in the optimization criteria, by restricting the search to specific regions of the experimental domain.

At this stage, it is expected that a satisfactory empirical model for each response of interest has been built and properly validated. Indeed, RSM owes its denomination to the way in which the experimental optimum is located, where the analyst "moves along" the function y of several variables, i.e., the response surface. Several mathematical and graphical strategies for the optimization of single and multiple responses have been proposed, which are briefly described in this section.

As mentioned in Sect. 4.2.3, in the case of parametric models (MLR), it is possible to obtain a graphical representation. If only two factors are involved, a 3D landscape (or 2D contour plot) will provide a complete description of the problem. Therefore, the optimum experimental conditions can be directly derived by visual inspection. In this regard, some of the most common situations can be appreciated in Fig. 4.5a-c. In particular, when the response surface is of the type of Fig. 4.5a, in a maximization problem, the location of the optimum region is evident, since the surface clearly exhibits a maximum point. Equivalently, the function displayed in Fig. 4.5b shows a point of minimum response. In both cases, extreme points are known as "stationary points." Conversely, in Fig. 4.5c, there is not a region of maximum or minimum response, and the center of the surface is called "saddle point." As evidenced, the geometry of the model will play a critical role in the location of the optimum.

On the other hand, when more than two factors are considered, 3D or 2D plots can still be obtained, but examining only two factors at a time and keeping the others at fixed values. Clearly, this represents only a limited part of the entire response surface in the experimental domain. For the particular case of three factors, a graphical strategy for the examination of the optimum consists in overlaying different contour plots, obtained at different combinations of factors, properly chosen. Through the superposition of these graphs, a compromise region that satisfies the response requirements can be visually determined. However, this strategy is limited when more than three factors are included in the model since the localization of the optimum from the superposition of several contour plots becomes a tricky task.

More formal and general approaches have been proposed for k-dimensional problems, based on traditional mathematical optimization methods. In calculus, maximum, minimum, and saddle points are generally referred to as "critical points" of a function. If the considered function depends on j independent variables, then a critical point is defined as a point where the partial derivatives are equal to zero or, at

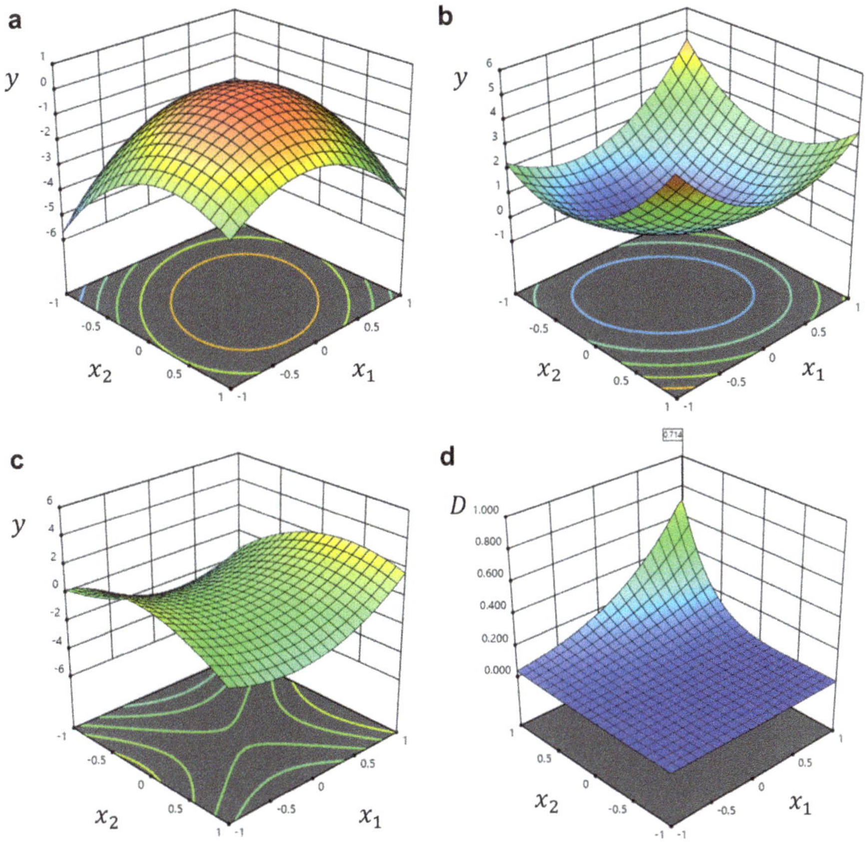

Fig. 4.5 (**a**) A second-order response surface with a maximum point. (**b**) A second-order response surface with a minimum point. (**c**) A second-order response surface with a saddle point. (**d**) Graphical representation of a global desirability D as a function of two experimental factors

least, one partial derivative does not exist. The k-vector that collects the partial derivatives of a function $y = f(x_1, x_2 \ldots, x_k)$, known as gradient vector ∇y, is defined as:

$$\nabla y = \left(\frac{\partial y}{\partial x_1}, \frac{\partial y}{\partial x_2}, \ldots, \frac{\partial y}{\partial x_k} \right) \quad (4.33)$$

Therefore, the philosophy of optimization methods consists in differentiating the adjusted model with respect to each independent variable and then finding the values of $\mathbf{x} = x_1, x_2 \ldots, x_k$ for which:

$$\nabla y = \mathbf{0} \quad (4.34)$$

The numerical method par excellence to solve Eq. (4.34) is the so-called gradient steepest ascent, for maximization problems, or, equivalently, gradient steepest descent when the goal is to find minimum points [2]. The main property of ∇y is that at any point $\mathbf{x}$ represents a direction of maximum increase of y. Thus, in a gradient-based maximization method, ∇y is evaluated at a given point $\mathbf{x}_t$, so that the values of the independent variables in the next iteration $\mathbf{x}_{t+1}$ are adjusted in the same direction of ∇y (or in the opposite direction for minimization), according to:

$$\mathbf{x}_{t+1} = \mathbf{x}_t + \eta \nabla y(\mathbf{x}_t) \tag{4.35}$$

where η is a small enough scalar. This procedure is iteratively repeated until convergence. It is important to remark that in RSM problems, $\mathbf{x}$ is bounded to the experimental domain. In addition, it is relevant to mention that in many situations, the solution of Eq. (4.34) is not unique, e.g., functions that exhibit saddle points. In such cases, the analyst will not find an optimal point but experimental regions for which the response reaches a desirable value. The selection of one particular region is generally based on experimental and/or economic criteria.

4.4.1 Multiple Response Optimization: The Desirability Function

On several occasions, the RSM problem involves the optimization of several responses. Therefore, it is necessary to find an overall compromise solution within the experimental domain that meets the optimization criteria for all the responses, simultaneously. In this regard, Derringer and Suich proposed in 1980 [10] an intuitive and straightforward method for multiple response optimization that became very popular. This approach is based on the so-called desirability function and constitutes the offered strategy in most software packages.

The desirability function is based on the idea that the quality of an experimental system that holds several responses is "undesirable" if one of them lies outside an acceptable boundary. Therefore, the function is used to seek working conditions to guarantee consistency among the criteria of all responses, providing the best compromise values for them.

The implementation of this multiple response approach consists of the following steps. Considering a problem of I responses of interest, firstly, an individual or partial desirability function d_i is computed for each response, taking the predicted responses for all combinations of the factors in the experimental domain, i.e., $y_i(\mathbf{x})$. The formulation of the desirability function will depend on the optimization criteria (maximization, minimization, target, or range of values). For instance, if the goal is to maximize the i-th response,

$$d_i(\hat{y}_i(\mathbf{x})) = \begin{cases} 0 & \hat{y}_i(\mathbf{x}) < L_i \\ \left(\frac{\hat{y}_i(\mathbf{x}) - L_i}{U_i - L_i}\right)^s & L_i \leq \hat{y}_i(\mathbf{x}) \leq U_i \\ 1 & \hat{y}_i(\mathbf{x}) > U_i \end{cases} \quad (4.36)$$

where the value of d_i will depend on the value of the response that is bounded to an acceptable range delimited by lower and upper values (L_i and U_i, respectively). The exponent s is a weight parameter that establishes how important it is for the response to reach the maximum. This parameter allows obtaining more restrictive solutions.

On the other hand, if a given response is to be kept at a target value T_i, the partial desirability function is defined as:

$$d_i(\hat{y}_i(\mathbf{x})) = \begin{cases} 0 & \hat{y}_i(\mathbf{x}) < L_i \\ \left(\frac{\hat{y}_i(\mathbf{x}) - L_i}{T_i - L_i}\right)^{s_1} & L_i \leq \hat{y}_i(\mathbf{x}) \leq T_i \\ \left(\frac{U_i - \hat{y}_i(\mathbf{x})}{U_i - T_i}\right)^{s_2} & T_i \leq \hat{y}_i(\mathbf{x}) \leq U_i \\ 0 & \hat{y}_i(\mathbf{x}) > U_i \end{cases} \quad (4.37)$$

where the exponents s_1 and s_2 are weight parameters.

As it can be noticed in Eqs. (4.36) and (4.37), partial desirability functions always vary between 0 and 1. This means that, if $d_i(y_i(\mathbf{x})) = 0$, then the response is unacceptable for the working conditions established by $\mathbf{x}$. On the contrary, if $d_i(y_i(\mathbf{x})) = 1$, the response is totally desirable. Intermediate values of $d_i(y_i(\mathbf{x}))$ indicate more or less desirable responses. It is interesting that desirability functions can also be calculated for experimental factors, allowing them to be incorporated into the optimization procedure.

Once individual desirability functions have been computed, they are combined in a global desirability function (D), which allows finding the combination of responses that best satisfies all the optimization criteria simultaneously. Global desirability function is defined as:

$$D = \left(d_1^{r_1} d_2^{r_2} \ldots d_I^{r_I}\right)^{\frac{1}{\sum r_i}} = \left(\prod_{i=1}^{I} d_i^{r_i}\right)^{\frac{1}{\sum r_i}} \quad (4.38)$$

where the exponents r_i indicate the relative importance attributed to each response. In particular, these parameters are useful in problems where the analyst is particularly interested in one of the responses or when a specific response is more critical than the others. Equation (4.38) implies that if D takes any value different from zero, all the responses are assumed to have desirable values. On the contrary, if $d_i(y_i(\mathbf{x})) = 0$ for at least one response, then the global desirability will also be zero. It is important to remark that the purpose of this optimization methodology is to find solutions that meet the requirements for the whole set of responses. This occurs when D is different from zero. Therefore, it is not mandatory to obtain solutions for which D is near 1. Any solution different from zero constitutes a desirable solution.

The global desirability function can also be graphically represented. In this sense, most software packages usually plot the values of D as a function of the experimental factors. As is the case for response surfaces, if more than two factors are involved in the problem, partial D surfaces can be generated by varying a pair of factors and keeping the others at fixed values. A generic example of a particular solution for D function in a two-factor problem is depicted in Fig. 4.5d. As observed, the desirability is maximum when both factors are kept at their upper levels in the experimental domain.

Finally, it is interesting to mention that RSM always offers the possibility for the analyst to perform new experiments to confirm or validate the predictions that arise from different models.

4.5 Software

The use of RSM is becoming more and more popular among the scientific community and industry. For this reason, it is not surprising that numerous computational tools have proliferated for the implementation of DOE and optimization methods.

On the one hand, there exists an interesting list of different software packages, provided by different companies. Some of the most traditional and well-known software are Design-Expert (http://www.statease.com/), Minitab (http://www.minitab.com/), SPSS (www.ibm.com/analytics/us/en/technology/spss/), JMP (https://www.jmp.com/), Six Sigma (https://www.isixsigma.com/six-sigma-software/), The Unscrambler (http://www.camo.com/), NCSS (http://www.ncss.com/), Cornerstone (www.camline.com/), and SAS (https://support.sas.com/rnd/app/qc/qc/qcdesign.html). MATLAB platform also offers some built-in functions for DOE. In particular, a free-access graphical user interface (GUI) written in MATLAB for optimization through ANN modeling has been published by Giordano et al. in 2017 [11].

On the other hand, it is relevant to mention that, more recently, numerous specific packages for the implementation of RMS in the open-source languages R (https://www.r-project.org/), such as *doe.base*, *frf2*, *rsm*, *mixexp*, and *desirability*, and Python (https://www.python.org/), such as *pyDOE* and *dexpy*, have been developed and are freely available.

References

1. Walpole RE, Myers RH, Myers SL, Ye K (2012) Probability and statistics for engineers and scientists, 9th edn. Pearson, Boston
2. Myers RH, Montgomery DC (2009) Response surface methodology: process and product optimization using designed experiments. Wiley, New York
3. James G, Witten D, Hastie T, Tibshirani R (2013) An introduction to statistical learning: with applications in R. Springer, New York
4. Cook RD, Weisberg S (1982) Residuals and influence in regression. Chapman and Hall, New York
5. Christensen R, Pearson LM, Johnson W (1992) Case-deletion diagnostics for mixed models. Technometrics 34:38–45
6. Miller D (1984) Reducing transformation bias in curve fitting. Am Stat 38:124–126
7. Scheffé H (1958) Experiments with mixtures. J R Stat Soc B 20:344–366
8. Haykin S (2009) Neural networks and learning machines, 3rd edn. Pearson, New York
9. Despagne F, Massart DL (1998) Neural networks in multivariate calibration. Analyst 123: 157R–178R
10. Derringer G, Suich R (1980) Simultaneous optimization of several response variables. J Quality Technol 12:214–219
11. Giordano PC, Goicoechea HC, Olivieri AC (2017) SRO_ANN: an integrated MatLab toolbox for multiple surface response optimization using radial basis functions. Chemom Intell Laborat Syst 171:198–206

Chapter 5
Quality by Design in Pharmaceutical Product and Process Development

Marília Medeiros and Márcia C. Breitkreitz

Abstract Chapter 1 provided an overview of the QbD strategy, including its goals and main elements as well as its relation to the lifecycle approach. This chapter will mainly focus on practical aspects of QbD for product and process pharmaceutical development. Important aspects related to raw materials will be discussed based on the most recent IPEC (International Pharmaceutical Excipients Council) guideline. Then we discuss how to structure DOE studies in order to uncover possible interactions among materials, formulation, and process variables. A practical guide of the input process variables, material and formulation attributes is provided, based on the most representative manufacturing processes as well as representative examples from literature. In the final section, examples of intermediate and final quality attibutes for the development of different dosages forms (solid, liquid, and topical) are presented.

Keywords Pharmaceutical development · Quality by design · Product optimization · Process optimization · Quality attributes

Pharmaceutical development is a complex and fascinating activity, controlled by multi-factorial relationships between materials, formulation, and process parameters. Pharmaceutical development can be carried out either by the traditional approach or by the enhanced QbD-based approach, the latter being highly recommended in the current quality scenario. As mentioned in Chap. 1, the QbD approach starts with the elaboration of the quality target product profile, followed by the identification of global critical quality attributes (CQAs) for that specific pharmaceutical dosage form. This is followed by the risk analysis of input materials, formulation, and

M. Medeiros
Ache Laboratórios Farmacêuticos, Guarulhos, SP, Brazil

M. C. Breitkreitz (✉)
Laboratory of Pharmaceutical Research and Chemometrics (LABFARQUI),
Institute of Chemistry, University of Campinas (UNICAMP), Campinas, SP, Brazil
e-mail: marciacb@unicamp.br

M. C. Breitkreitz, H. C. Goicoechea (eds.), *Introduction to Quality by Design in Pharmaceutical Manufacturing and Analytical Development*, AAPS Introductions in the Pharmaceutical Sciences 10, https://doi.org/10.1007/978-3-031-31505-3_5

process parameters and risk evaluation to identify critical material attributes (CMAs) and process parameters (CPPs) affecting the global and intermediate CQAs, if applicable. Then design spaces can be proposed as well as control strategies. A natural thought is how to structure studies in order to gather important information in a rational framework? We will try to bring some contributions about this matter in this chapter.

5.1 Raw Materials: A Quality by Design Point of View

It is reasonable to say that the entire pharmaceutical development starts with raw materials that are placed together in a given proportion (formulation) by a given manufacturing process. Raw material characteristics will impact the performance of the product in many aspects; therefore, it is of upmost importance to look at raw material attributes, and this has been highlighted within the QbD framework. The importance of studying the physicochemical properties of active pharmaceutical ingredients is well established. Fortunately, the role of excipients has also been recognized in the past years. In 2020, the International Pharmaceutical Excipients Council (IPEC) has launched the guideline "Incorporation of Pharmaceutical Excipients into Product Development using Quality by Design (QbD)," which was a very important update for suppliers and developers.

IPEC is an international association of regulatory authorities, excipient manufacturers/distributors, and end users (mostly pharma companies). There are regional pharmaceutical excipient industry associations located in the Americas, Europe, Japan, China, and India. IPEC's objective is to contribute to international excipient standard developments and harmonization, new excipient development and introduction, and best practice and guide development concerning excipients [1].

This guideline brought important discussions of excipients within QbD framework, from which we can highlight:

- CMAs should be defined ***according to the formulation and manufacturing process and not be considered as a general truth (regardless of the formulation and manufacturing process)***. For example, the flowability of a diluent that is a major component of a formulation in a direct compression manufacturing process is definitely a CMA. On the other hand, if the excipient is a minor component or the process involves its solubilization, it might not be considered a CMA.
- It is important to specify the ***grade of the excipients***; chemical name is not sufficient. Different specifications from the same supplier can have different impact on the product CQAs. The same holds true for the same excipient from different suppliers.
- ***Pharmacopeia compliance is not enough to justify the use of a given excipient***. The pharmacopoeia monographs provide a general guidance and cannot be solely used to justify the use of a given excipient in a specific product without evaluating its impact on the product/process CQAs. It is clearly stated in the IPEC guideline that *"Within*

this guide, a CMA is an excipient physical, chemical, or microbiological attribute, not necessarily reflected *in supplier specifications or monographs."*

- ***Variability of excipients*:** Excipients are inherently variable as they are not controlled at an individual parameter level. According to IPEC guide, it is expected that "both excipient *performance* and excipient *variability* impact on the CQAs should be a DoE output." The natural variability of excipients can be incorporated in the design spaces calculations by propagation of error (POE) [2] or Monte Carlo simulation [3] tools. For example, imagine one wants to study the influence of raw material particle size (x variable, measured by d_{90}) on the dissolution (y variable, the CQA). Suppose that the study includes the evaluation of the x variable at two levels (e.g., 50 μm and 150 μm) – which will be the input levels for this factor in the DOE study. However, these are average values, and there is a variation (Δx) associated with each of these levels. This variability can be taken into account by propagation of error or Monte Carlo simulations, after the establishment of regression models that relate the x and y variables. These approaches have been already incorporated in DOE commercial software and can also be used to incorporate variability of process variables. It is interesting to notice that the FDA simulated dossier brings the "excipient variability" in the process risk analysis [4] and attributed "low risk" for this input, since variability was well characterized and understood by the developers.

Another important tool to evaluate raw material variabilities within the QbD scenario is principal component analysis (PCA), often referred to as an MVA (multivariate analysis) tool. Stauffer and co-authors [5] studied the variability of eight API batches produced using different synthetic routes, crystallization, drying, delumping processes, and processing equipment. These batches were extensively characterized, and 17 properties from 7 characterization techniques were studied by principal component analysis (PCA). The authors concluded that conducting systematic studies on the batch-to-batch variability is a crucial aspect for robust formulation and process development and should be part of an integrated process involving preformulation, formulation, and process development teams.

Recently, some API and excipient manufacturers have provided to pharmaceutical companies "QbD samples," i.e., samples that encompass larger variability than encountered in commercial batches. This can be highly informative for the pharmaceutical researchers to evaluate the impact of raw material variabilities in their product CQAs. Portier and co-authors [6] assessed the impact of microcrystalline cellulose (MCC) batch-to-batch variability and process settings in a continuous twin screw wet granulation and semi-continuous drying line. Authors carried out the study enriching the commercial sample set with "QbD batches" provided by the manufacturer. Based on extensive raw material characterization and subsequent principal component analysis, raw material variability was quantitatively introduced in the design of experiments approach. For the studied formulation, it was concluded that MCC batches with low water binding capacity, low moisture content, and high bulk density generated granules with the most desirable quality attributes.

5.2 How to Structure DOE Studies Within the QbD Framework

As described in Chap. 1, DOE studies are carried out after risk analysis. Therefore, only critical material attributes (CMAs) and critical process parameters (CPPs) should be included in a DOE setup. Besides CMAs and CPPs, the formulation composition (percentage of each ingredient) should be studied and established. There is no "golden rule" for the use of DOE to study CMAs, CPPs, and formulation variables; nevertheless, a good suggestion might be as follows:

1. Establish "formulation DOEs," including critical material attributes, keeping the process variables fixed. The inclusion of CMAs in formulation development allows taking into account interactions between formulation amounts and material attributes, for example, API particle size vs the amount of a disintegrant in a formulation. It is intuitive to think that the amount of disintegrant required to achieve a given dissolution is dependent on the API particle size. Keeping the manufacturing process variables fixed at this point could be easier however, it could cause a "blind spot" in the development since possible interactions between formulation and process variables will not be taken into account. Therefore, in some situations, it could be interesting to carry out a DOE study for formulation and process variables that are likely to present interactions during this first step.
2. Split the process into its manufacturing units and define the global and intermediate (typically "in-process controls") CQAs. Intermediate CQAs are sometimes named ICQAs – some examples for solid dosage forms include granule sizes, granule shapes, moisture, flowability, etc. It should be noted that the critical material attributes of API and excipients are likely to mostly influence the first operation of a manufacturing process - if it has many. For further manufacturing operations, the material attributes are defined by the material characteristics of the previous operation.
3. "Process DOEs" can generate design spaces (DS) that can be transferred to Phases 2 and 3 of the product lifecycle, considering that the formulation is already developed and will be fixed. The purpose is to have an "everyday" DS that could be used to set process parameters and accommodate raw material variabilities during manufacturing. Even though it is possible to include variables of different manufacturing steps in a single study, it is recognized that developing separate studies for each unit operation is often simpler and more informative [7]. For this reason, the FDA simulated report [4] presents DoE studies for each manufacturing unit operation.

Figure 5.1 summarizes the discussion above. It is important to highlight that the suggestion presented in Fig. 5.1 allows uncovering all possible interactions between different types of variables, since interactions can only be seen if the variables under investigation are placed together in the same design.

As mentioned above, interactions are present between different variables throughout the pharmaceutical development and should be taken into consideration

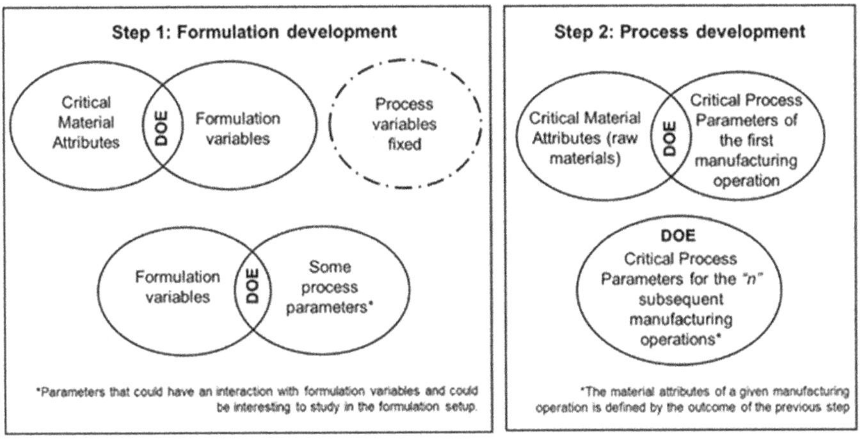

Fig. 5.1 A suggestion on how to structure DOE studies in order to gather information during formulation and process development

for a global understanding of the system. Univariate methods do not allow obtaining the whole picture, since all variables are fixed, except one that is individually varied.

It should be highlighted that the simple use of a DOE tool does not constitute the QbD approach; however, in some situations, we will keep the reference for guidance on the studied variables. DOE theory is not included in the scope of this chapter, since it will be addressed in details in other chapters. Nevertheless, we would like to highlight a very interesting paper that discusses major practical aspects of DOE within the QbD approach, presented by Mahesh Shivhare and Graham McCreath: "Practical implementation of DOE in QbD scenario" [8]. The authors discuss points such as the selection of the factors and their ranges (implications of ranges being too narrow or too wide), intrinsic variability, randomization, blocking, replication, Curvature evaluation, and statistical main aspects (ANOVA, analysis of residuals, goodness of prediction, etc.).

A systematic review of current status, challenges, and future perspectives of QbD has been recently published by Grangeia and co-authors [9]. The authors carried out a critical analysis of 60 research papers, whose contents were classified and compared. In their survey, they classified QbD applications involving formulation attributes (FA) and process parameters (PPs). It should be noted that in their classification, FAs are related to excipients' attributes (e.g., chemical type of excipient, particle size, solubility, grade) and also concentrations used in the drug product. In this text, we prefer to use separated terms, indicating the concentrations by "formulation variables" and keeping the CMAs' nomenclature to API and excipients' characteristics.

According to their classification, the combined study of FAs and PPs (FAs + PPs) represented 60% of total ($n = 36$), configuring the most prevalent category, followed by papers that focused only on PPs (28%, $n = 17$). The lowest percentage corresponded to the study of FAs alone (12%, $n = 7$). Within these studies that included both the FAs and PPs, it was further analyzed whether these factors (FAs and PPs) were included in the same DoE or in separate ones. In the pool of papers that analyzes both FAs and PPs ($n = 36$), 30% ($n = 11$) described an integrated DoE. These papers detected the presence of interactions between FAs and PPs, which highlights the benefits of using this integrated analysis. Regarding the papers that focus only on PPs ($n = 17$), it was noted that the majority presented a DoE with PPs from only one manufacturing unit operation. A different approach was reported by Tadashi Norioka et al. [10] that included PPs from different manufacturing unit operations (granulation, drying, milling, blending, and compression) in the same DoE. Even though this results in an overall evaluation of the process, details regarding intermediate CQAs can be missed.

5.3 Quality by Design in Pharmaceutical Product and Process Development

During the development of a pharmaceutical product, there are plenty of factors what may affect the final product's quality. As explained previously, the source of these factors might be present in the process parameters (PP), formulation attributes (FA), and/or material attributes (MA) involved in the product's manufacturing. Understanding the characteristics and particularities of the PPs, FAs and MAs of each unit operation that composes the manufacturing process is important to evaluate if they affect product's quality attributes (QA).

Some PPs, FAs and MAs do not have significant influence on product's QA, that is, a change in their values may never affect final product quality. On the other hand, there are PPs, FAs and MAs which have significant effects on the CQAs affecting the security and efficacy of drug products. A process parameter is considered critical (CPP) whenever its variation impacts at least one CQA. The same is applicable to material attributes; whenever there is impact on one CQA, it is considered then a critical material attribute (CMA). Knowing the details of each process step is also important to identify CPPs and CMAs that may influence the product CQAs.

The following pages are divided into sections according to the pharmaceutical dosage – solid, liquid, semisolids, and self-emulsifying drug delivery systems (SEEDS). The most relevant process parameters and material attributes will be indicated, as well as their potential impact on the product's quality attributes. Finally, a tentative list of the potential quality attributes for the mentioned pharmaceutical dosage forms will be described. It should be highlighted that the lists provided for PPs, FAs, MAs, and QAs mentioned in this chapter are not meant to be exhaustive.

5.4 Solid Dosage Forms

Solid dosage forms are the most popular among all pharmaceutical dosage forms. Oral route is preferred; tablets represent half of the solid dosage forms. Among the benefits of solid oral dosages are patient compliance and access to the treatment, cost reduction and efficiency in solid manufacturing process, and API physical and chemical stability in solid dosages [11, 12]. A solid dosage form can be obtained by different manufacturing processes, the most common are direct compression and wet (high shear; fluid bed) and dry (roller compaction) granulation.

5.4.1 Direct Compression

Direct compression is a manufacturing process largely used in the pharmaceutical industry for tablet production due to its reduced number of steps compared to other techniques, which reduces the cost and time for manufacturing [11, 12]. The direct compression process involves two main unit operations: dry mixing and compression. In the dry mixing process, the formulation ingredients are blended aiming at producing a uniform mixture. Table 5.1 presents the main material attributes which may cause impact on the product's quality attributes for direct compression, and Table 5.2 describes the main process parameters for blending and compression.

The PP and MA presented for direct compression may also be considered for the evaluation of the other manufacturing processes which present the unit operations of mixing and/or compression. An example is described in FDA report for IR dosage forms [4], where the impact of the number of revolutions (speed x time) (N_{rev}, Factor A, critical process parameter) and the drug particle size (d90) (Factor A, a critical material attribute) were evaluated using a two-factor, three-level full factorial design to investigate blend uniformity (an intermediate CQA) of the dry powder mixture. The analysis of variance (ANOVA) indicated that both linear coefficients were significant, as well as the interaction between them and the quadratic coefficient for Factor A (95% confidence level). The response surface is shown in Fig. 5.2a and the interaction graph in Fig. 5.2b.

The blend uniformity was measured by the relative standard deviation (% RSD). The goal was to minimize this response, and acceptable values were equal to or lower than ≤5%. The interaction is characterized by the non-parallel edges of the response surface: with a lower number of revolutions (100), the PSD has an important effect on the blend uniformity (blue arrow), whereas with the higher number of revolutions, PSD has no impact on this CQA (red arrow). Figure 5.2b shows the interaction plot where it can be clearly seen that with a lower number of revolutions, it is not possible to reach the 5% RSD (blue horizontal line) regardless of the particle size of the API. As far as the number of revolutions increases, the two lines approach, and, reaching 300 revolutions, both PSD values provide the same blend uniformity.

Table 5.1 Material attributes/formulation attributes (MA/FA) for direct compression

Process step	MA/FA	Comments
Mixing/ blending	Particle size	Fine particles tend to have a worst flow than large particles, which may affect blending process Blending ingredients with different particle sizes may lead to segregation and demixing [12, 13]
	Particle shape	Spherical and cubic particles presented greater mobility and tend to flow better than needle and plate particles, which are more likely to agglomerate [12, 13]
	Bulk/tapped density	Differences between the density of powders in the blend can lead to segregation [12]
	Flowability	Ingredients with poor flow may affect blend uniformity [4]
	Moisture content	The increase of moisture content may form agglomerates affecting blend uniformity [12, 13]
	Cohesive/adhesive properties	Cohesive powders may present poor flow and lead to processability problems. Cohesive powers may also cause agglomeration which may affect uniformity [12, 13]
Compression	Consolidation mechanism	Blending ingredients may present different consolidation/ deformation mechanisms such as elasticity, plasticity, and brittleness [14]. Material consolidation mechanisms may impact tablet's hardness and dissolution
	Particle shape	Contact area and bond formation between particles are influenced by particle shape, which may affect powder compactability. This property may have an influence on the presence of defects such as lamination and capping [13]
	Bulk/tapped density	The ratio between tap density and bulk density represents the propensity of the powder to be compressed and provides a dimension of volume reduction after compression. This ratio can also be related to powder's flow [13]
	Particle size distribution	Particle size and flowability are closely related, and both play a role in the compression process specially in the die filling stage.[15] Particle size distribution may affect tablet's tensile strength and porosity [16]
	Cohesive/adhesive properties	Highly cohesive/adhesive materials may disturb the compression process by sticking and film formation in the equipment [13]
	Moisture content	May affect tablet porosity and tensile strength, affecting bonding between particles [17]

Mishra and co-authors described a QbD approach for design, development, and optimization of orally disintegrating tablet formulation of carbamazepine [20]. The authors studied the compression force (x_1) together with two formulation variables (concentration of sublimating agent, x_2, and the disintegrant concentration, x_3) by a Box-Behnken response surface design. The responses were tablet crushing strength (hardness), tablet porosity, disintegration time, water absorption time, tablet friability, and drug dissolution. When considering several CQAs, each one might require a different model (linear, linear with interactions, quadratic, or higher order); therefore, it is important to consider a design that allows the calculation of all these potentially needed models. Box-Behnken is a design that provides degree of freedom

Table 5.2 Process parameters of direct compression

Process	Process parameters	Comments
Mixing/ blending	Fill volume	The increase of the fill level may decrease blending efficiency, affecting content uniformity and dissolution time. This parameter may have an interaction with mixing time, since for short mixing times the fill volume factor is more relevant [12, 18, 19]
	Mixing speed	Mixing speed influences powder's flow into the blender, affecting blending efficiency and consequently the product's content uniformity. Material attributes, such as cohesivity, may also have an interaction with this parameter [12, 18, 19]
	Mixing time	Pre-lubrification: Crucial to reach blend uniformity target. Excessive mixing time can lead to demixing Lubrification: Low mixing time for lubrification is recommended to avoid over-lubrication which may affect product's dissolution [12, 18, 19]
	Order of addition	May influence homogenization and the uniformity achievement [4]
Compression	Feeder mechanism and speed	An inadequate feed may cause filling problems and powder segregation [13, 14]
	Press speed (Dwell time)	Press speed higher than optimal may lead to filling problems, weight variability, and capping. Also press speed may affect tablet hardness and content uniformity [4, 14, 18]
	Pre/main compression force	Inadequate compression forces may affect tablet hardness and friability, consequently affecting disintegration and dissolution. It can also lead to lamination and capping [4, 14, 18, 19]

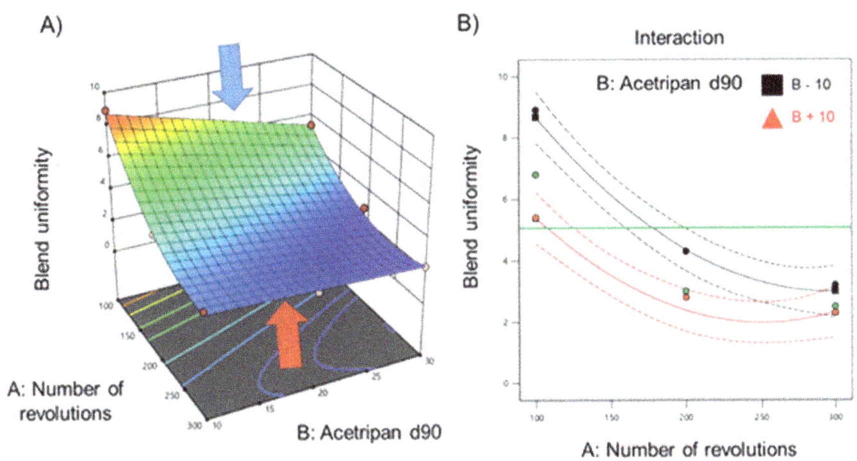

Fig. 5.2 Response surface (**a**) and interaction graph (**b**) of particle size distribution (d90) and number of revolutions. Figures generated by data available in FDA report for IR dosage forms [4]

to estimate linear, interaction, and high-order terms. In this study, tablet porosity and disintegration time were linear functions of the input variables, whereas crushing strength, water absorption time, and tablet friability required higher-order models. The formulation components and compression force to achieve acceptable values for all CQAs were determined based on desirability functions, using the established models for each CQA.

5.4.2 *Dry Granulation*

Dry granulation is a potential technique of choice for manufacturing solid dosage forms with cohesive, temperature–/moisture-sensitive, or poorly flowable drugs [12, 21]. There are different techniques, so that for dry granulation such as roll compaction and slugging, the first is preferred as it is more efficient and controllable [12, 22, 23]. Roll compaction is also very present and explored in literature and will be more detailed later. Figure 5.3 shows the roller compaction, milling, and sieving system.

The aim of dry granulation manufacturing is to promote a very intense compression of the powder blend, in order to mold it into a ribbon [12]. Dry granulation formulations demand the presence of materials which allows ribbon formation due to its deformation character, as plastic excipients, for example. The presence of a lubricant is also important in this kind of formulation to avoid process defects as adherence. Table 5.3 shows an overview of the main material attributes which are important to be considered in dry granulation.

Table 5.4 presents the process parameters for roller compaction granulation. Through this process, a ribbon is produced as an intermediate product, and the main quality attributes of the ribbon are thickness, density, and tensile strength [4, 24]. Ribbon relative density, also called ribbon solid fraction, is an indicator of ribbon quality [15, 16].

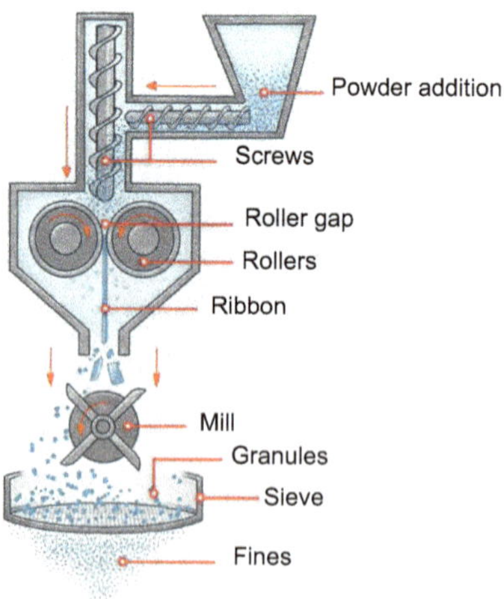

Fig. 5.3 Roller compaction, milling, and sieving system

Table 5.3 Material attributes/formulation attributes (MA/FA) for dry granulation

MA/FA	Comments
Type of excipients	*Diluents:* Plastic or brittle character may lead to ribbon formation/integrity. Brittle excipients can cause ribbon lamination. Deformation character of the vehicle may affect granule particle size. In general, plastic materials lead to the formation of coarser granules and less fine ones. Diluents may also influence tablet hardness, affecting consequently dissolution [21, 24] *Binder*: May affect size, compressibility, and friability of the granules and tablets [12, 24, 25] *Lubricants:* Prevent powder adhesion, affecting ribbon formation. May be applied in the blend or directly on the equipment surface [26]
Particle size	Material's particle size may impact ribbon properties, consequently impacting tablet hardness, content uniformity, and dissolution [12, 19, 27]
Propensity to adhere	The material may stick in the roll surface due to its propensity to adhere [25]

Table 5.4 Process parameters for roller compaction granulation

Process parameters	Comments
Roller surface design	The choice of roll surface (smooth, knurled, serrated, or pocketed, e.g.) depends on the compaction behavior of the material and its tendency to stick and may affect ribbon density [12, 19, 25]
Roller speed	May affect ribbon strength and density, leading to an impact on granule particle size, flow, and compressibility. Consequently, roller speed may affect tablet hardness and dissolution [4, 12, 19, 25]
Roller pressure/force	Promotes powder cohesion, affecting ribbon density and strength, leading to an impact on granule particle size, flow, and compressibility [4, 19, 28]. High pressures may cause ribbon lamination or breakage. It also leads to a reduction of the tablet tensile strength and may impact dissolution [12, 21]
Roller gap	May influence ribbon density and thickness, which can impact PSD, flow, compressibility, and compactability of milled granules. Consequently, may affect tablet hardness and dissolution [4, 25, 28]
Feed screw speed	The forced feed in a roller compaction is promoted by a screw. Feed screw speed influences the delivery of powder to the rolls. It may affect ribbon properties [4, 12, 25]
Fines recycled	The powder which was not compacted into coarse granules during roll compression represents the fine particles. The amounts of fine particles may be recycled in the roller compactor. This process can be repeated more than once if necessary. Controlling fine particles is important to improve flow and prevent powder segregation and weight variation during compression [21]

The ribbon will be the input of the following milling process. The milled ribbon will produce the granules, which will be sieved to separate the granules and the fine pieces, which can be recycled in the roller compaction process. The aim is obtaining granules with good flowability to allow compression or encapsulation. Granule main quality attributes are particle size, compressibility, and flow [6].

Milling technique is used in many processes of pharmaceutical manufacturing aiming at different objectives such as enhancing uniformity by particle size calibration, improving flow, producing bulk with better compressibility, and improving

Table 5.5 Process parameters for milling

Process parameters	Comments
Mill type	Affects granule particle size and consequently the density, flow, and compactability
Mill speed	Granules with poor flow and segregation may cause variability in tablet weight, assay, and content uniformity
Mill screen orifice type	Granule particle size also may affect hardness and dissolution [4, 21, 25, 28]
Mill screen orifice size	

dissolution [12]. Milling process parameters which are crucial to granule quality are presented in Table 5.5. These PPs can be considered for any unit operation of milling present in drug manufacturing, not only for roller compaction.

An integrated approach considering four process variables was described in an FDA report for IR dosage forms [4]. Two variables related to the roller compaction (roller pressure and roller gap) were studied together with two milling variables (mill speed and mill screen orifice size) using a fractional factorial design 2_{IV}^{4-1}. The responses were the intermediate CQAs: ribbon density (g/cc); granule d50 (μm); granule flowability (fcc); hardness (kP) at 10 kN) and the final CQA tablet content uniformity (% RSD).

The results indicated that the variable mill speed did not influence any of the CQAs; therefore, it was considered an inert ("slack") variable. By removing this variable from the design matrix, the 2_{IV}^{4-1} design has been changed to a 2^3 full factorial design, which allows the construction of linear models. Since the authors included center points in the design, it was possible to evaluate the suitability of linear models for each CQA. Analysis of variance (ANOVA), curvature-check test, and residual graphs indicated that all CQAs presented a linear dependence of the studied factors. Therefore, it was possible to use these models for prediction and to build the design space, either by the overlay or by using desirability functions (Fig. 5.4).

It should be highlighted that if one CQA did not follow a linear model, the design should be changed to allow the calculation of higher-order terms. Regarding the design space construction, a brief comparison of overlay and desirability procedures is suitable here. The overlay procedure requires a regular (rectangular) range of acceptable values of each output (CQA). The acceptable region for each CQA is sought by using the model equation, and then intersection region is determined for the global design space (yellow region in Fig. 5.4a, where all combinations are equally acceptable). This is a simpler, though limited, procedure. On the other hand, desirability procedure allows one to indicate the goals in a more specific manner, for example, to minimize/maximize or to find a specific value (target). This is illustrated in Fig. 5.4b where the criteria for granule flowability has been changed to "maximize," with a lower limit of 6, and the uniformity content was changed to "minimize" considering the upper limit of 5 (% RSD). This allows a differentiation within the design space, shown by the colors from green to orange, as indicated by the gray arrow. If the desirability criteria are set in a rectangular way for all CQAs, both procedures will lead to exactly the same results.

A)

Output Variables	Lower range	Upper Range
Ribbon density (g/cc)	1.0	1.2
Granule d50 (μm)	400	800
Granule flowability (ffc)	6	-
Hardness 10kN	9	-
Tablet CU (%RSD)	-	5

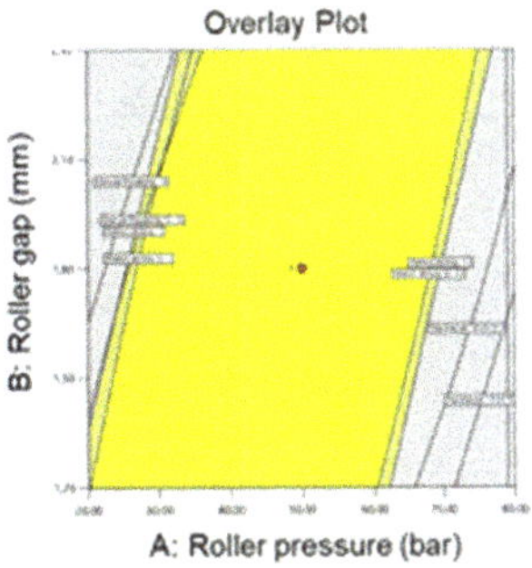

B)

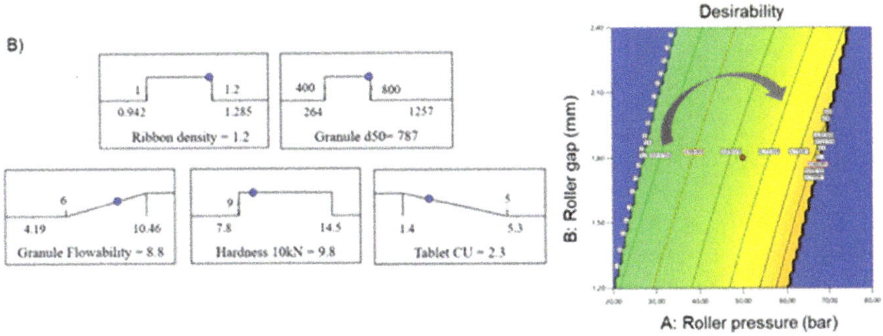

Fig. 5.4 (**a**) CQA criteria (rectangular ranges) in the overlay procedure and the corresponding design space (yellow region). (**b**) Desirability criteria for each CQA (rectangular for ribbon density, granule d50, and hardness; "maximize" for granule followability and "minimize" for content uniformity RSD) and the corresponding design space. The blue circles indicate the predicted CQA values for the maximum global desirability

5.4.3 *Wet Granulation*

Wet granulation is a technology explored by the pharmaceutical industry to improve powder flowability and compressibility. This goal is achieved by adding a solvent to the powder to transform small particles into granules [29, 30]. There are different types of wet granulation processes: low shear, high shear, fluidized bed granulation, foam granulation, freeze granulation, melt granulation among others [29]. The most common are high shear and fluidized bed granulation, which will be discussed in more details below. Due to the similarities between the different types of wet granulation, the material attributes which may present impact on the product's quality attributes were joined in Table 5.6.

High Shear Granulation

High shear granulation occurs in a granulator bowl with two blades, the impeller and the chopper. The impeller is responsible for mixing the particles and has an important role in nucleation and growth/consolidation steps.

The chopper breaks up the wet mass. The granulation starts when the granulating liquid is sprayed through a nozzle over the powder – as the droplets are mixed with the powder, granules are formed. Table 5.7 presents a list of process parameters of

Table 5.6 Material attributes/formulation attributes (MA/FA) for wet granulation

MA/FA	Comments
Type of solvent	Water is the most popular solvent used in wet granulation due to its low cost and safety. However, depending on the process/formulation needs and sensibility of the API to water, organic solvents may also be used [31]
Amount of solvent	The amount of solvent used in granulating liquid influences granule growth and density. The optimal density may be reached by the unspecified amount of solvent added An excess of solvent may cause over-granulation [31, 32]
Granulating liquid viscosity	Whenever the binder is mixed with the solvent, it influences the granulating liquid viscosity. Less viscous solutions spread more easily on the particles. Highly viscous solutions may increase the granule growth rate [29]
Concentration of binder	Binder concentration may influence tablet strength [29]
Solubility/ hydrophilicity	Hydrophilic materials tend to absorb water and lead to granule increase in the early stages of granulation process. Materials with high solubility may present uncontrolled granule growth. End-point determination might be more challenging [29, 32]
Particle size	Material particle size may affect granule growth and processability [29]
API-excipient ratio	For low API dosages, content uniformity may be critical due to difficulty of ensuring uniform distribution. For high API dosages, to ensure the uniform distribution of functional excipients such as lubricants and binders is critical [29]

high shear wet granulation which may cause impact on the product quality attributes, and Fig. 5.5 shows an illustrative representation.

A very detailed study was carried out by Kim and co-authors [28] describing the QbD approach for the pharmaceutical development comparing two important manufacturing processes: wet granulation (for metformin) and roller compaction (for dapagliflozin). The authors present the QTPP and risk assessment as the basis for the selection of CPPs for both manufacturing processes. Intermediate quality attributes (IQAs) and final CQAs were studied as responses for DoE. The CPPs in the high shear wet granulation were (1) impeller speed (50–100 rpm), (2) massing time (10–50 s), and (3) binder solvent volume (1–4 mL). The binding solvent spray speed was fixed at 100 mL/h. The selected IQAs were intrinsic dissolution rate, granule size, true density, bulk density, Carr's index, angle of repose, and granule strength. Final CQAs were studied as well: swelling property at 1 h, 3 h, and 5 h; weight gain at 1 h, 3 h, and 5 h; mass loss at 1 h, 3 h, and 5 h; gel strength at 1 h, 3 h, and 5 h; dissolution at 1 h, 3 h, and 10 h; and contact angle. For the roller compaction process, the CPPs were roller pressure (25–85 bar), roller gap (1.2–2.4 mm), and mill screen size (0.5–1.5 mm), whereas roll speed, feed screw, and roller pressure were fixed at 4 rpm, 10 rpm, and 55 bar, respectively. The IQAs were intrinsic dissolution rate, granule size, ribbon density, bulk density, tapped density, granule strength, and granule uniformity. Final CQAs were tablet content uniformity; dissolution at 5 min, 10 min, and 15 min; and contact angle. The models generated present linear, interaction, and quadratic significant coefficients, depending on the response considered. This highlights the importance of using a design that allows calculations of

Table 5.7 Process parameters of high shear granulation

Process parameters	Comments
Impeller speed	Impeller speed may influence granule size and density. Consequently, it may impact dissolution [28, 30, 33]
Chopper speed	Chopper speed may influence granule size. Consequently, it may impact dissolution [28, 33]
Solvent addition method and rate	The influence of method and rate addition depends on the affinity of the powder to the solvent and may impact uniformity and granule size [32]
Granulating liquid temperature	The temperature of granulating liquid may influence granule growth [32]
Binder addition method	Binders can be added in dry mixture or solubilized in the solvent. Binder addition method may influence granule shape and density and consequently the granule flow [29, 31, 33]
Wet massing time/kneading time	May influence granule densification, size, and flow. It may consequently affect drug product's dissolution [30, 34]
End-point determination	It is determined by the agglomerate cohesive force or tensile strength. End-point determination affects tablet's properties such as hardness, friability, and dissolution

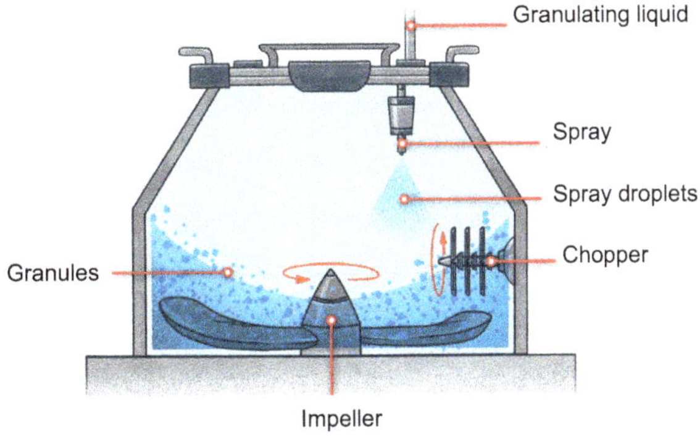

Fig. 5.5 High shear granulation

these coefficients, especially when several responses are being considered. A design space was established, and Monte Carlo simulations were conducted to include the uncertainties. The authors also used Pearson correlation and principal component analysis (PCA) to stablish relationships between IQAs and CQAs. In high shear wet granulation process, IQAs such as granule size, bulk density, and intrinsic dissolution rate were significantly correlated with CQAs such as dissolution profiles, swelling properties, and contact angle. In the roller compaction process, IQAs such as granule size, intrinsic dissolution rate, ribbon density, and granule strength were significantly correlated with CQAs such as dissolution profiles, tablet C.U., and contact angle.

In another detailed study by Han and co-authors [35], the IQAs, CQAs, and CPPs of a telmisartan tablet prepared by high shear wet granulation were determined and then analyzed by a similar integrated approach based on DOE and PCA to understand the relationship between intermediate quality attributes (IQAs), critical quality attributes (CQAs), and critical process parameters (CPPs).

Fluidized Bed Granulation

Fluidized bed is considered a process with “all-operations-in-one-equipment” since the mixing, granulating, drying, and coating can be carried out with the same equipment [36].

In fluidized bed granulation, heated air is injected into the chamber aiming to suspend the particles in a fluidized state. Next, a granulating liquid is sprayed through a nozzle, which atomizes the droplets that will wet the fluidized particles. Then, the powder humidified by the granulating liquid agglomerates into granules. Figure 5.6 presents wet granulation process in a fluidized bed. Table 5.8 presents the main process parameters of fluidized bed granulation which may cause impact on product quality attributes.

Gavan and co-authors [40] carried out a very comprehensive study on a fluidized bed granulation, including many QbD tools. The authors started by presenting an Ishikawa diagram to analyze which might be the process parameters and material

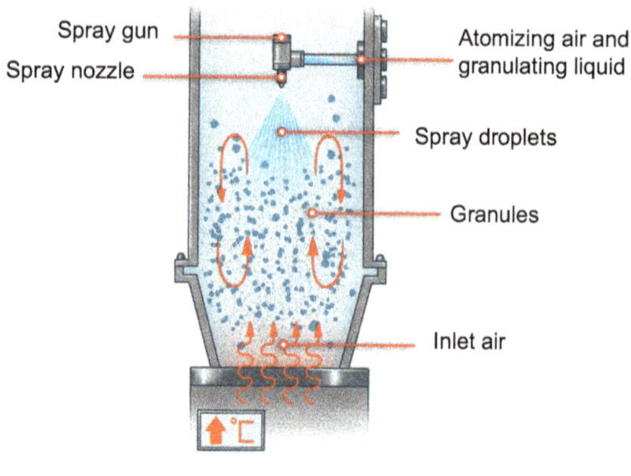

Fig. 5.6 Fluidized bed granulation

Table 5.8 Process parameters of fluidized bed granulation

Process parameters	Comments
Inlet air temperature	Granule size may be larger whenever the inlet air temperature is lower [37, 38]
Inlet air flow	May influence granule size High air flow may increase particle attrition and lead to small granule formation [38, 39]
Spray gun position	The spray gun may be positioned on the top, bottom, or tangential to the fluidized particles in the equipment. The spray gun position may influence bulk characteristics such as density, particle size, and morphology [36]
Atomizing air pressure	The atomizing air pressure affects material fluidization and influences granule size [38]
Granulating liquid spray rate	May impact granule size. Increasing the spray rate may lead to granule growth. Controlling this parameter may be important to prevent over- or under-granulation [29, 37]
Spray nozzle	Determines droplet size, which has a direct relation to granule size [39]

attributes with highest impact on product's quality attributes. The ones considered critical were evaluated through FMEA risk assessment. The PP and MA with the highest scores in FMEA present higher risk; that is the reason they were selected to be evaluated by a DOE. A D-optimal DOE was performed considering two quantitative PPs (binder spray rate, which corresponds to the granulating liquid spray rate, and atomizing pressure) and three qualitative MAs (paracetamol, ibuprofen, and microcrystalline cellulose of different sorts, suppliers, and characteristics). The DOE demonstrated that both PPs and the drug properties presented an effect on the granule size. This study also covered another QbD tool: an in-line NIR monitoring of fluid bed to quantify moisture content during the drying process.

Wang and co-authors [41] presented a study with a wet granulation technique named pulsed spray fluidized bed granulation. The aim of this technique is to provide better control of granule growth, which can be achieved by balancing the moisture content in the fluidized bed, controlling granulating liquid application. Through a Box-Behnken experimental design, three PPs (pulsed frequency, binder spray rate, and atomization pressure) were evaluated to investigate their influence on granule mean size, yield, relative width of granule distribution, Hausner ratio, and final granule moisture content. This work has shown that the PPs evaluated are interrelated and play a role in the granule's properties. Thus, a design space could be built for these PPs in order to achieve the granule target characteristics.

5.4.4 *Drying*

Drying process is explored in drug manufacturing whenever there is water or another solvent in the raw materials or final product that should be removed. It is often used after wet granulation. Drying process, as spray dryer technique, may also be applied for liquid intermediates which are atomized and dried by a gas into a powder. There are diverse types of drying processes, which may vary from cheap and long-lasting as tray to more expensive and faster as spray dryer. The main material attribute for drying process is the amount of volatile/moisture content present in the process intermediates of preview steps. The most common types of drying techniques and its process parameters are described in Table 5.9.

Table 5.9 Process parameters for drying

Type	Process parameters
Tray	Type of tray, tray depth, drying time, temperature, air flow, and amount of product [12, 13]
Fluidized bed	Inlet air flow, air temperature, product temperature, and drying time [12, 13]
Microwave	Frequency and drying time [12, 13]
Spray drying	Atomizer type, nozzle type, atomization pressure, feed flow rate, inlet air temperature, drying gas flow rate, particle collection [12, 13, 42]

A QbD-based approach for formulation development was described to prepare spray dried microparticles of erlotinib hydrochloride for sustained release [43]. After defining the quality target product profile (QTPP), the use of risk assessment by an Ishikawa diagram assisted in the identification of three risk factors: two formulation variables, the concentration of polycaprolactone (PCL) and polyvinyl alcohol (PVA), and one process variable, the inlet temperature of spray dryer. A Box-Behnken design was used to study the effect of these factors on particle size and drug encapsulation efficiency and sustained-release dissolution. A design space was built, and the optimized formulation provided promising IC50 (inhibitory concentration) values against the lung cancer cells (A549) indicating improved therapeutic efficacy and effective sustained release potential of the formulated microparticles.

5.5 Liquid and Semisolid Dosage Forms

There are different types of liquid and semisolid dosage forms as solutions, suspensions, and emulsions. The delivery route of these forms could vary among oral, parenteral, aero, and topical route. Liquid and semisolid dosage forms have in common the same main unit operation: mixing. Mixing has a different goal in the manufacturing of the different dosage forms. When it comes to solutions, the objective is promoting ingredient solubilization; in case of suspensions and semi-solids, homogenization of the raw materials; and in case of emulsions, to create co-solvent droplets and incorporate the API. Table 5.10 presents the main process parameters involved in the mixing process, and Table 5.11 presents the main material attributes and formulation attributes for liquid and semisolid dosage forms. The impact of the PP and MA/FA on the pharmaceutical product is also covered.

There are also other unit operations involved in liquid and semisolid dosage production, such as filtering, sterilization, and particle size reduction, but they will not be covered in this chapter.

Quality by Design has been used for the optimization of specialized liquid drug delivery systems such as SEDDS (self-emulsifying drug delivery system), solid lipid nanoparticles (SLN), nanostructured lipid carriers (NLC), nanoemulsions, and nanosuspensions [47]. An interesting research work was presented by Kuk and co-authors [50], who carried out the development of a nanosuspension of resveratrol using the anti-solvent process based on the QbD approach. After the construction of the QTTP, CQA identification, and risk analysis (RA), the authors carried out two sequential DOEs to study formulation and process variables. Initially, a Box-Behnken design was used for the optimization of three high-risk formulation parameters: polyvinylpyrrolidone vinyl acetate (PVP VA64), polyvinylpyrrolidone K12 (PVP K12), and sodium lauryl sulfate (SLS). Nine responses related to particle size measurement and zeta potential were selected. A design space for formulation variables was established by overlay plots (Design-Expert, Stat-Ease) and Monte Carlo simulation (Modde, Umetrics). Later, the process variable injection rates,

Table 5.10 Process parameters for mixing

Process parameters	Comments
Type of equipment	The equipment selection depends on the dosage form produced and its characteristics such as viscosity and rheological behavior of the drug product [44, 45]
Mixer type	There are plenty types of mixers, which have different blade forms, such as impellers, propellers, paddles, and helical ribbons, for example. The choice of mixer is made according to the dosage form, considering its viscosity and product quality requirements [44–46]
Mixing/homogenization speed and time	Homogenization speed depends on the formulation characteristics, such as viscosity. The time to reach homogenization must be studied during product development. These parameters might influence product uniformity, viscosity, stability, and depending on the formulation droplet size or also nanostructure [46, 47]
Temperature control	The temperature control during the process has multiple goals, such as melting/warming the phase, promoting the ingredient solubilization, and maintaining the adequate heat the process requires. Process temperature may influence product stability and viscosity [46, 48]
Cooling cycle	The cooling cycle is important for some formulations in order to ensure their uniformity, viscosity, and stability [46, 48]
Order of addiction	For formulations which have an oily and an aqueous phase, order of addition will determine the method the emulsion is formed, consequently on droplet size For APIs sensitive to heat, if the manufacturing process involves heating, a strategy must be designed for introducing it into the cooled system in order to avoid drug assay issues, instability, recrystallization, and precipitation The order of addiction of some excipients may also influence product viscosity [46, 48]

temperature, and mixing speed were studied by a factorial design in order to evaluate process robustness, under optimized composition conditions. The process was observed to be robust within the studied range of the variables. For more details on Quality by Design applications in nanoparticle formulations, the readers are encouraged to see Chap. 8 of this book.

It should be highlighted that formulation proportions can be optimized either by factorial/response surface designs [9, 47, 50] or by mixture designs [51]. In the former, the input variables are independent, whereas in the latter they are statistically dependent since the total amount of the studied excipients is fixed. Another characteristic of mixture designs is that the responses are not dependent on the total amount but only on the proportion of the mixture components. Factorial/response surface designs are often used when there is a diluent or q.s. excipient, which provides the degree of freedom to vary the other components in an independent manner and allows to maintain the total mass constant. Another possibility to use factorial/response surface designs is to study mixtures varying the ratio of excipients, keeping

Table 5.11 Material attributes /formulation attributes (MA/FA) for mixing

MA/FA	Comments
Solubility	Drug solubility evaluation is important before choosing the other excipients in the formulation depending on the final dosage form. If it is a solution, it is crucial to ensure the solubility in the vehicle and co-solvents in order to obtain a stable formulation For formulations which have two phases, knowing drug substance and excipient solubilities allows knowing in which phase (oily or aqueous) it is possible to achieve solubilization and may facilitate uniformization during the manufacturing process [46]
Particle size	In formulations in which the ingredients are not solubilized during the process, particle size may influence product appearance, uniformity, drug release, and distribution [48]
Excipients' proportions or concentrations	May impact products' viscosity, stability, uniformity, and depending on the formulation droplet size or nanostructure [46, 49]
Viscosity/rheology of the ingredients or thickening/rheological agents	The viscosity/rheology of the ingredients used in the formulation may impact final product viscosity and stability, as well as the use of a thickening/rheological agent in the formulation. This may lead to an impact on product's sensorial properties and depending on the dosage form also on the permeation [48]
Surfactants	They may increase wettability or stability of the formulations, avoiding precipitation/aggregation [44, 48]
Suspending agent	For suspensions, the presence of suspending agent improves product's stability

their total amount fixed, as described in the formulation development example in the FDA report [4].

Mixture designs can be straightforwardly used for the development of SEDDS formulations, which have as the basic composition a mixture of a solvent, a surfactant, and an oily component [52]. An approach to develop SEDDS formulation can be as follows: after the selection of the excipients, the preparation of different compositions within the ternary diagram can provide a guidance toward the most interesting region, considering the measured responses (e.g., particle size, polydispersity, zeta potential). Inside the selected region, experimental points are set according to a mixture design. If the region is regular, simplex lattice or simplex centroid designs can be used. If an irregular region is selected, optimal designs can be used [53]. The next step is to carry out the statistical data treatment, including desirability procedure for the overall optimization when more than one response is considered (Fig. 5.7).

Examples of the use of QbD for transdermal/topic formulations can be found in references [48, 54, 55].

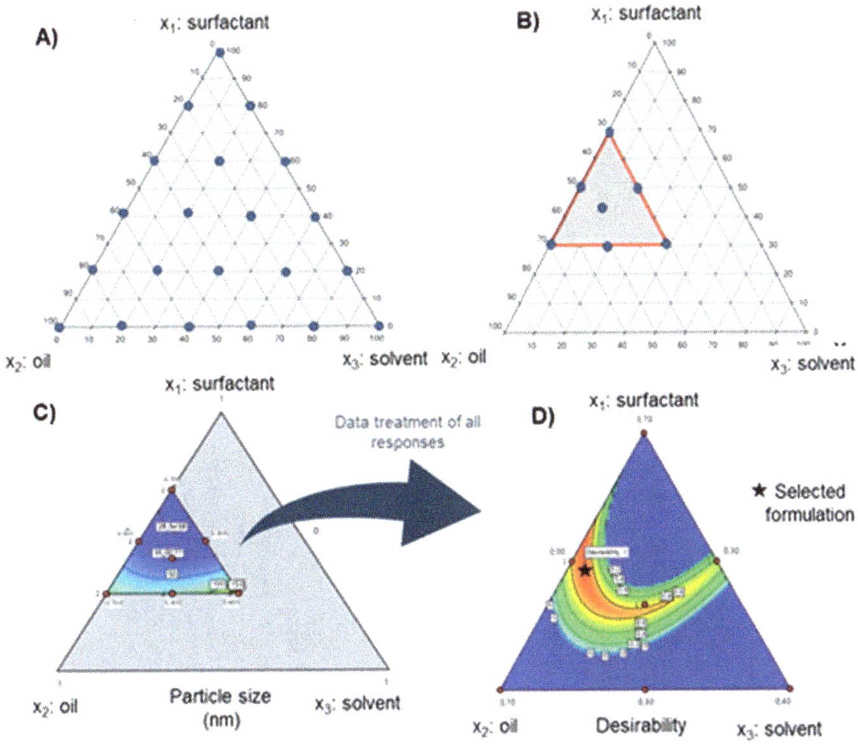

Fig. 5.7 A general representation of SEDDS development by mixture designs: (**a**) study of different compositions within the ternary diagram; (**b**) setup of a mixture design in the most promising region; (**c**) data treatment of all responses; and (**d**) application of desirability functions for the simultaneous optimization of all CQAs – the red region indicates the higher desirability values for this illustrative example

5.6 Quality Attributes

As previoulsy explained, the quality attributes are only considered CQAs if they have an impact on safety and efficiency of the pharmaceutical product and are likely to be affected by material attributes, formulation and process variables. These two criteria depend on the manufactring process, API and formulation characteristics. Therefore, it is not adequate to provide a general list of critical quality attributes for a given dosage form. Instead, we provide in Tables 5.12 and 5.13 the *potential* quality attributes for different pharmaceutical dosage forms. We highlight that the quality attributes described in these tables are not exhaustive, and the applicability of each one to a specific product must be individually evaluated.

Table 5.12 Potential quality attributes for solid dosage forms

Tablets [13, 18, 30, 33, 56, 57]	Capsules [56, 57]	Granules/powders for reconstitution [13, 56, 57]	
Assay Hardness Friability Disintegration Dissolution Content uniformity (uniformity of dosage units) Moisture content/water content Microbial limits Impurities Description Identification Tablet weight (mean or RSD [%]) Thickness Tablet porosity/density API polymorphic form (in final product) *For intermediate product (granules and/or ribbon)*: Particle size Flow Density Homogeneity/uniformity Porosity Moisture content Tensile strength Brittleness Elasticity	*Hard gelatin capsules* Assay Brittleness Disintegration Dissolution Content uniformity (uniformity of dosage units) Moisture content/water content Microbial limits Impurities Description Identification API polymorphic form (in final product) *Soft gelatin capsules* Assay Disintegration Dissolution Content uniformity (uniformity of dosage units) Moisture content/water content Microbial limits pH Leakage Pellicle formation Impurities Description Identification	*Granules/powder* Assay Uniformity of dosage units Moisture content/water content Drug polymorphic form Microbial limits Antimicrobial preservative content Antioxidant preservative content Extractables Solvent content Impurities Description Identification Particle size Flow Density Homogeneity/uniformity Porosity Strength (torque or power consumption) Brittleness Elasticity API polymorphic form (in final product)	*After reconstitution* Redispersibility Reconstitution time Dissolution Particle size distribution pH Viscosity Clarity Formation of precipitate Rheological properties

Table 5.13 Potential quality attributes for liquid and semisolid dosage forms

Solution [56, 57]	Parenteral [56, 57]	Suspensions [56, 57]	Topical [56, 57]	Lipid-based nanosystems [47, 49]
Assay pH Viscosity Microbial limits Clarity Formation of precipitate Antimicrobial preservative content Antioxidant preservative content Extractables Solvent content Rheological properties Impurities Description Identification	Assay pH Clarity Color Particulate matter Endotoxins Sterility Antimicrobial preservative content Antioxidant preservative content Extractables Functionality testing of delivery systems Osmolarity Particle size distribution Dissolution Particle size distribution Drug polymorphic form Rheological properties Redispersibility Reconstitution time Impurities Description Identification *Powders for reconstitution and suspensions* Uniformity of dosage units Reconstitution time Water content	Assay Dissolution Particle size distribution Drug polymorphic form pH Viscosity Microbial limits Clarity Formation of precipitate Antimicrobial preservative content Antioxidant preservative content Extractables Alcohol content Rheological properties Redispersibility Impurities Description Identification API polymorphic form (in final product)	Assay Clarity Homogeneity pH Suspendability (lotions) Consistency Viscosity Particle size distribution (for suspensions) Microbial limits Weight loss Impurities Description Identification API polymorphic form (in final product)	Assay Particle size Polydispersity index Zeta potential Encapsulation efficiency Drug release Morphology Surface characteristics Permeation properties Loading capacity Drug-lipid solubility Drug protection rheological properties Appearance Impurities Cytotoxicity Uniformity of dose API polymorphic form (in final product)

References

1. The International Pharmaceutical Excipients Council – IPEC (2020). Incorporation of pharmaceutical excipients into product development using quality-by-design (QbD). Available in https://www.ipec-europe.org/uploads/publications/ipec-qbd-guide-final-for-federation-f-20201030-1613984210.pdf. Accessed 4 Sep 2022.
2. Whitcomb P (2020) Overview of robust design, propagation of error, and tolerance analysis. Available in https://www.statease.com/webinars/overview-robust-design-propagation-error-tolerance-analysis/. Accessed 3 Jun 2022.
3. Muralidhar K (2003) Monte Carlo Simulation. In: Encyclopedia of information systems. Academic Press, pp 193–201
4. Food and Drug Administration (2012) Quality by Design for ANDAs: An Example for Immediate-Release Dosage Forms
5. Stauffer F, Vanhoorne V, Pilcer G et al (2018) Raw material variability of an active pharmaceutical ingredient and its relevance for Processability in secondary continuous pharmaceutical manufacturing. Eur J Pharm Biopharm 127:92–103
6. Portier C, Vigh T, Pretoro DG et al (2021) Continuous twin screw granulation: impact of microcrystalline cellulose batch-to-batch variability during granulation and drying – a QbD approach. Int J Pharm 3:100077. https://doi.org/10.1016/j.ijpx.2021.100077
7. Abraham J (2010) International conference on harmonization of technical requirements for registration of pharmaceuticals for human use. In: handbook of transnational economic governance regimes. Brill | Nijhoff, p 1041–1053
8. Shivhare M, McCreath G (2010) Practical considerations for DoE implementation in quality by design. BioProcess International. Available in https://bioprocessintl.com/manufacturing/information-technology/practical-considerations-for-doe-implementation-in-quality-by-design-297328/. Accessed 21 Aug 2022.
9. Grangeia HB, Silva C, Simões SP, Reis MS (2020) Quality by design in pharmaceutical manufacturing: a systematic review of current status, challenges and future perspectives. Eur J Pharm Biopharm 147:1–40
10. Norioka T, Kikuchi S, Onuki Y, Takayama K, Imai K (2011) Optimization of the manufacturing process for Oral formulations using multivariate statistical methods. J Pharm Innov 6:157–169
11. Iqubal M (2014) Recent advances in direct compression technique for pharmaceutical tablet formulation. IJPRD:49–57
12. Augsburger LL, Hoag SW (2008) Pharmaceutical dosage forms - tablets: unit operations and mechanical properties, 3rd edn. CRC Press, Boca Raton
13. Yu LX, Amidon G, Khan MA et al (2014) Understanding pharmaceutical quality by design. AAPS J 16:771–783
14. Natoli D, Levin M, Tsygan L, Liu L (2009) Development, optimization, and scale-up of process parameters: tablet compression. In: Developing solid oral dosage forms. Academic Press, p 725–759
15. Zhao J, Yin D, Rowe J et al (2018) Understanding the factors that control the quality of mini-tablet compression: flow, particle size, and tooling dimension. J Pharm Sci 107:1204–1208
16. Wünsch I, Finke JH, John E et al (2021) The influence of particle size on the application of compression and compaction models for tableting.c 1–17
17. Sun CC (2008) Mechanism of moisture induced variations in true density and compaction properties of microcrystalline cellulose. Int J Pharm 346:93–101
18. Kapsi S, Castro L, Muller F, Wrzosek T (2012) Development of a design space for a unit operation: illustration using compression-mix blending process for the manufacture of a tablet dosage form. J Pharm Innov 7:19–29
19. Fahmy R, Kona R, Dandu R et al (2012) Quality by design I: application of failure mode effect analysis (FMEA) and plackett-burman design of experiments in the identification of "Main

factors" in the formulation and process design space for roller-compacted ciprofloxacin hydrochloride immediate-release tablets. AAPS PharmSciTech 13:1243–1254
20. Mishra SM, Rohera BD (2017) An integrated, quality by design (QbD) approach for design, development and optimization of orally disintegrating tablet formulation of carbamazepine. Pharm Dev Technol 22:889–903
21. Perez-Gandarillas L, Perez-Gago A, Mazor A et al (2016) Effect of roll-compaction and milling conditions on granules and tablet properties. Eur J Pharm Biopharm 106:38–49
22. Shanmugam S (2015) Granulation techniques and technologies: recent progresses. Bioimpacts 5:55–63
23. Kleinebudde P (2004) Roll compaction/dry granulation: pharmaceutical applications. Eur J Pharm Biopharm 58:317–326
24. Chang CK, Alvarez-Nunez FA, Rinella JV Jr et al (2008) Roller compaction, granulation and capsule product dissolution of drug formulations containing a lactose or mannitol filler, starch, and talc. AAPS PharmSciTech 9:597–604
25. Rowe JM, Charlton ST, McCann RJ (2017) Development, scale-up, and optimization of process parameters: roller compaction theory and practice. In: Developing solid oral dosage forms, 2nd edn. Academic Press, pp 869–915
26. Dawes J, Allenspach C, Gamble JF et al (2013) Application of external lubrication during the roller compaction of adhesive pharmaceutical formulations. Pharm Dev Technol 18:246–256
27. Kushner J, Langdon BA, Hiller JI, Carlson GT (2011) Examining the impact of excipient material property variation on drug product quality attributes: a quality-by-design study for a roller compacted, immediate release tablet. J Pharm Sci 100:2222–2239
28. Kim JY, Chun MH, Choi DH (2021) Control strategy for process development of high-shear wet granulation and roller compaction to prepare a combination drug using integrated quality by design. Pharmaceutics 13:1–27
29. Hiremath P, Nuguru K, Agrahari V (2019) Material attributes and their impact on wet granulation process performance. Handbook of Pharmaceutical Wet Granulation, Academic Press, In, pp 263–315
30. Badawy SIF, Narang AS, LaMarche K et al (2012) Mechanistic basis for the effects of process parameters on quality attributes in high shear wet granulation. Int J Pharm 439:324–333
31. Morkhade DM (2017) Comparative impact of different binder addition methods, binders and diluents on resulting granule and tablet attributes via high shear wet granulation. Powder Technol 320:114–124
32. Liu L, Levin M, Sheskey P (2009) Process development and scale-up of wet granulation by the high shear process. Developing Solid Oral Dosage Forms, Academic Press, In, pp 667–699
33. Liu B, Wang J, Zeng J et al (2021) A review of high shear wet granulation for better process understanding, control and product development. Powder Technol 381:204–223
34. Verma R, Patil M, Paz CO (2019) Current practices in wet granulation-based generic product development. In: Handbook of pharmaceutical wet granulation. Academic Press, pp 203–259
35. Han JK, Shin BS, Choi DH (2019) Comprehensive study of intermediate and critical quality attributes for process control of high-shear wet granulation using multivariate analysis and the quality by design approach. Pharmaceutics 11:1–25
36. Patil P, Karnavat D, Chavan S (2018) Review on fluidized bed granulation and coating using fluidized bed coater. IAJPR 8:1794–1809
37. Gohel M, Parikh R, Baldaniya L et al (2007) Fluidized bed systems: a review. In: Pharmainfonet. Pharmainfonet, pp 1–37
38. Van Buskirk GA, Asotra S, Balducci C et al (2014) Best practices for the development, scale-up, and post-approval change control of IR and MR dosage forms in the current quality-by-design paradigm. AAPS PharmSciTech 15:665–693
39. SaintyCo A definitive guide to fluid bed granulation process (updated). Available in https://www.saintytec.com/fluid-bed-granulation-process/. Accessed 22 Aug 2022

40. Gavan A, Iurian S, Casian T et al (2019) Fluidised bed granulation of two APIs: QbD approach and development of a NIR in-line monitoring method. AJPS:1–12
41. Liu H, Wang K, Schlindwein W, Li M (2013) using the box–behnken experimental design to optimise operating parameters in pulsed spray fluidised bed granulation. Int J Pharm 448:329–338
42. Santos D, Maurício AC, Sencadas V et al (2017) Spray drying: an overview. Biomaterials - Physics and Chemistry, Intechopen, In, pp 9–35
43. Soni G, Yadav K, Gupta M (2020) QbD based approach for formulation development of spray dried microparticles of Erlotinib hydrochloride for sustained release. J Drug Deliv Sci Technol 57:1–15
44. Kulshreshtha A, Singh O, Wall G (2010) Pharmaceutical suspensions. From Formulation Development to Manufacturing. Springer
45. Block L (2006) Non-parenteral liquids and semisolids. In: Pharmaceutical process scale up, 2nd edn. CRC Press, North Carolina, pp 89–125
46. Simões A, Veiga F, Vitorino C, Figueiras A (2018) A tutorial for developing a topical cream formulation based on the quality by design approach. J Pharm Sci 107:1–10
47. Cunha S, Costa CP, Moreira JN et al (2020) Using the quality by design (QbD) approach to optimize formulations of lipid nanoparticles and nanoemulsions: a review. Nanomedicine 27:1–17
48. Namjoshi S, Dabbaghi M, Roberts MS et al (2020) Quality by design: development of the quality target product profile (QTPP) for semisolid topical products. Pharmaceutics 12:1–12
49. Zagalo D, Silva BMA, Silva C et al (2022) A quality by design (QbD) approach in pharmaceutical development of lipid-based nanosystems: a systematic review. J Drug Deliv Sci Technol 70:1–55
50. Kuk DH, Ha ES, Ha DH et al (2019) Development of a resveratrol nanosuspension using the antisolvent precipitation method without solvent removal, based on a quality by design (QbD) approach. Pharmaceutics 11:1–22
51. Cornell J (2011) A primer on experiments with mixtures. Wiley
52. Morozowich W, Gao P (2009) Improving the oral absorption of poorly soluble drugs using SEDDS and S-SEDDS formulations. In: Developing solid oral dosage forms. Academic Press, pp 443–468
53. Visetvichaporn V, Kim K-H, Jung K et al (2020) Formulation of self-microemulsifying drug delivery system (SMEDDS) by D-optimal mixture design to enhance the Oral bioavailability of a new Cathepsin K inhibitor (HL235). Int J Pharm 273:1–9
54. Savic SM, Cekic ND, Savic SR et al (2021) 'All-natural' anti-wrinkle emulsion serum with Acmella Oleracea extract: a design of experiments (DoE) formulation approach, rheology and in vivo skin performance/efficacy evaluation. Int J Cosmet Sci 00:1–16
55. Kim EJ, Choi DH (2012) Quality by design approach to the development of transdermal patch systems and regulatory perspective. J Pharm Investig 51:669–690
56. International Conference on Harmonisation (1999) Specifications: test procedures and acceptance criteria for new drugs substances and new drug products chemical substances Q6A. Available via https://www.ich.org/page/quality-guidelines. Accessed 04 Sept 2022
57. World Health Organization (2008) WHO expert committee on specifications for pharmaceutical preparations. Rev Inst Med. Trop, São Paulo 50:144

Chapter 6
A Quality by Design Update on Nano-drug Delivery Systems

Darío Leonardi, María Celina Lamas, and Claudio Javier Salomon

Abstract The production of nanoparticles is a very promising approach for improving the delivery of hydrophobic drugs due to their unique properties and advantages. Even though such nanostructures are very attractive for their applications in human medicine, the optimization of the formulations and manufacturing methodologies to obtain a product with the desired quality is a very difficult and time-consuming task. The conventional strategy applied for the optimization of these drug delivery systems is the study of one variable at a time while maintaining the others constant. In this regard, the application of the quality by design approach may avoid such drawbacks in addition to improving the efficiency and safety of the drug product. The design of experiments allows prediction of the properties of the pharmaceutical formulations even without preparing them as well as identifying potential interactions between the variables. Thus, the implementation of multivariate design of experiments and optimization of processes by means of the response surface methodology led to novel nano-scale drug delivery systems with desired properties. In this regard, this chapter reviews the latest advances on the development and optimization of drug nanocrystals, loaded polymeric nanoparticles, and loaded lipid vesicles.

Keywords Quality by design · Nanoparticles · Hydrophobic drugs · Dissolution · Bioavailability · Stability

D. Leonardi · M. C. Lamas · C. J. Salomon (✉)
Departamento de Farmacia, Facultad de Ciencias Bioquímicas y Farmacéuticas, Universidad Nacional de Rosario, Rosario, Argentina

Instituto de Química de Rosario, Consejo Nacional de Investigaciones Científicas y Tecnológicas, Rosario, Argentina
e-mail: leonardi@iquir-conicet.gov.ar; mlamas@fbioyf.unr.edu.ar; csalomon@fbioyf.unr.edu.ar

M. C. Breitkreitz, H. C. Goicoechea (eds.), *Introduction to Quality by Design in Pharmaceutical Manufacturing and Analytical Development*, AAPS Introductions in the Pharmaceutical Sciences 10, https://doi.org/10.1007/978-3-031-31505-3_6

6.1 Introduction

In the last decades, there has been a remarkable advance in the application of different technologies, including combinatorial chemistry and high-throughput screening to the discovery of new molecules with promising biological activities [1]. The majority of these new molecules contain different hydrophobic moieties that are capable to promote the binding to proteins and/or enzymes with the aim to improve their pharmacological performance. Thus, most of those substances may exhibit high molecular weight, high log *P*, and high melting point values and, therefore, display notable hydrophobic behavior resulting in very poor aqueous solubilities [2, 3]. Solubility is a fundamental physicochemical property of substances with therapeutic activity because it is essential in terms of absorption, distribution, metabolism, and excretion of an orally administered drug. It is a parameter mainly influenced by several factors including pH, pKa, polymorphism, salt formation, particle size, and hydrophilic/lipophilic balance [4, 5]. However, an increasing proportion of potential drug candidates are classified as slightly water-soluble, and the formulation of these molecules into viable dosage forms represents a great challenge to the pharmaceutical field [6, 7]. In agreement with the literature, an arsenal of alternatives was successfully developed lately for overcoming the concerns arising from the reduced solubility and dissolution rate [5, 6, 8]. However, the reduction of drug particle size to the nanometer range has been, probably, the most recent and attractive approach to overcome such drawbacks [9, 10].

6.2 Nanotechnology

Nanotechnology is the study, design, creation, synthesis, manipulation, and application of materials, devices, and functional systems through the control of matter at the nanometer scale (1–100 nm) [11]. When matter is manipulated at such a scale, the atoms and molecules demonstrate entirely new phenomena and properties, and its application has diversified exponentially in almost all areas of knowledge, creating new research and development disciplines [12]. The development of the nanotechnology is based on the proposals of Richard Feynman, who, in 1960, presented a lecture entitled “There is plenty of room at the bottom” during the meeting of the American Physical Society [13]. There, he discussed the concepts of manipulation of matter at the atomic level. In 1974, Japanese scientist Norio Taniguchi was the first who use the term “nanotechnology” to describe semiconductor processes that occur at 1 nm. According to his concepts, nanotechnology consisted of the procedure of separation, consolidation, and deformation of materials atom by atom or molecule by molecule [14]. Then, in 2011, the European Commission [15] defined nanomaterials as “natural, generated or manufactured material containing particles in the free or aggregate state, where 50% or more of the particles in the distribution in number have a size of 1–100 nm.”

Pharmaceutical nanotechnology is a key area related to the development of attractive and effective drug delivery systems because it is possible to obtain new products with individual functions that are integrated into an architecture that fulfills multiple functions [16, 17]. When facing the novelty of nanomedicines, knowledge of the physicochemical properties of the drug, the carrier, and the site at which the drug is administered are even more necessary, to ensure safety and efficacy [18]. Moreover, the administration route of medicines is in agreement with the criteria imposed by the type of pathology, the urgency of the treatment, and the selection of the dosage form [19, 20]. The oral route is the preferred and more convenient way for the delivery and controlled release of drugs, peptides, proteins, and nucleic acids. However, these nanostructures are also suitable for being administered through the intranasal, ocular, and parenteral routes [21–23].

6.3 Production of Nanostructures

In general, the selection of a suitable carrier is a key step on the development of drug delivery systems being the active molecule dissolved, entrapped, encapsulated, or attached to such carriers [24]. In this regard, there are two main methodologies applied to the production of nanostructures with improved biopharmaceutical properties. One of them starts from larger raw material that is divided into minor pieces by means of high-energy processes until particles of the desired size are obtained. The other one starts from single atoms and molecules that are assembled in order to obtain the nanoparticles. These two methodologies are, respectively, called top-down and bottom-up [25].

Specifically, top-down techniques are based on the application of wet milling techniques or homogenization using high pressure [26, 27]. In the case of milling, micronized drugs are subjected to collisions at a very high speed with pearls of different materials (stainless steel, zirconium oxide, resins, glass) of about 3 mm in diameter, which constitute the active element in the process. The milling process is performed loading the chamber with the drug, a stabilizing agent (polymer or surfactant), and water. In some cases, other auxiliary substances such as carbohydrates or salts may be added to assist the treatments on the nanonized product [28]. The reduction of the particle size of the solid takes place through a double mechanism: creation of high-energy shear forces associated with the rotation of the rotor and the impact of the beads on the particles of the solid. Nowadays, it is one of the most common techniques to prepare nanosuspensions, and one of its benefits is the appearance of very few polymorphic transitions because of the controlled temperature of the process [29, 30]. The other top-down method is the high-pressure homogenization that was developed by Muller more than 20 years ago. In the processes performed in aqueous medium, a drug micronized and dispersed in an aqueous solution containing a stabilizing agent passes under pressures, from 100 to 2000 bar, through a very narrow gap (about 25 μm) [31]. This process produces a sharp reduction in static pressure leading to the formation and growth of bubbles when the suspension leaves the homogenizer gap. In this case, the reduction of the size of the particles is mainly by cavitation forces, particle collision, and shear forces produced by the piston or jet stream homogenizer [32]. The recently formed nanosuspension may be subjected more than once to the high-pressure homogenizer depending on the type of material, the pressure of the equipment, and the desired properties of the resulting product. Moreover, this methodology requires the addition of stabilizers for avoiding further aggregation of the formed nanostructures [33]. The bottom-up process involves the dissolution of a drug in a solvent and the addition of this solution to a non-solvent solution with or without a stabilizer. Once added, a supersaturated solution is formed which leads to the origin of many small nuclei [34].

There are four main techniques to produce such nanostructures: liquid solvent/non-solvent precipitation, precipitation mediated by supercritical fluids, solvent evaporation and precipitation, and precipitation using high-energy processes [35]. Even though a precipitation process can lead to the formation of amorphous substances, a further recrystallization may occur in function of time and/or temperature. The main variables to modulate the properties of the precipitates are the temperature, the homogenization speed, and the concentration of drug dissolved in the organic phase. Thus, a reduction in temperature normally results in a lower coefficient of solubility of the solid and facilitates the supersaturation and further nucleation [36]. Similarly, a minor amount of drug reduces the potential aggregation of the formed particles and allows a more homogeneous supersaturation, yielding smaller particles [37, 38]. Regarding the homogenization speed, an increase of this parameter produces smaller particles by improving the homogeneity of supersaturated solution and increasing the number of nuclei [39]. Regardless of the selected methodology for the preparation of nanostructures, the reduction of the particle size produces a substantial increase of the specific surface area and surface energy (Gibbs free energy) of the drug particles. Therefore, the new nanostructures are thermodynamically unstable and may aggregate or agglomerate in order to decrease the total free energy [40]. Usually, the terms agglomerates and aggregates describe the same event because both define the assembly of nanoparticles and generation of clusters. The agglomeration phenomenon is defined as the formation of a collection of weakly joined particles or aggregates or a mixture of the two, where the resulting external surface area is similar to the sum of the surface areas of the individual components. The forces that hold an agglomerate together are weak forces, for example, van der Waals forces. The term aggregation comprises strongly bonded or fused particles where the resulting external surface area may be significantly less than the sum of the surface areas of the individual components. The forces that hold an aggregate together are strong, for example, covalent bonds, or those that result from sintering [41]. However, while the aggregation is a reversible event, the agglomeration is an irreversible one that takes place by collision of particles due to the Brownian motion [42]. Another important aspect of nanoparticle production is the Ostwald ripening theory, focused on the growth of the smallest particles of dispersed systems and the formation of microparticles. It is due to the variations in dissolution pressure/apparent saturation solubility between particles of different size [43]. Thus, if the dispersed systems contain particles of a high variable size, the smallest particles will diffuse to the surface of the larger ones and will dissolve faster than the larger structure. In this regard, the formation of a supersaturated solution in the proximity of larger particles will lead to the growth of these particles. This phenomenon will increase the mean diameter of a particle size distribution [44].

Taking into account these facts, Ostwald ripening could be avoided or reduced using a stabilizer in order to formulate more homogeneous nanostructures [45]. However, it is not simple to elucidate the selection of the most appropriate stabilizer or a particular drug, because the type and concentration of this excipient possess a remarkable influence on the stability and therapeutic performance of the final product [46]. Depending on the properties of the molecules used as stabilizer, they can create a steric or electrostatic barrier preventing the attraction between the newly formed nanoparticles. It is important to note that the amount of the stabilizer is crucial for decreasing the Ostwald ripening. An insufficient concentration of such substances could partially cover the surface of the nanoparticles, and, thus, the attractive forces could cause the aggregation/agglomeration of the particles [47].

The selection of a stabilizer is based, mainly, on its type of stabilization (steric vs. electrostatic) and the route of administration of the nanosized product. Different polymers including polyvinyl pyrrolidone, polyvinyl alcohol, polyethylene glycols, and cellulose derivatives as well as non-ionic surfactants, including poloxamers and polyoxyethylene sorbitan monooleates, are the most widely substances for providing steric hindrance around the particles. On the other hand, ionic surfactants, including sodium lauryl sulfate, lecithin, and docusate sodium, are able to stabilize the formation of nanoparticles by means of the electrical double layer theory [48]. In addition, it is possible to combine both types of stabilizations through the mixture of ionic and non-ionic stabilizers to prevent the formation of aggregates/agglomerates. An attractive alternative is to combine ionic stabilizers (electrostatic stabilization) and non-ionic stabilizers (steric stabilization) in order to avoid the issues related to the stability of the formed nanoparticles because the addition of neutral stabilizers may decrease the self-repulsion between the charged stabilizers [49].

The zeta potential measurement is a valuable technique to calculate the effective electric charge on the surface of the nanoparticles and gives an idea of the potential physical stability of the systems because it indicates the degree of repulsion between charged particles in such systems. It is a property of materials that measure the electrical potential in colloidal solutions and represents the potential difference between the dispersion medium and the stationary layer attached to the dispersed particle [50]. For an electrostatically stabilized nanostructure, a minimum zeta potential of ± 30 mV is required, while in the case of combined electrostatically and sterically stabilized nanostructures, a minimum zeta potential of ± 20 mV is required. However, if the zeta potential is lower than those values, the existence of van der Waals or any other attractive forces would lead to a physical instability due to the aggregation or agglomeration phenomenon [51]. Each particular process and/or variable involved in the formulation of nanostructures will influence the properties of the final product including the particle size, polydispersity index, zeta potential, shape, drug loading, stability, permeability across barriers, bioavailability, and therapeutic efficacy [52]. Thus, the identification and optimization of each variable as well as the reproducibility and scaling up processes are crucial for ensuring the quality, reliability, and safety of the final product. However, to reach those goals by means of conventional procedures is very complicated and time-consuming [53]. For those reasons, the quality by design (QbD) approach is a systematic way to a rational pharmaceutical development, which starts from predefined objectives, emphasizing product knowledge and understanding and control of processes [54].

As is described in previous chapters, QbD is based on the use of verifiable science and decision-making based on Quality Risk Management (QRM), and this approach is giving new opportunities and challenges in terms of productivity, leading to more efficient pathways to pharmaceutical manufacturing [55]. The aims of the QbD applied to drug delivery systems include the achievement of significant product quality specifications based on clinical performance; the improvement of process capacity and reduction of product variability and defects by enhancing the product and process design, understanding, and control; and the improvement of the product development and manufacturing efficiency [56].

6.4 Nanocrystals

Drug nanocrystals are nanoparticles or drug nanoparticles constituted by a solid and usually crystalline substance covered by a stabilizer (surfactant or polymer). Drug nanocrystals exhibit many advantages including high efficiency of drug loading, easy scale-up for manufacture, relatively low cost for preparation, and applicability to various administration routes, such as oral, parenteral, ocular, and pulmonary delivery [57].

Although the statistical design of experiments (DoE) is a successful and widely applied strategy in different research fields and industries, its application to the development and design of drug nanocrystals has attained significant attention in the last decade. Pinar et al. described the formulation and optimization of nanocrystals of cyclosporine A (CsA), a huge molecule composed of 11 amino acids used for immunosuppression and pediatric skin diseases, such as psoriasis and atopic dermatitis [58]. The nanosuspension for oral administration of CsA was prepared using the wet milling, a top-down method, and optimized by analyzing the effects of critical process parameters, bead volume, bead diameter, and milling speed, on the particle size, particle size distribution, and zeta potential of nanosuspensions. The optimum formulation was decided from the response surface graphs obtained at the end of DoE, and further studies were carried out with this optimum formulation. After the preformulation study, 54 formulations (33 factorial designs and 2 replicates) with low, medium, and high levels of independent variables were prepared. Following the initial screening, the bead diameter and milling speed were found to be significant factors on particle size. Also, the same variables affected the polydispersity index, but milling speed was the only factor that influenced the zeta potential. According to the DoE data, the optimized CsA formulation could be obtained using a milling speed of 600 rpm during 1 h with a bead size of 0.1 mm and a bead volume of 25 mL. Under these conditions, the resulting nanoformulation showed a particle size of 561.2 ± 31.3 nm, a polydispersity index of 0.374 ± 0.037, and a zeta potential of −24.4 ± 0.9 mV. It is important to note that several replications of this experiment using the optimized parameters yielded the CsA nanosuspension with very similar particle size, polydispersity index, and zeta potential. The stability study (25 °C/30 days) suggested the formulation was stable because the evaluation of these parameters shows very similar values to those of the fresh formulation. This result confirmed that the mixture of the stabilizers, hydroxypropyl methylcellulose and sodium dodecyl sulfate, was able to avoid the aggregation/agglomeration of the particles. Dissolution studies of the optimized formulation indicated important differences in terms of the dissolved drug. Hence, depending on the dissolution media, between 20% and 100% of drug was in solution in 120 min. In vivo evaluation of the optimized CsA system performed in fasted conditions showed higher C_{max} (2.12-fold) and $AUC_{0–24}$ (2.09-fold) values than the raw drug powder. This is in agreement with the in vitro dissolution results obtained in the Fasted State Simulated Intestinal Fluid (FaSSIF) media. In fed conditions, CsA nanoformulation exhibited higher C_{max} (3.99-fold) and AUC0–24 (5.51-fold) values than the raw drug powder, and these findings were in agreement with those obtained in the Fed State Simulated Intestinal Fluid (FeSSIF) media. Overall, these data confirmed the suitability of the optimized nanocrystals for increasing the bioavailability of CsA.

Shaikh et al. described the optimization and evaluation of cilnidipine nanosuspension for enhancing the drug dissolution rate and bioavailability [59]. Cilnidipine is a dihydropyridine calcium channel blocker that inhibits cellular influx of calcium and causes vasodilatation, but its oral bioavailability is very poor (~13%) in humans. In order to improve it, drug-loaded nanosuspensions were prepared employing a precipitation-ultrasonication method in combination with DoE to optimize the obtained results. Cilnidipine was dissolved in methanol (solvent phase) at different drug concentrations (10–30 mg/mL), while the anti-solvent aqueous phase contained a polymer (poloxamer 188, 1–2% w/v) and a surfactant (Tween® 80, 0.2–0.4% w/w). The organic phase was mixed with the aqueous phase using a propeller mixer at different speeds (1000–3000 rpm). Through a Plackett–Burman design, seven different parameters (polymer amount, solvent/anti-solvent ratio, drug amount, propeller speed, sonication amplitude, sonication time, and surfactant amount) were selected within their respective ranges for preparing the nanosuspension and to analyze their influence on the dependent variables: average particle size (nm) and drug release (%) at 30 min. Application of the initial screening (12 runs) found that the most significant variables for average particle size were drug amount, solvent/anti-solvent ratio, and sonication time. In the case of drug release at 30 min, the most significant variables were drug amount, propeller speed, and surfactant amount. Then, a Box-Behnken design (17 runs) was applied for the nanoformulation optimization considering the drug amount (mg/mL in methanol), sonication time (min), and surfactant amount (% w/v) as independent variables and particle size, release at 30 min, and encapsulation efficiency as dependent variables. The four other factors included in the Plackett-Burman design were set at a fixed level. The batches of this optimization showed particle sizes in a range from 230 to 940 nm. An increase of the drug amount leads to smaller particle sizes. It could be due to higher rates of nucleation, which increases the number of nuclei reducing the sizes of new particles. The drug in solution at 30 min was in a range between 57% and 99%, while the encapsulation efficiency was in a range from 51% to 98%. The dissolution assay indicated that all the batches showed higher drug dissolution rates than the raw drug that could be due to the smaller particle sizes of those samples, as described by the Noyes-Whitney equation where the dissolution is inversely proportional to particle size. Particularly, 50% of raw drug was dissolved in 60 min, while up to 99% of drug-loaded nanosuspensions was dissolved at the same time. The differences in the dissolution profiles of different batches are influenced by both the particle size of each nanoformulation and the amount of surfactant. Finally, the validation of the experimental design was performed by preparing and evaluating the batch for all the dependent variables. The predicted and experimental values for particle size were 278.86 and 280.10 nm, while these values for drug release at 30 min were 85% and 86.32%, and for percent encapsulation efficiency, 93.28% and 92.14%, respectively. As seen, the predicted and experimental values are very similar, which corroborates the validity of the model. After the optimization, the in vivo performance of cilnidipine from nanosuspensions was evaluated in an animal model. Pharmacokinetic parameters indicated that $AUC_{(0-t)}$ and C_{max} of the drug nanocrystals are higher than both the raw drug and marketed formulation. Oral bioavailability of drug nanocrystals in rats was 2.44-fold higher than the raw drug, which confirms that the improved in vivo performance of nanoformulated cilnidipine is directly related to the increasing solubility and dissolution rate achieved by decreasing the drug particle size.

Arrua et al. described the nanoformulation of benznidazole, a drug widely used for Chagas disease [60]. Benznidazole, discovered more than 50 years ago, exhibits a low aqueous solubility and permeability that produces an erratic bioavailability. Herein, drug nanocrystals employing liquid anti-solvent precipitation process were prepared for avoiding those disadvantages. Benznidazole was dissolved in ethanol, and the solution was added to an aqueous solution with poloxamer 188 (stabilizer) for the formation of drug nanocrystals. A Plackett-Burman design (12 runs) was applied to identify the effect of 5 factors over 4 responses that affect the desirable properties of the new particles. In this initial screening, the factors were evaluated at two levels. The selected parameters were drug and stabilizer concentrations, organic/water phase ratio, propeller speed during the first 60 min of stirring, and precipitation temperature, while the studied responses were yield, nanoprecipitation efficiency, particle size, and drug solubility. The applied model suggested that yield of the process was not affected by the factors, while the efficiency of the nanoprecipitation was affected by the drug amount. Both the drug and the stabilizer amounts influenced the particle size, while both the stabilizer amount and the organic/water ratio influenced the drug solubility. Considering the results of the initial screening, a central composite design (20 runs) was used for the optimization phase, which indicated that the nanoprecipitation and particle size were significantly affected by the drug amount. Indeed, an increased amount of drug in the organic phase could yield a major saturation degree in contact with the aqueous phase leading to a larger proportion of nuclei and further formation of smaller drug nanocrystals. In addition, higher amounts of poloxamer 188 (surfactant) might increase the solubility of small particles and its deposition over the surface of large particles. As a result, increase in the mean particle size with time would occur (Ostwald ripening phenomenon). Moreover, higher amounts of stabilizer might also increase the viscosity of the aqueous solution allowing the generation of larger droplets when the ethanolic and the aqueous phases are mixed. Thus, larger drug particles could precipitate. To confirm the suitability of the experimental design, drug nanocrystals were prepared and characterized and, then, compared with the results of the theoretical results. As seen, small modifications of the selected factors led to important modifications in the responses, specifically in the benznidazole solubility and particle size. The experimental conditions for the formulation of nanocrystals were 25 mg/mL of drug, 0.01 g/mL of stabilizer, and 25% of organic phase. The result of this experiment yielded optimized drug nanocrystals with a size of 57.40 ± 0.30 nm, a polydispersity index of 0.11 ± 0.005, and an aqueous solubility of 3.75 mg/mL, which is more than tenfold higher than the raw drug. Taking into account the oral administration of benznidazole, the dissolution studies were performed in different biorelevant media. Accordingly, 60% and 95% of drug were in solution after 1 and 6 h, respectively, in a Fasted State Simulated Gastric Fluid (FaSSGF) media at pH 1.2. In the presence of two biorelevant media (FeSSIF, pH 5.0, and FaSSIF, pH 6.8) that simulate the pH of the small intestine, 40% and 100% of drug were released after 1 and 6 h, respectively. The higher drug dissolution at low pH could be due to the protonation of the molecule that assists its solubilization in water, as already described. After these results, the permeability of optimized nanocrystals was analyzed in a Caco-2 cell model, and the findings showed that the drug permeability was improved by reducing its particle size as compared to the raw drug powder, confirming the suitability of this nanoformulation in terms of drug dissolution and permeability, two requisites for an improved absorption and better therapeutic efficacy.

6.5 Polymeric Nanoparticles

Polymeric nanoparticles exhibit several advantages over other drug delivery systems including a controlled drug release, a targeted specificity, and an improved therapeutic efficacy. The most widely used carriers for preparing these nanostructures are natural polymers like chitosan and gelatin and synthetic polymers including poly (lactic acid), poly(lactic acid-co-glycolic acid) copolymers, polymethacrylate-based copolymers, and poly(ε-caprolactone) (PCL) [61]. de Mattos et al. developed and optimized poly(lactic acid) (PLA) or poly(lactic acid)-poly(ethylene glycol) (PLA-PEG) blend nanoparticles loaded with 5-fluorouracil, a widely used agent for the treatment of actinic keratoses and cutaneous carcinomas (skin cancer) [62]. The nanoparticles were obtained by the double emulsion-solvent evaporation technique using methylene chloride as organic solvent. A 2^3 factorial experimental design was developed to analyze three different factors and their effect on the physicochemical properties of the loaded polymeric nanoparticles. The variables were set at two levels based on previous experiments. Polyvinyl alcohol (PVA) concentration in the internal aqueous phase (from 0.1 to 0.2% w/v), the ratio of organic to aqueous phase (from 1:3 to 1:5), and the surfactant type in the external aqueous phase (Tween® 80 3% and PVA 2%) were optimized. The results indicated that 0.2% of PVA increased the drug loading in comparison with 0.1% of PVA. Moreover, the optimization demonstrated that 0.2% of PVA concentrations in the inner aqueous phase and 2% of PVA in the outer aqueous phase were the most acceptable conditions for higher drug encapsulation. The particle sizes of the systems were between 283 and 350 nm, and the presence of PEG in the formulation of the nanoparticles did not modify these parameters, as confirmed by analysis of variance (ANOVA). Dissolution experiments suggested that drug release from PLA and PLA-PEG nanoparticles was similar ($p > 0.05$). A biphasic release was observed, and a maximum amount of 58% of the drug was dissolved after 320 h. The optimized nanoparticles presented prolonged 5-fluorouracil release keeping the drug cytotoxicity over Hep-2 cells, which suggest the usefulness of 5-fluorouracil-loaded nanoparticles in tumor growth inhibition. Additionally, all pharmacokinetic parameters were improved by the 5-fluorouracil-loaded nanoparticles after single oral administration in rats, and the bioavailability of the drug was increased fourfold in nanoparticles compared to free 5-fluorouracil.

Ravi et al. presented the design, optimization, and evaluation of PCL-based polymeric nanoparticles for oral delivery of lopinavir, an antiretroviral protease inhibitor used in combination with other antiretrovirals to treat HIV-1 infection [63]. The authors prepared polymeric nanoparticles by an O/W emulsion-solvent evaporation technique, and a Plackett-Burman design was used, as initial screening, to select the conditions of the procedures and the range of the excipients. Eight factors were evaluated at two levels to determine their influence on two responses (particle size and drug encapsulation). The factors studied were polymer amount, type and amount of surfactant (Tween® 80 or Pluronic® F127), type of organic solvent (methylene chloride or chloroform), volume of external phase, homogenization speed, homogenization time, and rate of external phase addition (rapid or slow). In agreement with the initial screening, Tween® 80 and methylene chloride were selected as excipients, and three critical factors including surfactant and polymer amounts and homogenization time were identified for the optimization process. Then, a three-level-three-factor Box-Behnken design (17 runs) was conducted to study main influence, interaction effects, and quadratic effects of identified initial critical factors: surfactant concentration (from 0.5% to 1.5% w/v), polymer amount (from 100 to 300 mg), and time of homogenization (from 15 to 45 min). The results indicated that particle sizes of loaded nanoparticles were between 138.4 and 232.6 nm, while the drug loading was between 73.5% and 94.4%. In order to obtain an optimized system, a Design-Expert software was used, and the conditions for preparing optimized loaded nanoparticles were Tween® 80 (1.36 mg/mL), PCL

(270.74 mg), and homogenization time (40.15 min). As seen, optimized nanoparticles exhibited a size of 195.3 nm and a drug loading of 93.9%. In vitro drug release study showed biphasic sustained-release behavior of lopinavir from nanoparticles. The dissolution of the non-encapsulated drug was complete after 5 h, while the encapsulated drug showed a biphasic profile, characterized by a fast release after 10 h (60%) followed by a sustained release during 75 h. The optimized loaded formulation was also characterized at the solid state. The system exhibited a spherical shape, and the thermal evaluation suggested the amorphization of the drug after the encapsulation procedure. Pharmacokinetic studies in male Wistar rats indicated an increase in oral bioavailability of lopinavir by fourfold after incorporation into nanoparticles. The encapsulated drug was significantly accumulated in the liver and spleen tissues as compared to the free drug post-oral administration. Such accumulation could lead to a higher drug concentration in these reservoirs, which could result in a higher therapeutic efficacy in comparison to the current treatment.

Deepika et al. developed a novel delivery system of frovatriptan succinate, a triptan derivative that acts on the three-level 5-HT 1B/1D receptor, narrows blood vessels around the brain, and reduces the migraine headaches. Loaded polymeric nanoparticles for brain targeting by nasal route were prepared by means of a double emulsion methodology [64]. A central composite design was employed for optimizing the particle size, zeta potential, and drug entrapment. A total of 27 experiments were performed to reach a quality target product profile. Four factors were selected in this research: homogenization speed (from 10,000 to 15,000 rpm), homogenization time (from 10 to 15 min), surfactant concentration (from 1% to 1.5%), and polymer concentration (from 20 to 25 mg). In agreement with the results of the central composite design, higher amounts of PLGA yielded particles of higher size, probably due to the formation of large coacervates. In contrast, higher amounts of surfactant decreased both the zeta potential and size of the particles, which could be due to a more efficient stabilization decreasing the potential particle aggregation or agglomeration phenomenon. In addition, an increase in the speed of homogenization could create higher shear stress in the solution and might produce coacervates of small size and further smaller particles. The formulation obtained by using a homogenization speed of 15,000 rpm, homogenization time of 15 min, surfactant concentration of 1.25%, and amount of polymer 20 mg was the optimum, showing average particle size of 264.4 ± 0.04 nm, zeta potential of −35.17 ± 0.07 mV, and encapsulation efficiency of 65.2 ± 0.06%. The physicochemical characterization indicated that the optimized nanoparticles were spherical in shape, uniform, and non-aggregated. The encapsulated drug was, mainly, in an amorphous state, which could affect its solubility. Release studies suggested an initial burst phenomenon, around 25% of drug release in 30 min, followed by a prolonged release of 75% of drug in 72 h related, probably, to a drug diffusion from the PLGA matrix. The biopharmaceutical performance of optimized nanostructures demonstrated that the drug permeated through the nasal cavity reaching directly the brain in high concentration without entering the systemic circulation.

Nair et al. reported the preparation by polymeric nanoparticles loaded with ansamycin, a lipophilic macrocyclic antibiotic that easily diffuses through membranes. The nanostructures were prepared by means of the emulsification-solvent diffusion technique and optimized following a response surface methodology [65]. According to the initial assays, the amount of polymer (polylactic-co-glycolic acid) and surfactant (poloxamer) as well as the ratio of organic/aqueous phase were the variables that affected the particle size, zeta potential, and encapsulation efficiency. Thus, a central composite rotatable design was applied for the optimization of these three independent variables consisting of polymer concentration (from 5% to 20% w/v), surfactant concentration (from 0.25% to 1% w/v), and organic-to-aqueous phase ratio (from 5 to 15 v/v). The optimization diagram indicated that higher amounts of polymer lead to higher particle size and encapsulation efficiency. On the other hand, higher amounts of surfactant produced a lower drug encapsulation because the drug could prefer to be part of the micelles formed by the surfactant instead of being encapsulated by the polymer. In contrast, a higher ratio of organic solvent led to higher encapsulation efficiency. It is important to note that the ratio of the organic/aqueous phase modifies the viscosity of the medium and, therefore, influences both the particle size and drug encapsulation. Then, the ideal formulation parameters for the nanoparticles were based on polymer concentration of 5% w/v, surfactant concentration of 1% w/v, and 1:10 v/v organic/aqueous phase. Optimized drug-loaded nanoparticles showed an encapsulation efficiency of 89%, mean particle size of 121 nm, and zeta potential value of −25 mV. Release assay demonstrated a burst drug release from the optimized nanoparticles, and 24%, 52%, and 96% of drug in solution were found at 2 h, 12 h, and 36 h, respectively.

Sharma et al. described the formulation of nanoparticles for the intranasal delivery of lorazepam, a benzodiazepine that acts on the brain to treat anxiety, active seizure episodes, and sleeping issues [66]. The loaded nanoparticles were prepared by the emulsion-solvent evaporation (nanoprecipitation) method using PLGA as polymer and poloxamer 407 as surfactant (stabilizer). An optimization approach, through a Box-Behnken design, was selected to maximize the percentage of drug entrapment in the nanoparticles and to minimize the mean particle size. The factors, polymer concentration (w/v), surfactant concentration (w/v), aqueous/organic phase ratio (v/v), and drug concentration (lorazepam, w/v) were evaluated at three levels (26 runs). Following this set of experiments, the optimized formulation was obtained using a combination of polymer concentration (10 mg/mL), surfactant concentration (9.42 mg/mL), aqueous/organic phase ratio (10/1 v/v), and drug concentration (4.5 mg/mL). The effects of the independent variables over the particle size indicated that an increase on PLGA amount gives an increase in particle size, which could be due to the higher viscosity of the organic phase in the presence of higher polymer concentrations. Thus, the generation of larger droplets, first step on the nanoprecipitation phenomenon, could occur, leading to the appearance of larger particles. In contrast, an increase on surfactant concentration led to loaded nanoparticles of lower size. The presence of higher amounts of surfactant could improve the interfacial stability, which may decrease the nucleation rate and the formation of aggregates/agglomerates, yielding smaller particles. Regarding the other factors, an increase in drug amount did not influence the size of the loaded nanoparticles, but an increase in the amount of water of the aqueous/organic ratio produced particles of larger size, probably due to this ratio decreasing the amount of organic solvent for the formation of a primary emulsion, before the nanoprecipitation process. Concerning the other dependent variable, the drug entrapment increased with increasing values of both PLGA and lorazepam concentrations. These findings could be due to a higher viscosity of the organic phase by increasing the polymer concentration that could diminish the diffusion of lorazepam to the water leading to a higher loading of the drug into the nanostructure. In the case of higher drug amounts, it could be possible to obtain more drug solubilized in the organic solvent allowing, as a consequence, a higher entrapment into the nanoparticles. In contrast, an increase in the amount of poloxamer 407 led to lower drug encapsulation, because it may induce a higher diffusion of the solubilized drug into the aqueous phase and, as a consequence, a smaller amount of drug available for encapsulation in the organic phase. As a result, the formulated nanoparticles exhibit a particle size of 168.2 nm, a polydispersity index of 0.08, and a drug encapsulation of 83.8% that was in good agreement with the predicted values obtained from the response surface methodology model. Once optimized, the dissolution rate of the loaded formulation exhibited an initial burst release of nearly 22% from the nanoparticles after 2 h, whereas nearly 70% was released after 24 h, suggesting that a sustained drug release was obtained from the optimized nanoparticles. These in vitro results were in agreement with the drug release of the nanoparticles in nasal mucosa of sheep confirming the sustained release of lorazepam.

Rafiei and Haddadi reported a robust systematic design to formulate and optimize polymeric nanoparticles of PLGA for docetaxel intravenous delivery [67]. Docetaxel is a taxane derivative prescribed for treating breast, lung, prostate, and stomach cancer. The nanostructures prepared by a modified emulsification-solvent evaporation procedure used PLGA as carrier in ethyl acetate (solvent). A L16 Taguchi design (a "fractional" factorial design) was employed for the identification of influencing factors in order to obtain docetaxel nanoparticles with desired properties. The analyzed factors were docetaxel concentration (from 0.25 to 1.5 mg/mL), organic/aqueous phase ratio (from 1:2 to 1:5 v/v), polymer molecular weight (from 0.15–0.25 to 0.55–0.75 dL/g), polymer terminus (acid or ester), lactide/glycolide ratio (from 50:50 to 75:25), and PVA concentration (from 2.2% to 9%). Using this "fractional" factorial design, two factors were analyzed at four levels, and four factors were analyzed at two levels. By the application of this model, the combination of six different factors and their corresponding levels included the optimization of particle size, polydispersity index, zeta potential, drug loading, and drug loading efficiency. Thus, the preparation of nanoparticles following the 16 runs of factors/levels leads to an average size between 93.5 ± 6.9 and 191.4 ± 21.1 nm, a range of zeta potential between −17.8 ± 1.2 and −36.5 ± 0.6 mV, and a range of drug loading efficiency between 25.2 ± 1.1% and 96.2 ± 4.0%. As observed, the linear regression analysis indicated that both PVA (surfactant) amount and organic/aqueous phase ratio influenced the particle size ($p < 0.05$). The response surface approach showed that maximum response (i.e., particle size) was obtained lowering PVA amount (level 1). Probably, the location of the surfactant on the particle surface may avoid the interactions and further aggregation of particles resulting in the formation of smaller particles. The ratio of the solvents (level 1) also influenced the particle size because larger amounts of water would dispersed homogeneously the organic droplets with less chances of further aggregation. In addition, the diffusion rate of the organic phase to the aqueous environment would be higher in case of larger amount of aqueous phase, which produces a faster formation of particles and less adhesion. Regarding the PVA effect over the particle size, higher amounts of the surfactant may produce smaller particles because it is possible to avoid the aggregation of particles by stabilizing the system. In addition, the polymer molecular weight (MW) has a direct impact over the average particle size. In the presence of PLGA of low MW, the organic phase tends to be less viscous, which finally will produce droplets of small size during sonication. According to this, smaller particles would be provided. On the other hand, PLGA of high MW would increase the viscosity of the organic phase, and, therefore, both the droplets and new particles would be bigger. The analysis of the zeta potential of the particles after the 16 runs was in a range from -16 to -37 mV. In order to see how each factor affects this parameter, the ranges between the lower and higher values of zeta potential were calculated, and the results indicated that polymer terminus and polymer MW were the factors that influenced the zeta potential in higher and lower degrees, respectively. This finding could be due to the functional group of the polymer (acid) that may be negatively charged increasing the zeta potential value. The evaluation of the drug loading efficiency was in a range from 25% to 96%, according to the Taguchi design. The organic/aqueous phase ratio affects the loading more than the other factors, followed by the amount of drug. Hence, lower drug loading was yielded when the organic/aqueous phase ratio was 1:5 and drug amount was the highest. In contrast, the highest drug loading was seen when the organic/aqueous phase ratio was 1:2. These findings could be attributed to the fact that less amount of water allows the drug molecules to remain in the organic phase where it dissolves the polymer. Thus, during the formation of polymeric nanoparticles, it is possible to encapsulate larger amounts of drugs. On the other hand, higher amounts of drug may saturate the polymeric matrix leading, probably, to a minor entrapment of drug. Overall, the optimal nanoparticles presented an average diameter of 123.6 ± 9.5 nm, a polydispersity index of 0.245 ± 0.041, and a zeta potential of -28.3 ± 1.25 mV. Once optimized, the PLGA nanoparticles were PEGylated (surface modification) in order to reduce drug degradation as well as in vivo elimination. Then, a dissolution assay of both non-modified and modified nanoparticles showed different drug release rates. First, a burst release of 20% and near 40% at 1 day for the non-modified and modified formulations was observed, respectively, which may be due to the dissolution of drug molecules located at the surface of the nanoparticles. However, at 5 days, the non-modified nanoparticles release around 25%, while the other nanoparticles released 50% of drug. It may be due to the PEGylation that facilitates the wettability of the nanoparticles increasing the drug release rate. In any case, the optimized nanoparticle formulation demonstrated suitable properties for intravenous sustained drug release.

6.6 Liposomes

Liposomes are colloidal vesicles of size between 50 nm and some microns, consisting of one or more aqueous spaces, enclosed in an equal number of phospholipid lamellae or bilayers. Their compositions are very similar to that of the cell membranes [16]. These vesicles have the ability of encapsulating hydrophilic, lipophilic, or even amphiphilic drugs reducing their degradability in addition to controlling their release into different biological environments improving their bioavailability and their therapeutic performance [23]. The formulation of liposomes implies several steps as well as a wide range of excipients, and, therefore, the application of the QbD strategy to find the most critical factors and responses during the production steps of the development process is very attractive.

Refaat et al. published the optimization and evaluation of propolis liposomes prepared by spraying technique as a promising therapeutic approach for COVID-19 [68]. Propolis extract has a high potential for the inhibition of angiotensin-converting enzyme (ACE), a functional receptor for SARS-CoV-2, pathogenic coronavirus that cause a pandemic of acute respiratory disease (COVID-19). Propolis extract, cholesterol, and lipoid S75 (phospholipids with 70% phosphatidylcholine) were dissolved in absolute ethanol and sprayed on an aqueous solution containing sucrose. After solvent evaporation, the loaded liposomes were formed in the aqueous solution. The optimization of the formulation parameters was done by means of the response surface methodology. Lipid amount, percentage of cholesterol, and propolis extract loading were the independent variables analyzed, while the responses were encapsulation efficiency and release of the encapsulated extract. The variables were analyzed at three levels: lipid (80, 60, 40 mM), cholesterol (66%, 43%, 20%), and extract loading (5, 3.25, 1.5 mg). A Box-Behnken design (17 runs) and the response surface diagram were applied to optimize a liposomal formulation. According to the DoE results, the encapsulation efficiency of the liposomes was in a range from 35 ± 1.73% to 79.2 ± 0.9%. The highest percentage was obtained when both the amounts of lipid and propolis extract were used. The proposed models herein suggested a good agreement between the predicted and the experimental values confirming the significance of the model for the prediction of the encapsulation efficiency within the levels of the factors considered for running the experiments. The correlation coefficient showed the similarities between those values, and the ANOVA analysis confirmed that the amount of lipoid S75 and drug loading affect directly the drug entrapment, because an increase in both factors leads to a higher encapsulation efficiency. As reported, a higher lipid amount may increase the thickness of the lipid bilayer membrane producing larger loading of hydrophobic molecules (propolis extract). In contrast, cholesterol amount did not influence this response. Regarding the other response, the 17 runs indicated that percentage of propolis extract released was between 27.5 ± 2.3% and 84.15 ± 1.17%, being the highest value obtained with a formulation developed with maximum level of lipid, medium level of drug loading, and minimum level of cholesterol. In particular, 40 mM of lipid and 20% of cholesterol released more than 75% of extract, but in the presence of 60% of cholesterol, the released amount decreased to 27%. It could be due to a minor fluidity of the lipid membrane (less drug released) by increasing the amount of cholesterol. Thus, using a smaller amount of cholesterol could result in higher propolis release and improved biological activity. Following the application of this approach, the optimized formulation parameters were 60 mM of lipid amount, 20% of cholesterol amount, and propolis loading of 5 mg/mL. Under these conditions, the encapsulation efficiency was 70.11% and a propolis release of 81.80%. In order to confirm the applied model, five experiments using the same parameters were performed, and the results showed an encapsulation efficiency of 68 ± 2.4% and a release of 76 ± 3.2%, which is in good agreement with the predicted values. The antiviral effects of the optimized loaded liposomes against COVID-19 virus, done by RT-PCR, showed that this lipid nanostructure inhibited the virus replication much more than the non-encapsulated extract. It could be due to an improved permeability of the propolis extract when it was encapsulated in a lipid bilayer membrane, very similar to the bilayer of cell membrane, which may facilitate the uptake processes of both the hydrophilic and hydrophobic substances of the extract. The loaded liposomes exhibit very similar in vitro inhibitory activity against COVID-19 virus than the commercial antiviral drug remdesivir suggesting the relevance of this optimization.

Zhu et al. described the development and optimization of galangin-loaded, liver-targeting oral liposomes and evaluated their hepatoprotective efficacy [69]. Galangin is a natural flavonoid that may protect cells from being damaged by free radicals because of its antioxidant capacity and might help to reduce/avoid the growth of cancer cells. Liposomes were prepared through the thin-film hydration method using soybean lecithin, cholesterol, isopropyl myristate, and galangin dissolved in absolute ethanol. After ultrasonication, the solvent was evaporated, and sodium cholate, dissolved in ethanol, was added to the lipid mixture. The procedure was optimized via response surface methodology-Box-Behnken design for improving the encapsulation efficiency. Three parameters were optimized (3 levels/17 runs). The independent variables were different molar ratios of phospholipid to galangin, isopropyl myristate to galangin, and sodium cholate to galangin, while the encapsulation efficiency was the dependent variable. Following the matrix design, the encapsulation efficiency of the flavonoid was between 64.12% and 91.54%. The response surface methodology was used to see the influence of each factor on the response. The phospholipid/galangin and sodium cholate/galangin ratios influenced positively the encapsulation efficiency, but isopropyl myristate/galangin and sodium cholate/galangin ratios influenced negatively the response. Therefore, phospholipid/galangin ratio exhibited the major effect on the encapsulation efficiency. Then, a batch of loaded liposomes was prepared for validation of the model, and a remarkable similarity on the encapsulation efficiency was observed. The theoretical value of encapsulation efficiency was 91.23%, while the experimental encapsulation efficiency from the prepared sample was 92.36 ± 2.71%, confirming the validity of the optimization process. The ability of phospholipid to encapsulate galangin in high yield was also confirmed. The optimized loaded formulation showed a spherical shape, a homogeneous surface, and droplets of less than 100 nm. The liposomes exhibited a particle size of 74.67 ± 7.09 nm, a polydispersity index of 0.14 ± 0.06, and a zeta potential of −47.61 ± 3.65 mV. Such values confer an acceptable stability to the colloidal system. Then, in vivo studies of the optimized loaded liposomes showed a difference in T_{max} between raw galangin (2 h) and loaded liposomes (4 h) caused by, possibly, the slower release of drug from the lipid vesicles. Moreover, the oral bioavailability in rats of galangin liposomes was more than 450-fold higher than the non-encapsulated drug, probably due to an increase of galangin solubility in the lipid nanostructure. The presence of sodium cholate, a bile salt, in the lipid bilayers of the liposomes might assist the transport through the paracellular pathway increasing the drug permeability and bioavailability, as well. In vivo studies in mice demonstrated that these loaded liposomes were very effective for decreasing the levels of two aminotransferases used to indicate liver damage in animal models. In conclusion, loaded optimized liposomes revealed interesting hepatoprotective effects in comparison with the free drug and control reference.

Kutbi et al. reported the optimization of hyaluronate-based liposomes to augment the oral delivery and the bioavailability, in rats, of berberine [70]. It is a natural alkaloid widely used to treat several gastrointestinal disorders and is also a potential effective substance for the treatment of type 2 diabetes mellitus. In order to improve its permeability and bioavailability, liposomes were prepared by the film hydration methodology using soya bean phosphatidylcholine and cholesterol as lipid matrix. A full factorial design was applied, and the selected factors were the total lipid amount and berberine and hyaluronic acid amounts (two levels). The optimized responses were the encapsulation efficiency (%) and the mean particle size. Three-dimensional response surface diagrams showed that the encapsulation efficiency was in a range between 29.91% and 78.12% and hyaluronic acid presented a significant influence on this response, even more than lipid and drug amounts. This finding could be attributable to potential ionic interactions between the anionic carboxy-group of hyaluronic acid and cationic ammonium group of berberine, which could incorporate higher amounts of drug into the lipid vesicle. Hyaluronic acid may also increase both the viscosity of the inner space of the liposomes and the aqueous medium nearby to the lipid structure that could help to avoid the leaking of drug from the liposome. Regarding the other response, the plots indicated that particle size was in range between 389.7 and 825.7 nm and that berberine amount produced the highest influence on liposome particle size. On the other hand, lipid amount and the addition of hyaluronic acid did not affect the response. Hence, the optimal formulation obtained with 1:20 berberine/lipid ratio and 30 mg of sodium hyaluronate presented an encapsulation efficiency of 78.1 ± 0.1% with mean size 520.7 ± 19.9 nm. A morphological evaluation revealed a multi-lamellar vesicle and a clear deformation of these systems. In agreement with the electronic images, it was possible to see that the color of the inner phase is due to the hyaluronic acid, whereas the deformation of the vesicles is because of the viscosity of hyaluronic acid at 25 °C. The release assay showed that free drug was quantitatively in solution after 6 h. In contrast, encapsulated berberine exhibited a first burst release followed by a sustained dissolution. Such prolonged effect could be due to the already mentioned ionic interactions between the drug and the hyaluronic acid. On the other hand, the solubilization of hyaluronic acid in the external aqueous phase of the vesicles may result on a higher viscosity of the internal organic phase. As a result, a lower drug diffusion through both phases could occur extending the drug release as a function of time. In vivo studies in rats demonstrated that the loaded liposomal with hyaluronic acid formulation showed a C_{max} higher than the plain loaded liposomes (without hyaluronic acid) and the berberine solution. In addition, the AUC_{0-24} of the loaded liposomal with hyaluronic acid, plain loaded liposomes, and the berberine solution were 854, 439.9 ng h/mL, and 194.8 ng h/mL, respectively. The analysis of both parameters confirmed that the drug absorption from the optimized liposomes was greater than that from the solution. Relative bioavailability of the hyaluronic liposomes and plain liposomes was 4.38 times and 2.26 times higher than raw drug confirming the ability of the lipid vesicles for improving the delivery of berberine.

6.7 Conclusions

The successful development of different nanostructures for improving drug delivery is one of the most recent and important achievements on the road to efficient therapeutic tools with better biopharmaceutical properties. The reduction of drug particle sizes to the nanometer range leads to the formation of particles with higher surface area, which results in higher drug solubility and dissolution. The choice of a proper procedure and a careful selection of polymers and surfactants are fundamental steps toward the final product with the appropriate properties. However, in most of the cases, these methodologies require a lot of time and efforts as well as waste of material, particularly in case of complex formulations. As a consequence, the application of design of experiments strategy is an attractive tool for increasing the capability of the formulation procedure while lowering the variability of the final pharmaceutical product by improving the design of both the process and its product. Thus, the versatility of these principles for achieving a crucial understanding of formulation and processing variables influencing the preparation of drug nanocrystals, loaded polymeric nanoparticles, and loaded lipid vesicles was herein summarized. Taking into account the evidence reported, it is fundamental to implement these concepts because more and more nanomedicines will be commercially available in the near future.

References

1. Suay-García B, Bueso-Bordils JI, Falcó A, Antón-Fos GM, Alemán-López PA (2022) Virtual combinatorial chemistry and pharmacological screening: a short guide to drug design. Int J Mol Sci 23:1620. https://doi.org/10.3390/ijms23031620
2. Xiao F, Chen Z, Wei Z, Tian L (2020) Hydrophobic interaction: a promising driving force for the biomedical applications of nucleic acids. Adv Sci (Weinh) 7:2001048. https://doi.org/10.1002/advs.202001048

3. Tsantili-Kakoulidou A, Demopoulos VJ (2021) Drug-like properties and fraction lipophilicity index as a combined metric. ADMET DMPK 9:177–190. https://doi.org/10.5599/admet.1022
4. Lipinski CA, Lombardo F, Dominy BW, Feeney PJ (2001) Experimental and computational approaches to estimate solubility and permeability in drug discovery and development settings. Adv Drug Deliv Rev 46:3–26. https://doi.org/10.1016/s0169-409x(00)00129-0
5. Williams HD, Trevaskis NL, Charman SA, Shanker RM, Charman WN, Pouton CW, Porter CJ (2013) Strategies to address low drug solubility in discovery and development. Pharmacol Rev 65:315–499. https://doi.org/10.1124/pr.112.005660
6. O'Donnell KP, Williams RO (2012) Optimizing the formulation of poorly water-soluble drugs. In: Williams R III, Watts A, Miller D (eds) Formulating poorly water soluble drugs, AAPS advances in the pharmaceutical sciences series, vol 3. Springer, New York. https://doi.org/10.1007/978-1-4614-1144-4_2
7. Bae YH, Park K (2020) Advanced drug delivery 2020 and beyond: perspectives on the future. Adv Drug Deliv Rev 158:4–16. https://doi.org/10.1016/j.addr.2020.06.018
8. Khadka P, Ro J, Kim H, Kim I, Kim JT, Kim H, Cho JM, Yun G, Lee J (2014) Pharmaceutical particle technologies: an approach to improve drug solubility, dissolution and bioavailability. Asian J Pharm Sci 9:304–316. https://doi.org/10.1016/j.ajps.2014.05.005
9. Gómez-Aguado I, Rodríguez-Castejón J, Vicente-Pascual M, Rodríguez-Gascón A, Solinís MÁ, Del Pozo-Rodríguez A (2020) Nanomedicines to deliver mRNA: state of the art and future perspectives. Nanomaterials (Basel) 10:364. https://doi.org/10.3390/nano10020364
10. Hejmady S, Pradhan R, Alexander A, Agrawal M, Singhvi G, Gorain B, Tiwari S, Kesharwani P, Dubey SK (2020) Recent advances in targeted nanomedicine as promising antitumor therapeutics. Drug Discov Today 25:2227–2244. https://doi.org/10.1016/j.drudis.2020.09.031
11. Drexler KE (1989) Nanosystems: molecular machinery, manufacturing, and computation, vol 25. Wiley, New York. https://doi.org/10.3390/molecules25010112
12. Poole CP, Owens FJ (2003) Introduction to nanotechnology. Wiley, New York
13. Feynman R (1960) There's plenty of room at the bottom. Eng Sci 23:22–36
14. Taniguchi N (1974) On the basic concept of nano-technology. In: Proceedings of the International Conference on Production Engineering, Tokyo
15. https://ec.europa.eu/environment/chemicals/nanotech/faq/definition_en.htm
16. Lombardo D, Kiselev MA, Caccamo MT (2019) Smart nanoparticles for drug delivery application: development of versatile nanocarrier platforms in biotechnology and nanomedicine. J Nanomater 12:1–26. https://doi.org/10.1155/2019/3702518
17. Yuan Y, Gu Z, Yao C, Luo D, Yang D (2019) Nucleic acid-based functional nanomaterials as advanced cancer therapeutics. Small 15:e1900172. https://doi.org/10.1002/smll.201900172
18. Merisko-Liversidge EM, Liversidge GG (2008) Drug nanoparticles: formulating poorly water-soluble compounds. Toxicol Pathol 36:43–48. https://doi.org/10.1177/0192623307310946
19. Martinez MN, Amidon GL (2002) A mechanistic approach to understanding the factors affecting drug absorption: a review of fundamentals. J Clin Pharmacol 42:620–643. https://doi.org/10.1177/00970002042006005
20. Chow SC (2014) Bioavailability and bioequivalence in drug development. Wiley Interdiscip Rev Comput Stat 6:304–312. https://doi.org/10.1002/wics.1310
21. Chakraborty C, Pal S, Doss GP, Wen ZH, Lin CS (2013) Nanoparticles as 'smart' pharmaceutical delivery. Front Biosci (Landmark Ed) 18:1030–1050. https://doi.org/10.2741/4161
22. Chavanpatil MD, Khdair A, Panyam J (2006) Nanoparticles for cellular drug delivery: mechanisms and factors influencing delivery. J Nanosci Nanotechnol 6:2651–2663. https://doi.org/10.1166/jnn.2006.443
23. Tyagi P, Subramony JA (2018) Nanotherapeutics in oral and parenteral drug delivery: key learnings and future outlooks as we think small. J Control Release 272:159–168. https://doi.org/10.1016/j.jconrel.2018.01.009

24. Biswas A, Bayer IS, Biris AS, Wang T, Dervishi E, Faupel F (2012) Advances in top-down and bottom-up surface nanofabrication: techniques, applications & future prospects. Adv Colloid Interf Sci 170:2–27. https://doi.org/10.1016/j.cis.2011.11.001
25. Verma S, Gokhale R, Burgess DJ (2009) A comparative study of top-down and bottom-up approaches for the preparation of micro/nanosuspensions. Int J Pharm 380:216–222. https://doi.org/10.1016/j.ijpharm.2009.07.005
26. Aryal S, Park H, Leary JF, Key J (2019) Top-down fabrication-based nano/microparticles for molecular imaging and drug delivery. Int J Nanomedicine 14:6631–6644. https://doi.org/10.2147/IJN.S212037
27. Van Eerdenbrugh B, Van den Mooter G, Augustijns P (2008) Top-down production of drug nanocrystals: nanosuspension stabilization, miniaturization and transformation into solid products. Int J Pharm 364:64–75. https://doi.org/10.1016/j.ijpharm.2008.07.023
28. Li M, Azad M, Davé R, Bilgili E (2016) Nanomilling of drugs for bioavailability enhancement: a holistic formulation-process perspective. Pharmaceutics 8:17. https://doi.org/10.3390/pharmaceutics8020017
29. Hagedorn M, Bögershausen A, Rischer M, Schubert R, Massing U (2017) Dual centrifugation – a new technique for nanomilling of poorly soluble drugs and formulation screening by an DoE-approach. Int J Pharm 530:79–88. https://doi.org/10.1016/j.ijpharm.2017.07.047
30. Moribe K, Ueda K, Limwikrant W, Higashi K, Yamamoto K (2013) Nano-sized crystalline drug production by milling technology. Curr Pharm Des 19:6246–6258. https://doi.org/10.2174/1381612811319350003
31. Keck CM, Müller RH (2006) Drug nanocrystals of poorly soluble drugs produced by high pressure homogenisation. Eur J Pharm Biopharm 62:3–16. https://doi.org/10.1016/j.ejpb.2005.05.009
32. Kluge J, Muhrer G, Mazzotti M (2012) High pressure homogenization of pharmaceutical solids. J Supercrit Fluids 66:380–388. https://doi.org/10.1016/j.supflu.2012.01.009
33. Karadag A, Ozcelik B, Huang Q (2014) Quercetin nanosuspensions produced by high-pressure homogenization. J Agric Food Chem 62:1852–1859. https://doi.org/10.1021/jf404065p
34. de Waard H, Frijlink HW, Hinrichs WL (2011) Bottom-up preparation techniques for nanocrystals of lipophilic drugs. Pharm Res 28:1220–1223. https://doi.org/10.1007/s11095-010-0323-3
35. Sinha B, Müller RH, Möschwitzer JP (2013) Bottom-up approaches for preparing drug nanocrystals: formulations and factors affecting particle size. Int J Pharm 453:126–141. https://doi.org/10.1016/j.ijpharm.2013.01.019
36. Chan HK, Kwok PC (2011) Production methods for nanodrug particles using the bottom-up approach. Adv Drug Deliv Rev 63:406–416. https://doi.org/10.1016/j.addr.2011.03.011
37. Huang F, Zhang H, Banfield JF (2003) Two-stage crystal-growth kinetics observed during hydrothermal coarsening of nanocrystalline ZnS. Nano Lett 3:373–378. https://doi.org/10.1021/nl025836+
38. Yadav SK, Mishra S, Mishra B (2012) Eudragit-based nanosuspension of poorly water-soluble-drug: formulation and in vitro-in vivo evaluation. AAPS PharmSciTech 13:1031–1044. https://doi.org/10.1208/s12249-012-9833-0
39. Mehrotra A, Pandit JK (2012) Critical process parameters evaluation of modified nanoprecipitation method on lomustine nanoparticles and cytostatic activity study on L132 human cancer cell line. J Nanomed Nanotechnol 3:149. https://doi.org/10.4172/2157-7439.1000149
40. Arrua EC, Hartwig O, Ho DK, Loretz B, Murgia X, Salomon CJ, Lehr CM (2021) Surfactant-free glibenclamide nanoparticles: formulation, characterization and evaluation of interactions with biological barriers. Pharm Res 38:1081–1092. https://doi.org/10.1007/s11095-021-03056-2
41. Luo W, Hu W (2013) Gibbs free energy, surface stress and melting point of nanoparticle. Physica B 425:90–94. https://doi.org/10.1016/j.physb.2013.05.025

42. Zare Y (2016) Study of nanoparticles aggregation/agglomeration in polymer particulate nanocomposites by mechanical properties. Composites A 84:158–164. https://doi.org/10.1016/j.compositesa.2016.01.020
43. Ashraf MA, Peng W, Zare Y, Rhee KY (2018) Effects of size and aggregation/agglomeration of nanoparticles on the interfacial/interphase properties and tensile strength of polymer nanocomposites. Nanoscale Res Lett 13:214. https://doi.org/10.1186/s11671-018-2624-0
44. Dagtepe P, Chikan V (2010) Quantized Ostwald ripening of colloidal nanoparticles. J Phys Chem C 114:16263–16269. https://doi.org/10.1021/jp105071a
45. Thessing J, Qian J, Chen H, Pradhan N, Peng X (2007) Interparticle influence on size/size distribution evolution of nanocrystals. J Am Chem Soc 129:2736–2737. https://doi.org/10.1021/ja068072z
46. Lindfors L, Skantze P, Skantze U, Rasmusson M, Zackrisson A, Olsson U (2006) Amorphous drug nanosuspensions. 1. Inhibition of Ostwald ripening. Langmuir 22:906–910. https://doi.org/10.1021/la0523661
47. Lou H, Gao L, Wei X, Zhang Z, Zheng D, Zhang D, Zhang X, Li Y, Zhang Q (2011) Oridonin nanosuspension enhances anti-tumor efficacy in SMMC-7721 cells and H22 tumor bearing mice. Colloids Surf B Biointerfaces 87:319–325. https://doi.org/10.1016/j.colsurfb.2011.05.037
48. Madkour M, Bumajdad A, Al-Sagheer F (2019) To what extent do polymeric stabilizers affect nanoparticles characteristics? Adv Colloid Interf Sci 270:38–53. https://doi.org/10.1016/j.cis.2019.05.004
49. Tuomela A, Hirvonen J, Peltonen L (2016) Stabilizing agents for drug nanocrystals: effect on bioavailability. Pharmaceutics 8:16. https://doi.org/10.3390/pharmaceutics8020016
50. Inam W, Bhadane R, Akpolat RN, Taiseer RA, Filippov SK, Salo-Ahen OMH, Rosenholm JM, Zhang H (2022) Interactions between polymeric nanoparticles and different buffers as investigated by zeta potential measurements and molecular dynamics simulations. VIEW 3:20210009. https://doi.org/10.1002/VIW.20210009
51. Reis CP, Neufeld RJ, Ribeiro AJ, Veiga F, Nanoencapsulation I (2006) Methods for preparation of drug-loaded polymeric nanoparticles. Nanomedicine 2:8–21. https://doi.org/10.1016/j.nano.2005.12.003
52. Gaumet M, Vargas A, Gurny R, Delie F (2008) Nanoparticles for drug delivery: the need for precision in reporting particle size parameters. Eur J Pharm Biopharm 69:1–9. https://doi.org/10.1016/J.EJPB.2007.08.001
53. Paliwal R, Babu RJ, Palakurthi S (2014) Nanomedicine scale-up technologies: feasibilities and challenges. AAPS PharmSciTech 15:1527–1534. https://doi.org/10.1208/s12249-014-0177-9
54. Lionberger RA, Lee SL, Lee L, Raw A, Yu LX (2008) Quality by design: concepts for ANDAs. AAPS J 10:268–276. https://doi.org/10.1208/s12248-008-9026-7
55. Yu LX (2008) Pharmaceutical quality by design: product and process development, understanding, and control. Pharm Res 25:781–791. https://doi.org/10.1007/s11095-007-9511-1
56. Raw AS, Lionberger R, Yu LX (2011) Pharmaceutical equivalence by design for generic drugs: modified-release products. Pharm Res 28:1445–1453. https://doi.org/10.1007/s11095-011-0397-6
57. Müller RH, Gohla S, Keck CM (2011) State of the art of nanocrystals – special features, production, nanotoxicology aspects and intracellular delivery. Eur J Pharm Biopharm 78:1–9. https://doi.org/10.1016/j.ejpb.2011.01.007
58. Pınar SG, Canpınar H, Tan Ç, Çelebi N (2022) A new nanosuspension prepared with wet milling method for oral delivery of highly variable drug cyclosporine A: development, optimization and in vivo evaluation. Eur J Pharm Sci 171:106123. https://doi.org/10.1016/j.ejps.2022.106123
59. Shaikh F, Patel M, Patel V, Patel A, Shinde G, Shelke S, Pathan I (2022) Formulation and optimization of cilnidipine loaded nanosuspension for the enhancement of solubility, dissolution and bioavailability. J Drug Deliv Sci Technol 69:103066. https://doi.org/10.1016/j.jddst.2021.103066

60. Arrua EC, Hartwig O, Loretz B, Goicoechea H, Murgia X, Lehr CM, Salomon CJ (2022) Improving the oral delivery of benznidazole nanoparticles by optimizing the formulation parameters through a design of experiment and optimization strategy. Colloids Surf B Biointerfaces 217:112678. https://doi.org/10.1016/j.colsurfb.2022.112678
61. Banik BL, Fattahi P, Brown JL (2016) Polymeric nanoparticles: the future of nanomedicine. Wiley Interdiscip Rev Nanomed Nanobiotechnol 8:271–299. https://doi.org/10.1002/wnan.1364
62. Mattos AC, Altmeyer C, Tominaga TT, Khalil NM, Mainardes RM (2016) Polymeric nanoparticles for oral delivery of 5-fluorouracil: formulation optimization, cytotoxicity assay and pre-clinical pharmacokinetics study. Eur J Pharm Sci 84:83–91. https://doi.org/10.1016/j.ejps.2016.01.012
63. Ravi PR, Vats R, Dalal V, Gadekar N, Aditya N (2015) Design, optimization and evaluation of poly-ε-caprolactone (PCL) based polymeric nanoparticles for oral delivery of lopinavir. Drug Dev Ind Pharm 41:131–140. https://doi.org/10.3109/03639045.2013.850710
64. Deepika D, Dewangan HK, Maurya L, Singh S (2019) Intranasal drug delivery of frovatriptan succinate-loaded polymeric nanoparticles for brain targeting. J Pharm Sci 108:851–859. https://doi.org/10.1016/j.xphs.2018.07.013
65. Nair KGS, Velmurugan R, Sukumaran SK (2020) Formulation and optimization of ansamycin-loaded polymeric nanoparticles using response surface methodology for bacterial meningitis. BioNanoSci 10:279–291. https://doi.org/10.1007/s12668-019-00713-0
66. Sharma D, Maheshwari D, Philip G, Rana R, Bhatia S, Singh M, Gabrani R, Sharma SK, Ali J, Sharma RK, Dang S (2014) Formulation and optimization of polymeric nanoparticles for intranasal delivery of lorazepam using Box-Behnken design: in vitro and in vivo evaluation. Biomed Res Int 2014:156010. https://doi.org/10.1155/2014/156010
67. Rafiei P, Haddadi A (2019) A robust systematic design: optimization and preparation of polymeric nanoparticles of PLGA for docetaxel intravenous delivery. Mater Sci Eng C Mater Biol Appl 104:109950. https://doi.org/10.1016/j.msec.2019.109950
68. Refaat H, Mady FM, Sarhan HA, Rateb HS, Alaaeldin E (2021) Optimization and evaluation of propolis liposomes as a promising therapeutic approach for COVID-19. Int J Pharm 592:120028. https://doi.org/10.1016/j.ijpharm.2020.120028
69. Zhu J, Wang Q, Li H, Zhang H, Zhu Y, Omari-Siaw E, Sun C, Wei Q, Deng W, Yu J, Xu X (2018) Galangin-loaded, liver targeting liposomes: optimization and hepatoprotective efficacy. J Drug Deliv Sci Technol 46:339–347. https://doi.org/10.1016/j.jddst.2018.05.034
70. Kutbi HI, Asfour HZ, Kammoun AK, Sirwi A, Cavalu S, Gad HA (2021) Optimization of hyaluronate-based liposomes to augment the oral delivery and the bioavailability of berberine. Materials (Basel) 14:5759. https://doi.org/10.3390/ma14195759

Chapter 7
Control Strategies of Solid Dosage Forms by PAT Tools

Pierre-Yves Sacré, Charlotte De Bleye, Philippe Hubert, and Eric Ziemons

Abstract As the pharmaceutical manufacturing industry is moving from batch-wise production to continuous production, concepts such as quality by design, continuous process verification, and real-time release testing are gaining importance. These approaches need real-time measurements of critical quality attributes (CQAs) of the processed matter. The technologies enabling these measurements are grouped under the name of "process analytical technologies" (PAT). The present chapter covers some of the most used PAT tools and the general strategy to analyze their outputs.

7.1 Introduction

Whenever patients come to a pharmacy with a prescription, they expect to return home with high-quality medicines. This quality is ensured by a mastered manufacturing process. To verify that the process is under control and that the final product will have the desired quality, the deployment of a control strategy is mandatory.

Pharmaceutical manufacturing traditionally consists in a sequence of separated steps leading to the production of large batches of product. However, any small failure in one of these steps may lead to the quality control failure of the whole batch and eventually to possible drug shortages [1]. This is especially true in the actual globalized pharmaceutical environment with the active ingredient, the excipients,

P.-Y. Sacré (✉)
CIRM, Research Support Unit in Chemometrics, Department of Pharmacy, University of Liege (ULiege), Liege, Belgium
e-mail: pysacre@uliege.be

C. De Bleye · P. Hubert · E. Ziemons
CIRM, VibraSante Hub, Department of Pharmacy, Laboratory of Pharmaceutical Analytical Chemistry, University of Liege (ULiege), Liege, Belgium
e-mail: cdebleye@uliege.be; ph.hubert@uliege.be; eziemons@uliege.be

M. C. Breitkreitz, H. C. Goicoechea (eds.), *Introduction to Quality by Design in Pharmaceutical Manufacturing and Analytical Development*, AAPS Introductions in the Pharmaceutical Sciences 10, https://doi.org/10.1007/978-3-031-31505-3_7

and the finished product that are manufactured in different (sometimes very far) locations [2]. Therefore, the pharmaceutical manufacturing industry started a transition to be more agile, flexible, and robust. One of the proposed solutions is the shift from batch manufacturing to continuous manufacturing [3]. This shift of paradigm comprises several challenges in science and engineering, among which a better understanding, monitoring, and control of the process are mandatory.

In that perspective, the US Food and Drug Administration (US FDA) launched the guidance for process analytical technology (PAT) in 2004 to promote the establishment of advanced analytical strategies by the pharmaceutical industry [4]. This guidance defines PAT as "a system for designing, analysing, and controlling manufacturing through timely measurements (i.e., during processing) of critical quality and performance attributes of raw and in-process materials and processes, with the goal of ensuring final product quality." It was one of the early moves toward the control of the quality of the final drug product through the design and control of the manufacturing process. It has been shortly followed by the International Council for Harmonisation of Technical Requirements for Pharmaceuticals for Human Use (ICH) quality guidelines Q8, Q9, and Q10 describing the Quality by Design (QbD) framework [5–7]. In a fully validated QbD-compliant process, the quality of the final product is ensured if all the critical process parameters are under control.

The development of a QbD-compliant process follows a common scheme. First, the quality target product profile (QTPP) is defined regarding the quality, safety, and efficacy of the drug product. Then, the critical quality attributes (CQAs) are identified. The CQAs are the product properties having an impact on the product quality that can be studied and controlled (e.g., hardness, friability, weight uniformity, content uniformity, etc. for a tableting process). Once the CQAs are identified, the process is analyzed preferably via a design of experiments (DoE) approach to identify the critical process parameters (CPPs) and critical material attributes (CMAs) that influence the CQAs (Fig. 7.1). Generally, the DoE results enable the modelling of the relationship between CMAs, CPPs, and CQAs and their interactions providing a deeper understanding of the process and the establishment of limits for these CPPs and CMAs within which the quality of the final product is assured [7, 8]. This process understanding is intended to build a process that could be more resilient to variability of the input materials.

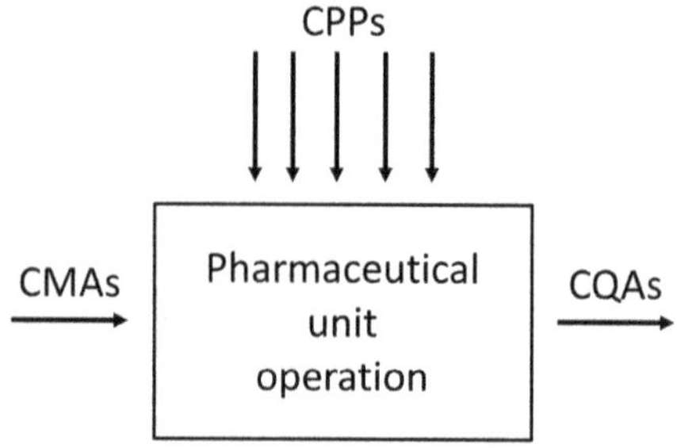

$$\text{CQAs} = f(CPP_1, CPP_2, \ldots, CMA_1, CMA_2, \ldots)$$

Fig. 7.1 Linking input critical material attributes (CMAs) and critical process parameters (CPPs) to output critical quality attributes (CQAs) for a pharmaceutical unit operation. (Adapted from Yu et al. [8])

When the different CMAs and CPPs are correctly identified and modelled, it is necessary to establish an effective control strategy to ensure that the process is working as expected and therefore that the final product quality requirements are fulfilled. This control strategy generally includes the control of the input material attributes, the product specifications, the unit operations identified as critical for the final product quality, and the CQAs during the processing. A key parameter of a continuous manufacturing process being the implementation of a real-time release testing, it is crucial to implement a monitoring program to verify and maintain the different chemometric models used.

The present chapter will present two major aspects of the control strategy, namely, the PAT tools involved in the at-, on-, or in-line measurements of the CQAs [4, 9] and the general chemometric workflow used to model and interpret the outputs of these PAT tools.

7.2 Process Analytical Technologies

Regarding the complexity of the manufacturing processes, a full understanding and mechanistic modelling of the process are not always possible at least during the process development stage. Therefore, it is necessary to perform frequent in-process control measurements of the CPPs and the CQAs of the intermediate and final product. The control strategy should include a combination of soft and hard sensors being able to monitor the several CQAs ideally in a non-destructive, minimally invasive, accurate, and fast way. Even if the CQAs are the same for batch and continuous manufacturing, the choice of the sensor type, the point of testing, and the sampling plan vary between both manufacturing modes. Indeed, the sensors used in continuous manufacturing may be used as in-process controls to re-adjust the process parameters and possibly divert the non-conform samples, whereas the measurements are generally used as process end-point detection in batch manufacturing [10]. The different sensors that will be presented measure either directly or indirectly the CQAs. The latter is especially the case for vibrational spectroscopy tools such as near-infrared (NIR) and Raman spectroscopy that may be used to monitor a wide amount of CQAs but that require the construction of a predictive multivariate chemometric model using the measurements from a reference method.

The majority of medicines are oral solid dosage forms (tablets and capsules) where active pharmaceutical ingredients (APIs) are mixed with excipients via three main processing routes: direct compression, wet granulation, and dry granulation. While direct compression is still used in specific cases, wet and dry granulation routes are the most common ones to improve the manufacturability of incoming materials (flowability and compressibility) but also the homogeneity of mixing [11]. The most common CQAs for tablet manufacturing are presented in Fig. 7.2. They may be divided as chemical properties, identity, moisture content, solid-state chemistry, homogeneity, and content uniformity, or physical properties, density,

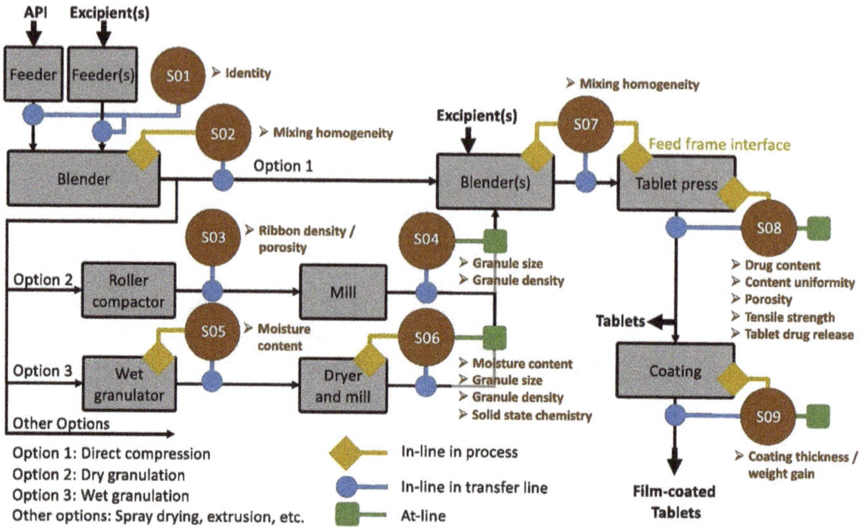

Fig. 7.2 Process flow diagram for direct compression (option 1), dry granulation (option 2), and wet granulation (option 3) highlighting the points of testing and properties of interest. S01–S09 refer to points of testing (sensor positions) in the process line. Red text refers to CQAs. (Reproduced with permission from ref. [10])

particle size, porosity, tensile strength, coating thickness, and coating integrity. These CQAs may be monitored by multiple PAT tools that will be presented below at different stages of the tablet manufacturing.

7.3 Near-Infrared Spectroscopy

NIR spectroscopy is certainly the most widely used PAT tool. Indeed, because of its nature and properties, it may be used to monitor the majority of the CQAs of oral solid dosage forms [12]. In addition, the recent technological developments, notably with the microelectromechanical systems (MEMS), enabled the conception of small, wireless, and relatively inexpensive sensors [13].

The near-infrared spectral region is situated in the 800–2500 nm wavelength range (12,500–4000 cm^{-1} wavenumbers). NIR spectroscopy is based on the absorption of light related to the overtones and combinations of fundamental vibrations of –CH, –NH, –OH, and –SH functional groups induced by the change in the molecular dipole moment [14]. These absorption bands are, by nature, broad and highly overlapped hindering the visual or univariate analysis of NIR spectra. Nevertheless, the weaker absorption intensity of NIR absorption bands compared to fundamental absorption bands in mid-infrared enables the analysis of samples without preparation.

The very first step in the manufacturing of medicines is the processing of incoming materials; it is natural that the first CQA to be verified is the identity of these raw materials. NIR, being sensitive to both the physical and the chemical properties of the sample, provides a very specific fingerprint of the raw materials (if they absorb the NIR light) [15]. Therefore, it has been used as primary control of incoming materials to ensure their identity [16–18].

When the tablets are produced by wet granulation, a liquid (notably pure water) is added to the incoming materials during mixing to improve the subsequent processing steps. Once the granules are produced, the liquid is removed by drying. However, the residual moisture should be tightly controlled since it will eventually impact the dissolution, hardness, and porosity of the produced tablets among others [19, 20]. Thanks to the two very intense absorption bands of water (1450 and 1940 nm), NIR spectroscopy is a technique of choice and has been used in process lines for moisture monitoring [21–23]. It has also been used to monitor the water uptake of theophylline during storage implying its conversion from anhydrous theophylline to theophylline monohydrate [24]. Other solid-state properties may be measured by NIR spectroscopy such as the polymorphic transformation [25] or to quantify a polymorphic purity [26, 27]. The control of the polymorphs in pharmaceutical products is very important since it may impact the product's processability, bioavailability, efficacy, and safety [28].

The measured NIR spectrum is highly influenced by the optical path length of the light in the sample. This, in turn, is influenced by the particle size and the density of the matter for the diffuse reflectance measurement modes. It is therefore possible to infer the particle size and the bulk density of the analyzed matter from the measured NIR spectra via a predictive multivariate modelling. Pauli et al. monitored the particle size of granules in a continuous granulation process [29]. They were able to predict the particle size distribution with root mean squared error of prediction of 17, 97, and 174 μm, for particle size fractions 20–234, 98–1017, and 748–2297 μm, respectively. These performances were sufficient to keep the process under control. The powder density is a critical parameter of solid dosage form manufacturing because several unit operations are based on a volumetric operation (feeding, capsule filling, tableting, etc.). Therefore, a variation in the material density might result in drug product quality issues [30]. NIR spectroscopy was used to monitor in real time the powder bulk density in a pilot plant facility [31]. The measured signal was used to feed-forward control the process line and take corrective actions in a real-time fashion. The authors concluded that this monitoring led to improvements in the final product quality despite process and raw material variations.

One of the most important CQAs of pharmaceutical manufacturing is the homogeneity of the blend. Indeed, this is a mandatory step to ensure that the whole production, from the first to the last tablet, has an API content and properties within specifications. Of course, depending on the production route (direct compression or granulation), this should be monitored at different steps (see Fig. 7.2). NIR spectroscopy may be used to ensure a homogeneous mixing. Two main approaches may be envisaged in this regard: a qualitative or a quantitative one. In the qualitative approach, the mix is supposedly homogeneous once a stable or desired spectral signature is obtained [32, 33]. Indeed, a spectral variation below a defined threshold indicates a stable chemical composition and can be linked to a homogeneous mixing. The quantitative approach necessitates the development of a predictive model to determine the strength of the blend [34, 35]. This approach demands more efforts but allows a better process control via feed-forward actions and process parameter adjustments.

NIR spectroscopy may also be used for the prediction of final drug product CQAs such as its assay [36], its dissolution [37, 38], or the coating thickness and integrity [39].

Recent developments enabled the control of 100% of the produced tablets with conventional NIR probes [40] or with the SAM-spec technology that provides also information regarding the homogeneity of the API distribution within the tablets [41]. Besides the probes, NIR spectroscopy may be coupled to an imaging device [42]. NIR chemical imaging has been advantageously used to monitor several CQAs in new production routes such as inkjet-printed formulations [43] or 3D printed pharmaceuticals [44].

7.4 Raman Spectroscopy

Raman spectroscopy is the second most used vibrational spectroscopy as PAT tool. It is based on the inelastic scattering of light caused by an energy transfer between the incident radiation and molecular vibrations (Stokes and anti-Stokes radiations). Contrary to NIR spectroscopy where absorption of light occurs when there is a change in the dipole moment, the Raman scattering occurs when there is a change in the molecular polarizability during vibration. The technique share some advantages with NIR spectroscopy such as its non-destructive character, the opportunity to analyze samples without preparation steps, and a wide variety of probes enabling the direct in-line monitoring of a wide variety of pharmaceutical processes [45]. Compared to NIR spectroscopy, it exhibits sharp and specific peaks enabling the univariate or visual inspection of spectra. As mentioned above, water is a very weak scatterer making Raman spectroscopy a PAT tool of choice for the monitoring of bioprocesses [46].

The major drawbacks of the technique are its relatively high instrumental costs, the weakness of the Raman effect that may imply long acquisition times reducing its applicability for real-time process monitoring, and the occurrence of sample auto-fluorescence. The latter is possibly the main limitation of Raman applications. Indeed, the weakness of the Raman effect necessitates the use of high-intensity light sources (i.e., lasers) and low-wavelength lights (e.g., 532 nm) because of their more energetic nature. However, auto-fluorescence of the analyzed matter is inversely proportional to the incident radiation wavelength being almost inexistent for near-infrared laser sources (e.g., 1064 nm), but the Raman signal is also very weak because of the λ^{-4} dependence of the scattering process. Therefore, a compromise needs to be found between fluorescence and Raman signal intensity [47]. For most pharmaceutical applications, this compromise is found with the 785 nm laser source.

Because the low-frequency Raman spectroscopy (10–200 cm^{-1}) is sensitive to intermolecular and lattice vibrations, it is commonly used to monitor the polymorphic state of pharmaceuticals [48–50]. It has also been used to monitor the cocrystals and their conversions during a wet manufacturing and a fluidized bed granulation process [51]. In addition, it allowed the observation of a sequence of conversion behaviors of the cocrystals during the process increasing its understanding and possibly its design.

Besides the monitoring of solid-state properties, Raman spectroscopy is also used for the monitoring of the blend homogeneity [52, 53] and the assay of produced tablets [54]. It has also shown a great interest for the monitoring of coating processes [55–57].

7.5 Microwave Resonance Technology (MRT)

Microwave resonance technologies (MRT) measure the dielectric (non-conductive) properties of a sample by the interaction between the molecules and a resonant electromagnetic field. When a dielectric object (tablet, granule, etc.) passes through a microwave resonator, the resonance frequency of the sensor decreases, and the resonance curve broadens. The magnitude of these changes can be correlated to the water content and the mass of the sample [41]. The system must be calibrated for each specific sample composition with a multiple linear regression. The very high wavelength of microwave radiations (1 mm to 1 m) allows the radiation to penetrate deep in the sensed material: several centimeters compared to a few millimeters for NIR spectroscopy. Therefore, MRT can advantageously replace NIR spectroscopy for determining the water content of samples, but it lacks its versatility [58–60]. A 3D MRT detector has been used by Brouckaert et al. to measure the mass of the produced tablets at very high speed in a non-destructive way [41]. This 3D configuration allowed the analysts to increase the accuracy of the method and remove the sensitivity of the measurement to the object shape, orientation, and movement. MRT could also be used to determine the bulk density of the powdered pharmaceutical materials [30, 61].

7.6 Acoustic Emission (AE)

Acoustic emission (AE) is based on the detection of the sounds produced by a process or a system. Particles may enter in collision with other particles or with the process elements (e.g., blender, fluidized bed chamber). The particles may also undergo changes at the atomic scale (plastic deformation or cracking) because of stress induced by the process. These phenomena generate transient elastic waves (vibrations) that embed physical and chemical information. These vibrations of high frequency (70–500 kHz) are measured by AE piezoelectric sensors. The signals of an amplitude greater than a predefined threshold are recorded. The AE sensors are generally placed on the outside of the process elements with a couplant material (e.g., grease or ultrasonic couplant) to improve the sensor sensitivity and avoid the sound attenuation by air.

AE has been used in-line to monitor the change in particle flowability due to the addition of lubricants in a V-blender [62]. AE has also been used to monitor the particle size and water content in granules during the fluidization process [63]. The results were encouraging, but several environmental parameters may jeopardize the applicability of AE such as the ambient air humidity. Another team used AE to monitor the granulation process state and several attributes such as the bed movement, liquid feed concentration, and reflux of fines to granulator [64].

7.7 Terahertz Spectroscopy

The terahertz (THz) region is situated between the infrared and the microwave regions of the electromagnetic spectrum corresponding to frequencies from 0.1 to 10 THz (3 to 133 cm^{-1}). Compared to infrared spectroscopy that is mainly related to intramolecular vibrations, the THz region is mainly dominated by weak intermolecular vibrations. It is therefore more sensitive to the physical properties of the bulk phase (conformation, packing geometry, etc.) than to its chemical nature [30]. The technique is based on a mixture of absorptions of a THz light by the weak intermolecular interactions (van der Waals forces, hydrogen bonding, dipolar, ionic, and other weak covalent interactions) [65]. Recent advances in the instrumentation allowed the generation of THz radiation by the irradiation of a terahertz emitter (generally a photoconductive antenna) by a pulsed NIR laser.

Like the microwaves, it may penetrate a broad range of non-conductive materials (e.g., cardboard, paper, plastic, ceramics) but, however, is blocked by liquid, water, or metals [66]. Two main THz-based techniques are used in the pharmaceutical sector: terahertz time-domain spectroscopy (THz-TDS) and terahertz pulsed imaging (TPI).

7.7.1 Terahertz Time-Domain Spectroscopy (THz-TDS)

THz-TDS has been largely used to study the crystallography and polymorphism of drug molecules, hydrates, and cocrystals, thanks to its capacity to directly probe intermolecular interactions. In addition, the fact that amorphous state materials do not exhibit a specific fingerprint in the THz domain attracted much interest to detect residual trace of crystallinity in material [67].

Besides its application to characterize the physical state of powdered materials, THz-TDS knows a renewed interest for the fast and non-destructive measure of tablet porosity [68]. Indeed, several studies showed that there is a correlation between the refractive index (sensed by transmission THz-TDS) and the total porosity of tablets that was in turn correlated with the disintegration and dissolution properties of tablets [69–71]. Nevertheless, some limitations remain such as the pulse broadening by the high absorption of most APIs that might lead to errors in the measured refractive index. In addition, high absorbing APIs and relatively thick tablets may completely block the THz radiation making the transmission measurements impossible. Moreover, the thickness and the geometry (curvature and possible lensing effect of the biconvex tablets) can cause further refraction of the THz beam and hence introduce some error in the results [72].

7.7.2 *Terahertz Pulsed Imaging (TPI)*

TPI is based on THz-TDS measurements at specific locations of the sample. As the measurements must be performed perpendicular to the sample surface at the same focal position, a prior step of topography measurement of the tablet is mandatory. Although TPI has been used to quantitatively map the tablet density [73], it is generally used to check the coating thickness and integrity.

In the pharmaceutical manufacturing of solid dosage forms, and especially tablets, a coating may be applied. The coating may have diverse functions: (i) film coatings may be used to ensure color uniformity and light protection of the API, (ii) active coating contains an API in the coat, and (iii) functional coating may be used to mask the taste or smell of a product, to protect the API against the acidity of the stomach, or to prolong API release. Therefore, the coating thickness and integrity are very important since it is directly linked to the API content for active coatings and to the medicine's efficacy and safety in the case of functional coating. When applied onto a tablet, the THz will penetrate the sample, and, at each coating interface or abrupt change in refractive index between drug particles and excipient matrix, a portion of THz will reflect to the detector. The film coating thickness is eventually determined by the separation of the adjacent reflexion peaks [74]. TPI has been used in several studies to measure the coating thickness, and it has been established that it was particularly powerful for thick coatings (40–200 μm) [75, 76].

7.8 Data Analysis and Process Modelling

The modelling of the process is of paramount importance to quantify all relevant process variables in real time. This allows the detection of process deviation and possibly the prediction of the future evolution of the process which in turns allows the analyst to take the appropriate measures in a proactive way to ensure the final product quality. Three main modelling properties exist [77]:

- The mechanistic modelling whose model equations are derived from prior knowledge of the phenomena of the process. In this case, the model parameters have a physical, chemical, or biological meaning, and the number of parameters is fixed. It usually uses ordinary or partial differential equations and represents the process knowledge. Another advantage is that they support dynamic changes (e.g., the bypass of an equipment under maintenance, etc.) and they can deliver good results even for situations that have never been seen before. Mechanistic models have a higher extrapolation power and a large application range. However, they require a deep knowledge of the validity space and the inner process structure and need more time to be set up.
- The data-driven modelling fits general mathematical models to predict variables from data measured in the process. In this case, the model parameters have no physical, chemical, or biological meaning since only correlations (possibly without causation) are established. These models need a huge amount of process data to be set up and do not allow any extrapolation. They instantly estimate unmeasured variables, can detect when the process deviates from its normal trajectory, and can determine unsuspected correlations between process variables. In that sense, they allow a better process understanding. Since they are easy and fast to calculate and set up, they are generally preferred over mechanistic models for new process not well understood yet or for processes that are too complex to be modelled. The data-driven models generally use multivariate statistical algorithms and are commonly named chemometrics [78].
- The hybrid modelling combines mechanistic and data-driven models. They are generally used to cope with the restrictions of each of the abovementioned approaches such as the correction of the outputs of a mechanistic model by a data-driven model or the incorporation of prior knowledge (e.g., reaction equation) in a data-driven model. In the end, what is important is to have a model that is sufficiently accurate fast and robust for the intended purpose.

The development of a data-driven or chemometric model is an iterative process (Fig. 7.3) that involves the selection of a representative sample set to develop the model, the tuning of the model parameters, and the evaluation of the model performances. The objective of the chemometric modelling is to link the variables measured by the abovementioned PAT technologies (among others) and the response variables (e.g., process parameters, concentrations, etc.) that are generally more difficult or costly to obtain.

7.8.1 Model Development

Sample Selection

Chemometric models are, as stated above, data-driven models. This implies that all the modelling such as the choice of the algorithm, the tuning of the model parameters, and the model performances are influenced by the data. Therefore, the

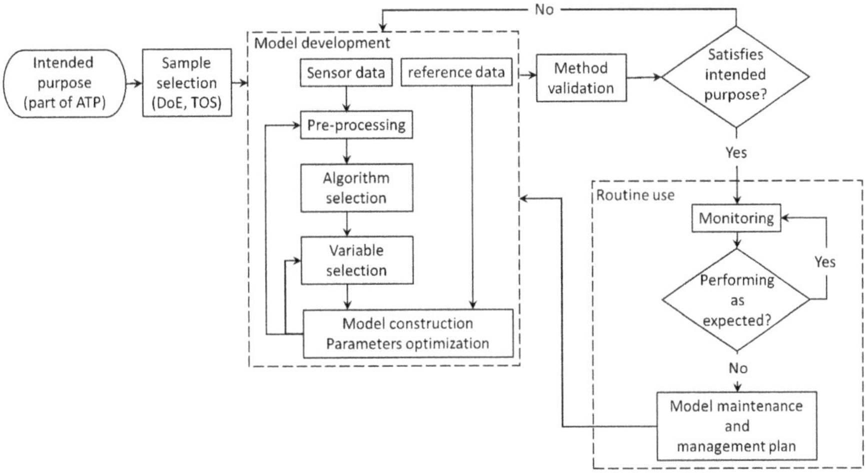

Fig. 7.3 Representation of the lifecycle workflow of a chemometric model. (Adapted from ref. [85])

acquisition of a representative set of high-quality data is crucial and may be performed following the theory of sampling (TOS) and design of experiments (DoE) [79, 80]. This set of data should include all possible process states and intra- and inter-batch variability. Another important point that might seem trivial is that the data must contain the expected information. It is therefore the responsibility of the analyst to choose the correct sensor for the correct application. The positioning of the sensor has also its importance [81].

Before collecting the data, the relevant range of variability and the applications must be clearly defined. Once this is done, a sampling plan should be envisaged depending on the purpose of the analysis. Indeed, if the aim of the method is to build a multivariate statistical process control, the model should be able to detect abnormal conditions. Thereupon, samples representing the normal variability of a process under control should be collected. On the one hand, for this kind of application, samples from a DoE should not be used since they represent the extremes of the deviation. On the other hand, DoE samples are excellent samples to build a good regression model [82, 83]. For both cases, historical data, when available, may be useful to increase the amount of relevant data [84].

Pre-processing

The data acquired by the different sensors are generally not directly exploitable. They typically need to be pre-processed to decrease the non-relevant interfering variance and to increase the interesting variance. Several kinds of pre-processing techniques may be applied such as denoising, baseline correction, alignment, normalization, and scaling [86]. These pre-processing methods have an impact on the data and consequently on the models. The optimal pre-processing strategy involves a selection of pre-processing techniques or combinations. The initial choice of pre-processing methods should be guided by a prior knowledge of the samples or a theoretical knowledge of the technique (e.g., correction of a multiplicative scattering effect) but is generally chosen based on the experience of the chemometrician. Once the initial selection is performed, it is refined and optimized through an iterative process.

There are no clear guidelines on when to use or avoid certain pre-processing techniques. Therefore, attempts have been made to build a systematic screening and selection of the best pre-processing methods or combinations [87–90]. This pre-processing step is also crucial when combining data of different sources and formats and in different time of samplings. In all cases, precautions must be taken to avoid overprocessing that can attenuate the signal, and one must be aware that some instruments already pre-process the data with proprietary techniques or common techniques named differently.

Algorithm Selection

Many different algorithms may be used to model the data depending on the nature of the data and the final objective of the method. These algorithms may be grouped into two main categories: unsupervised and supervised multivariate models. On the one side, unsupervised algorithms are essentially qualitative; they may be used as exploratory methods (e.g., principal component analysis, PCA) to acquire knowledge on the process, as outlier detection method, or as clustering method (k-means algorithm, hierarchical clustering) to group samples based on their measured properties. On the other side, supervised algorithms may be used for qualitative purposes such as class modelling (e.g., soft independent modelling of class analogy, SIMCA) to model the normal conditions of a process or identify raw materials or for quantitative purposes such as regression (e.g., partial least squares, PLS). The regression algorithms may handle linear and non-linear relationships but always rely on reference measurements. Indeed, since most PAT technologies are indirect techniques (e.g., vibrational spectroscopy), it is mandatory to provide the algorithm the reference values of the measured samples. Therefore, a cautious sampling should be performed to collect and annotate the processed matter corresponding to the measured data and eventually analyze it with a reference analytical method. This reference method (generally liquid chromatography) should be previously developed and validated. It is important to note that the performances of the developed chemometric model are de facto limited to the reference method performances. The latter should therefore be as performant as possible to keep the final error as low as possible. This step of acquisition of the calibration data with their reference measurements is the most consuming step of the model generation.

Variable Selection

The size and amount of data measured (e.g., spectral range and spectral vector length) are generally defined by the spectrometer design. Therefore, it is very likely that a large part of the measured variables is not bearing any relevant information. These non-informative variables should be minimized or removed to improve the model's performances and robustness. This has been demonstrated for the PLS algorithm [91]. These feature selection methods may be grouped into three main categories:

- Filter methods that use the output of the fitted model to select the relevant variables by defining a threshold on a variable ranking technique (e.g., variable importance in projection, VIP). This approach is simple to implement, fast, and easy to compute.
- Wrapper methods that use the filter method in an iterative way where the variables selected by the filter method are used to fit a new model. This procedure involves an iterative model fitting and variable selection procedure. Several variations exist among which the well-known genetic algorithm (GA) and interval PLS (iPLS).
- Embedded methods that incorporate the variable selection as an integrated part of the algorithm such as the random forests or gradient boosted algorithms.

A wide variety of feature selection methods exist. More details may be found elsewhere [92–94].

Model Optimization

Once the calibration data are acquired and pre-processed, the algorithm is chosen, and the relevant variables are selected, the algorithm's parameters must be tuned. This may be performed as an iterative process together with the variable selection step. In any cases, this is performed through the evaluation of the model's performances for the different tested algorithm parameter values. Therefore, the chemometrician needs a dataset to be predicted by the computed model. This dataset is generally, in a first step, a selected subset from the calibration set. Obviously, this subset is not representative of the full calibration set and even less of the future measurements. One way to circumvent this limitation is to repeat the subset selection several times. This process is called cross-validation (CV), and a wide variety of cross-validation strategies exist depending on the nature of the data, the amount of available data, and the objective of the method. A popular strategy is the K-fold CV where K holds for the number of splits. The two extremes to this strategy are the leave-one-out CV (LOOCV) where K corresponds to the number of samples in the dataset and the twofold CV where the dataset is split into two segments. Between these two extremes, the number of splits should be carefully selected. Indeed, there is a relationship between the number of splits and the measured error of cross-validation with LOOCV having the lowest error and twofold CV the highest. It is therefore important to note that the objective of the CV strategy is not to achieve the lowest error but the most realistic one compared to the routine use of the method. Another important aspect is to keep replicate measurements together during the splitting to avoid having different measurements of the same sample in both calibration and validation set that would lead to over-optimistic results [95, 96].

Other CV strategies involve the bootstrapping where, in each iteration, the validation subset is randomly selected with replacement. This procedure is repeated many times, generally more than 100 cycles. An extreme version of cross-validation is the Monte Carlo simulation where thousands or millions of re-samplings are carried out [97].

It is important to keep in mind that cross-validation is, by definition, different from a formal validation dataset and should never be used to perform the validation of a model but only to tune its parameters [98].

Method Validation

The objective of validation is to demonstrate that the performances of the method are suitable for its intended purpose. The wording here has its importance since method validation is different from model validation. On the one hand, model validation mainly focusses on model parameter tuning and has been addressed in the above paragraph. On the other hand, method validation is performed to give the analyst and the authorities the sufficient guarantees that correct decisions may be taken from the predicted results. Validation necessitates the use of *independent* and *external* data representative of the future routine measurements. The ICH Q2(R1) is the reference text treating this subject [99], and it addresses the different validation criteria to be fulfilled: specificity, linearity, accuracy, precision, and robustness among others. Regarding the regulatory expectations, information may be found in the relevant guidelines [100, 101].

A comprehensive review of the method validation process is out of the scope of this chapter; nevertheless, it is important to insist on its cyclic and dynamic character. Once validated, the majority of analytical procedures (e.g., chromatographic techniques) may work well in routine without little need for adjustments or modifications. This is not the case for chemometric models and spectroscopic technique applications. Indeed, the spectra measured are highly influenced by a series of properties such as the physicochemical properties of the sample, the dynamics of the analyzed matter, etc. All these properties are considered and handled at the time of model development and validation. But they may vary through time, and model maintenance and adjustments such as model update and calibration transfer are frequently required. Therefore, it is crucial to envisage model monitoring and maintenance concomitantly to its development [102, 103].

A control strategy that checks the model performances in routine must be developed and documented and should consider the risks associated with the model. Typically, multivariate spectroscopic models are considered as medium-impact ("models can be useful in assuring quality of the product but are not sole indicators of the product quality") or high-impact ("prediction from the model is a significant indicator of quality of the product") [104]. The changes to these methods are consequently critical and must be reviewed by authorities. This reviewing process complicates the application of changes since even if technical and documentary activities linked to the maintenance may take a few weeks to be completed, the reviewing process by the authorities may take several months [105]. This management plan could be included and applied via the Pharmaceutical Quality System and the change managements described in the ICH Q12 [106]. This is an actual discussion, and several new guidelines are also emanating from the relevant institutions such as the new ICH Q14 and Q2(R2) and the USP and the European Pharmacopeia chapters on chemometrics [85, 107, 108]. These regulatory and quality control aspects are particularly important in the actual context of agility increase of the pharmaceutical manufacturing [109].

7.9 Conclusion

The actual shift of the pharmaceutical manufacturing from batch to continuous manufacturing requires the development of process control strategies. This control strategy implies the need for accurate sensors at strategic places of the process to monitor the different CQAs and CPPs. The raw measurements are then converted into information with chemometric models. The deployment of this control strategy and more generally of PAT requires a close collaboration of a multidisciplinary team with backgrounds in many fields of science and engineering. This is important for the conception of the control strategy itself but also to interpret the outputs of the chemometric models and ensure that the latter sound reasoning. In addition, the knowledge gained by the implementation of the chemometric models may eventually lead to the development of mechanistic models of the process. Another advantage of the multidisciplinary nature of PAT is the diversity of point of views that may lead to innovative solutions.

Actual sensors and modelling strategies are well established and may constitute a basis for new developments. Indeed, new challenges arise such as the maintenance of the sensors and model performances over long periods (years), data storage, and handling together with new galenic forms (e.g., additive manufacturing, inkjet printing, etc.) [110]. Another active field of research that will gain importance in the future of PAT is the fusion of data. Indeed, advanced modelling of data from sensors (of different size and nature) together with process data will enable a deeper understanding and management of the manufacturing process. This fusion of data may be handled by different approaches such as multiblock algorithms or deep learning [38, 111, 112]. The huge amount of data, sensors, and knowledge gained of the process constitutes an unprecedented opportunity to increase the manufacturing agility and increase the product quality. This is achieved notably through the digitalization of the manufacturing process called the Pharma 4.0 and includes the use of digital twins among others [113–115].

Finally, it is very likely that PAT will continue to grow with the development of new sensors and innovative data modelling solutions. This implies new challenges to be tackled from a scientific and regulatory point of view. Nevertheless, the deep understanding of the process, the increased agility in the manufacturing, and in the end the improved quality of the newly produced pharmaceutical products will ensure that the patients get the best treatments.

References

1. De Weerdt E, Simoens S, Hombroeckx L et al (2015) Causes of drug shortages in the legal pharmaceutical framework. Regul Toxicol Pharmacol 71:251–258. https://doi.org/10.1016/j.yrtph.2015.01.005
2. USP (2022) Geographic concentration of pharmaceutical manufacturing: USP Medicine Supply Map analysis. https://qualitymatters.usp.org/geographic-concentration-pharmaceutical-manufacturing

3. Lee SL, O'Connor TF, Yang X et al (2015) Modernizing pharmaceutical manufacturing: from batch to continuous production. J Pharm Innov 10:191–199. https://doi.org/10.1007/s12247-015-9215-8
4. U.S. Food and Drug Administration (2004) Guidance for industry, PAT – a framework for innovative pharmaceutical development, manufacturing and quality assurance. FDA-2003-D-0032
5. International Conference on Harmonization (2005) ICH Q9: quality risk management
6. International Conference on Harmonization (2008) ICH Q10: pharmaceutical quality systems
7. International Conference on Harmonization (2009) ICH Q8 (R2): pharmaceutical development
8. Yu LX, Amidon G, Khan MA et al (2014) Understanding pharmaceutical quality by design. AAPS J 16:771–783. https://doi.org/10.1208/s12248-014-9598-3
9. Laske S, Paudel A, Scheibelhofer O et al (2017) A review of PAT strategies in secondary solid oral dosage manufacturing of small molecules. J Pharm Sci 106:667–712. https://doi.org/10.1016/j.xphs.2016.11.011
10. Markl D, Warman M, Dumarey M et al (2020) Review of real-time release testing of pharmaceutical tablets: state-of-the art, challenges and future perspective. Int J Pharm 582:119353. https://doi.org/10.1016/j.ijpharm.2020.119353
11. Wang B, Sun X, Xiang J et al (2022) A critical review on granulation of pharmaceuticals and excipients: principle, analysis and typical applications. Powder Technol 401:117329. https://doi.org/10.1016/j.powtec.2022.117329
12. De Beer T, Burggraeve A, Fonteyne M et al (2011) Near infrared and Raman spectroscopy for the in-process monitoring of pharmaceutical production processes. Int J Pharm 417:32–47. https://doi.org/10.1016/j.ijpharm.2010.12.012
13. Crocombe RA (2018) Portable spectroscopy. Applied spectroscopy. 72(12):1701–1751. https://doi.org/10.1177/0003702818809719
14. Ciurczak EW, Igne B, Workman J, Burns DA (eds) (2021) Handbook of near-infrared analysis, 4th edn. CRC Press, Boca Raton
15. Razuc M, Grafia A, Gallo L et al (2019) Near-infrared spectroscopic applications in pharmaceutical particle technology. Drug Dev Ind Pharm 45:1565–1589. https://doi.org/10.1080/03639045.2019.1641510
16. Sun L, Hsiung C, Pederson CG et al (2016) Pharmaceutical raw material identification using miniature near-infrared (MicroNIR) spectroscopy and supervised pattern recognition using support vector machine. Appl Spectrosc 70:816–825. https://doi.org/10.1177/0003702816638281
17. Pasquini C (2018) Near infrared spectroscopy: a mature analytical technique with new perspectives – a review. Anal Chim Acta 1026:8–36. https://doi.org/10.1016/j.aca.2018.04.004
18. Deidda R, Sacre P-Y, Clavaud M et al (2019) Vibrational spectroscopy in analysis of pharmaceuticals: critical review of innovative portable and handheld NIR and Raman spectrophotometers. TrAC Trends Anal Chem 114:251–259. https://doi.org/10.1016/j.trac.2019.02.035
19. Gabbott IP, Al Husban F, Reynolds GK (2016) The combined effect of wet granulation process parameters and dried granule moisture content on tablet quality attributes. Eur J Pharm Biopharm 106:70–78. https://doi.org/10.1016/j.ejpb.2016.03.022
20. Kestur U, Desai D, Sharif S et al (2017) Impact of moisture and magnesium stearate functionality on manufacturability of wet granulated metformin tablets. Pharm Dev Technol 22:715–723. https://doi.org/10.3109/10837450.2015.1116562
21. Chablani L, Taylor MK, Mehrotra A et al (2011) Inline real-time near-infrared granule moisture measurements of a continuous granulation-drying-milling process. AAPS PharmSciTech 12:1050–1055. https://doi.org/10.1208/s12249-011-9669-z
22. Fonteyne M, Arruabarrena J, de Beer J et al (2014) NIR spectroscopic method for the in-line moisture assessment during drying in a six-segmented fluid bed dryer of a continuous tablet production line: validation of quantifying abilities and uncertainty assessment. J Pharm Biomed Anal 100:21–27. https://doi.org/10.1016/j.jpba.2014.07.012

23. Avila CR, Ferré J, de Oliveira RR et al (2020) Process monitoring of moisture content and mass transfer rate in a fluidised bed with a low cost inline MEMS NIR sensor. Pharm Res 37: 84. https://doi.org/10.1007/s11095-020-02787-y
24. Otsuka M, Kanai Y, Hattori Y (2014) Real-time monitoring of changes of adsorbed and crystalline water contents in tablet formulation powder containing theophylline anhydrate at various temperatures during agitated granulation by near-infrared spectroscopy. J Pharm Sci 103:2924–2936. https://doi.org/10.1002/jps.24006
25. Calvo NL, Maggio RM, Kaufman TS (2018) Characterization of pharmaceutically relevant materials at the solid state employing chemometrics methods. J Pharm Biomed Anal 147:538–564. https://doi.org/10.1016/j.jpba.2017.06.017
26. Antonio M, Maggio RM (2018) Assessment of mefenamic acid polymorphs in commercial tablets using chemometric coupled to MIR and NIR spectroscopies. Prediction of dissolution performance. J Pharm Biomed Anal 149:603–611. https://doi.org/10.1016/j.jpba.2017.11.053
27. Alaoui Mansouri M, Ziemons E, Sacré P-Y et al (2021) Classification of polymorphic forms of fluconazole in pharmaceuticals by FT-IR and FT-NIR spectroscopy. J Pharm Biomed Anal 196:113922. https://doi.org/10.1016/j.jpba.2021.113922
28. Brittain HG (ed) (2018) Polymorphism in pharmaceutical solids, 2nd edn. CRC Press, Boca Raton
29. Pauli V, Roggo Y, Kleinebudde P, Krumme M (2019) Real-time monitoring of particle size distribution in a continuous granulation and drying process by near infrared spectroscopy. Eur J Pharm Biopharm 141:90–99. https://doi.org/10.1016/j.ejpb.2019.05.007
30. Stranzinger S, Markl D, Khinast JG, Paudel A (2021) Review of sensing technologies for measuring powder density variations during pharmaceutical solid dosage form manufacturing. TrAC Trends Anal Chem 135:116147. https://doi.org/10.1016/j.trac.2020.116147
31. Singh R, Román-Ospino AD, Romañach RJ et al (2015) Real time monitoring of powder blend bulk density for coupled feed-forward/feed-back control of a continuous direct compaction tablet manufacturing process. Int J Pharm 495:612–625. https://doi.org/10.1016/j.ijpharm.2015.09.029
32. Blanco M, Cueva-Mestanza R, Cruz J (2012) Critical evaluation of methods for end-point determination in pharmaceutical blending processes. Anal Methods 4:2694. https://doi.org/10.1039/c2ay25379h
33. Khorasani M, Amigo JM, Bertelsen P et al (2015) Detecting blending end-point using mean squares successive difference test and near-infrared spectroscopy. J Pharm Sci 104:2541–2549. https://doi.org/10.1002/jps.24533
34. Jaumot J, Igne B, Anderson CA et al (2013) Blending process modeling and control by multivariate curve resolution. Talanta 117:492–504. https://doi.org/10.1016/j.talanta.2013.09.037
35. Hetrick EM, Shi Z, Barnes LE et al (2017) Development of near infrared spectroscopy-based process monitoring methodology for pharmaceutical continuous manufacturing using an offline calibration approach. Anal Chem 89:9175–9183. https://doi.org/10.1021/acs.analchem.7b01907
36. Román-Ospino AD, Cárdenas V, Ortega-Zuñiga C, Singh R (2018) PAT for pharmaceutical manufacturing process involving solid dosages forms. Comput Aided Chem Eng 41:293–315. https://doi.org/10.1016/B978-0-444-63963-9.00012-9
37. Pawar P, Wang Y, Keyvan G et al (2016) Enabling real time release testing by NIR prediction of dissolution of tablets made by continuous direct compression (CDC). Int J Pharm 512:96–107. https://doi.org/10.1016/j.ijpharm.2016.08.033
38. Nagy B, Petra D, Galata DL et al (2019) Application of artificial neural networks for Process Analytical Technology-based dissolution testing. Int J Pharm 567. https://doi.org/10.1016/j.ijpharm.2019.118464
39. Gendre C, Boiret M, Genty M et al (2011) Real-time predictions of drug release and end point detection of a coating operation by in-line near infrared measurements. Int J Pharm 421:237–243. https://doi.org/10.1016/j.ijpharm.2011.09.036

40. Järvinen K, Hoehe W, Järvinen M et al (2013) In-line monitoring of the drug content of powder mixtures and tablets by near-infrared spectroscopy during the continuous direct compression tableting process. Eur J Pharm Sci 48:680–688. https://doi.org/10.1016/j.ejps.2012.12.032
41. Brouckaert D, Vandenbroucke F, Chauchard F et al (2022) Towards real-time release of pharmaceutical tablets: 100% in-line control via near-infrared spatially resolved spectroscopy and 3D microwave resonance technology. J Pharm Biomed Anal 209:114491. https://doi.org/10.1016/j.jpba.2021.114491
42. Ravn C, Skibsted E, Bro R (2008) Near-infrared chemical imaging (NIR-CI) on pharmaceutical solid dosage forms-comparing common calibration approaches. J Pharm Biomed Anal 48:554–561. https://doi.org/10.1016/j.jpba.2008.07.019
43. Stranzinger S, Wolfgang M, Klotz E et al (2021) Near-infrared hyperspectral imaging as a monitoring tool for on-demand manufacturing of inkjet-printed formulations. AAPS PharmSciTech 22:211. https://doi.org/10.1208/s12249-021-02091-x
44. Khorasani M, Edinger M, Raijada D et al (2016) Near-infrared chemical imaging (NIR-CI) of 3D printed pharmaceuticals. Int J Pharm 515:324–330. https://doi.org/10.1016/j.ijpharm.2016.09.075
45. Esmonde-White KA, Cuellar M, Uerpmann C et al (2017) Raman spectroscopy as a process analytical technology for pharmaceutical manufacturing and bioprocessing. Anal Bioanal Chem 409:637–649. https://doi.org/10.1007/s00216-016-9824-1
46. Esmonde-White KA, Cuellar M, Lewis IR (2022) The role of Raman spectroscopy in biopharmaceuticals from development to manufacturing. Anal Bioanal Chem 414:969–991. https://doi.org/10.1007/s00216-021-03727-4
47. Lewis IR, Edwards H (eds) (2001) Handbook of Raman spectroscopy, 1st edn. CRC Press, Boca Raton
48. Wang H, Boraey MA, Williams L et al (2014) Low-frequency shift dispersive Raman spectroscopy for the analysis of respirable dosage forms. Int J Pharm 469:197–205. https://doi.org/10.1016/j.ijpharm.2014.04.058
49. Feng H, Bondi RW, Anderson CA et al (2017) Investigation of the sensitivity of transmission Raman spectroscopy for polymorph detection in pharmaceutical tablets. Appl Spectrosc 71:1856–1867. https://doi.org/10.1177/0003702817690407
50. Nomura K, Titapiwatanakun V, Hisada H et al (2020) In situ monitoring of the crystalline state of active pharmaceutical ingredients during high-shear wet granulation using a low-frequency Raman probe. Eur J Pharm Biopharm 147:1–9. https://doi.org/10.1016/j.ejpb.2019.12.004
51. Otaki T, Tanabe Y, Kojima T et al (2018) In situ monitoring of cocrystals in formulation development using low-frequency Raman spectroscopy. Int J Pharm 542:56–65. https://doi.org/10.1016/j.ijpharm.2018.03.008
52. Nagy B, Farkas A, Gyürkés M et al (2017) In-line Raman spectroscopic monitoring and feedback control of a continuous twin-screw pharmaceutical powder blending and tableting process. Int J Pharm 530:21–29. https://doi.org/10.1016/j.ijpharm.2017.07.041
53. Riolo D, Piazza A, Cottini C et al (2018) Raman spectroscopy as a PAT for pharmaceutical blending: advantages and disadvantages. J Pharm Biomed Anal 149:329–334. https://doi.org/10.1016/j.jpba.2017.11.030
54. Casian T, Reznek A, Vonica-Gligor AL et al (2017) Development, validation and comparison of near infrared and Raman spectroscopic methods for fast characterization of tablets with amlodipine and valsartan. Talanta 167:333–343. https://doi.org/10.1016/j.talanta.2017.01.092
55. Bogomolov A, Engler M, Melichar M, Wigmore A (2010) In-line analysis of a fluid bed pellet coating process using a combination of near infrared and Raman spectroscopy. J Chemom 24:544–557. https://doi.org/10.1002/cem.1329
56. Müller J, Knop K, Wirges M, Kleinebudde P (2010) Validation of Raman spectroscopic procedures in agreement with ICH guideline Q2 with considering the transfer to real time monitoring of an active coating process. J Pharm Biomed Anal 53:884–894. https://doi.org/10.1016/j.jpba.2010.06.016

57. Feng H, Mohan S (2020) Application of process analytical technology for pharmaceutical coating: challenges, pitfalls, and trends. AAPS PharmSciTech 21:179. https://doi.org/10.1208/s12249-020-01727-8
58. Corredor CC, Bu D, Both D (2011) Comparison of near infrared and microwave resonance sensors for at-line moisture determination in powders and tablets. Anal Chim Acta 696:84–93. https://doi.org/10.1016/j.aca.2011.03.048
59. Austin J, Gupta A, Mcdonnell R et al (2013) The use of near-infrared and microwave resonance sensing to monitor a continuous roller compaction process. J Pharm Sci 102: 1895–1904. https://doi.org/10.1002/jps.23536
60. Peters J, Taute W, Bartscher K et al (2017) Design, development and method validation of a novel multi-resonance microwave sensor for moisture measurement. Anal Chim Acta 961: 119–127. https://doi.org/10.1016/j.aca.2017.01.021
61. Lourenço V, Herdling T, Reich G et al (2011) Combining microwave resonance technology to multivariate data analysis as a novel PAT tool to improve process understanding in fluid bed granulation. Eur J Pharm Biopharm 78:513–521. https://doi.org/10.1016/j.ejpb.2011.02.008
62. Crouter A, Briens L (2016) Monitoring lubricant addition using passive acoustic emissions in a V-blender. Powder Technol 301:1119–1129. https://doi.org/10.1016/j.powtec.2016.07.051
63. Matero S, Poutiainen S, Leskinen J et al (2009) The feasibility of using acoustic emissions for monitoring of fluidized bed granulation. Chemom Intell Lab Syst 97:75–81. https://doi.org/10.1016/j.chemolab.2008.11.001
64. Halstensen M, de Bakker P, Esbensen KH (2006) Acoustic chemometric monitoring of an industrial granulation production process—a PAT feasibility study. Chemom Intell Lab Syst 84:88–97. https://doi.org/10.1016/j.chemolab.2006.05.012
65. Ruggiero MT (2020) Invited review: modern methods for accurately simulating the terahertz spectra of solids. J Infrared Milli Terahz Waves 41:491–528. https://doi.org/10.1007/s10762-019-00648-3
66. Patil MR, Ganorkar SB, Patil AS, Shirkhedkar AA (2022) Terahertz spectroscopy: encoding the discovery, instrumentation, and applications toward pharmaceutical prospectives. Crit Rev Anal Chem 52:343–355. https://doi.org/10.1080/10408347.2020.1802219
67. Zeitler JA, Taday PF, Newnham DA et al (2010) Terahertz pulsed spectroscopy and imaging in the pharmaceutical setting – a review. J Pharm Pharmacol 59:209–223. https://doi.org/10.1211/jpp.59.2.0008
68. Bawuah P, Zeitler JA (2021) Advances in terahertz time-domain spectroscopy of pharmaceutical solids: a review. TrAC Trends Anal Chem 139:116272. https://doi.org/10.1016/j.trac.2021.116272
69. Juuti M, Tuononen H, Prykäri T et al (2009) Optical and terahertz measurement techniques for flat-faced pharmaceutical tablets: a case study of gloss, surface roughness and bulk properties of starch acetate tablets. Meas Sci Technol 20:015301. https://doi.org/10.1088/0957-0233/20/1/015301
70. Markl D, Sauerwein J, Goodwin DJ et al (2017) Non-destructive determination of disintegration time and dissolution in immediate release tablets by terahertz transmission measurements. Pharm Res 34:1012–1022. https://doi.org/10.1007/s11095-017-2108-4
71. Markl D, Bawuah P, Ridgway C et al (2018) Fast and non-destructive pore structure analysis using terahertz time-domain spectroscopy. Int J Pharm 537:102–110. https://doi.org/10.1016/j.ijpharm.2017.12.029
72. Bawuah P, Markl D, Farrell D et al (2020) Terahertz-based porosity measurement of pharmaceutical tablets: a tutorial. J Infrared Milli Terahz Waves 41:450–469. https://doi.org/10.1007/s10762-019-00659-0
73. Palermo R, Cogdill RP, Short SM et al (2008) Density mapping and chemical component calibration development of four-component compacts via terahertz pulsed imaging. J Pharm Biomed Anal 46:36–44. https://doi.org/10.1016/j.jpba.2007.10.023
74. Alves-Lima D, Song J, Li X et al (2020) Review of terahertz pulsed imaging for pharmaceutical film coating analysis. Sensors 20:1441. https://doi.org/10.3390/s20051441

75. Lin H, Dong Y, Markl D et al (2017) Measurement of the intertablet coating uniformity of a pharmaceutical pan coating process with combined terahertz and optical coherence tomography in-line sensing. J Pharm Sci 106:1075–1084. https://doi.org/10.1016/j.xphs.2016.12.012
76. Lin H, Dong Y, Shen Y, Zeitler JA (2015) Quantifying pharmaceutical film coating with optical coherence tomography and terahertz pulsed imaging: an evaluation. J Pharm Sci 104: 3377–3385. https://doi.org/10.1002/jps.24535
77. Solle D, Hitzmann B, Herwig C et al (2017) Between the poles of data-driven and mechanistic modeling for process operation. Chemie Ing Tech 89:542–561. https://doi.org/10.1002/cite.201600175
78. Pomerantsev AL, Rodionova OY (2012) Process analytical technology: a critical view of the chemometricians. J Chemom 26:299–310. https://doi.org/10.1002/cem.2445
79. Esbensen KH, Friis-Petersen HH, Petersen L et al (2007) Representative process sampling – in practice: variographic analysis and estimation of total sampling errors (TSE). Chemom Intell Lab Syst 88:41–59. https://doi.org/10.1016/j.chemolab.2006.09.011
80. Esbensen KH, Román-Ospino AD, Sanchez A, Romañach RJ (2016) Adequacy and verifiability of pharmaceutical mixtures and dose units by variographic analysis (theory of sampling) – a call for a regulatory paradigm shift. Int J Pharm 499:156–174. https://doi.org/10.1016/j.ijpharm.2015.12.038
81. De Leersnyder F, Vanhoorne V, Kumar A et al (2019) Evaluation of an in-line NIR spectroscopic method for the determination of the residence time in a tablet press. Int J Pharm 565: 358–366. https://doi.org/10.1016/j.ijpharm.2019.05.006
82. Kjeldahl K, Bro R (2010) Some common misunderstandings in chemometrics. J Chemom 24: 558–564. https://doi.org/10.1002/cem.1346
83. Scheibelhofer O, Grabner B, Bondi RW et al (2015) Designed blending for near infrared calibration. J Pharm Sci 104:2312–2322. https://doi.org/10.1002/jps.24488
84. Galvis L, Offermans T, Bertinetto CG et al (2022) Retrospective quality by design r(QbD) for lactose production using historical process data and design of experiments. Comput Ind 141: 103696. https://doi.org/10.1016/j.compind.2022.103696
85. United States Pharmacopeia (2022) General chapter, ⟨1039⟩ chemometrics. In: United States Pharmacopeia (ed) USP-NF, Rockville
86. Rinnan Å, van den Berg F, Engelsen SB (2009) Review of the most common pre-processing techniques for near-infrared spectra. TrAC Trends Anal Chem 28:1201–1222. https://doi.org/10.1016/j.trac.2009.07.007
87. Engel J, Gerretzen J, Szymańska E et al (2013) Breaking with trends in pre-processing? TrAC Trends Anal Chem 50:96–106. https://doi.org/10.1016/j.trac.2013.04.015
88. Roger J-M, Biancolillo A, Marini F (2020) Sequential preprocessing through ORThogonalization (SPORT) and its application to near infrared spectroscopy. Chemom Intell Lab Syst 199:103975. https://doi.org/10.1016/j.chemolab.2020.103975
89. Mishra P, Roger JM, Marini F et al (2021) Parallel pre-processing through orthogonalization (PORTO) and its application to near-infrared spectroscopy. Chemom Intell Lab Syst 212: 104190. https://doi.org/10.1016/j.chemolab.2020.104190
90. Mishra P, Roger JM, Marini F et al (2022) Pre-processing ensembles with response oriented sequential alternation calibration (PROSAC): a step towards ending the pre-processing search and optimization quest for near-infrared spectral modelling. Chemom Intell Lab Syst 222: 104497. https://doi.org/10.1016/j.chemolab.2022.104497
91. Nadler B, Coifman RR (2005) The prediction error in CLS and PLS: the importance of feature selection prior to multivariate calibration. J Chemom 19:107–118. https://doi.org/10.1002/cem.915
92. Mehmood T, Liland KH, Snipen L, Sæbø S (2012) A review of variable selection methods in partial least squares regression. Chemom Intell Lab Syst 118:62–69. https://doi.org/10.1016/j.chemolab.2012.07.010
93. Chandrashekar G, Sahin F (2014) A survey on feature selection methods. Comput Electr Eng 40:16–28. https://doi.org/10.1016/j.compeleceng.2013.11.024

94. Yun Y-H, Li H-D, Deng B-C, Cao D-S (2019) An overview of variable selection methods in multivariate analysis of near-infrared spectra. TrAC Trends Anal Chem 113:102–115. https://doi.org/10.1016/j.trac.2019.01.018
95. Jonathan P, Krzanowski WJ, McCarthy WV (2000) On the use of cross-validation to assess performance in multivariate prediction. Stat Comput 10:209–229. https://doi.org/10.1023/A:1008987426876
96. Westad F, Marini F (2015) Validation of chemometric models – a tutorial. Anal Chim Acta 893:14–24. https://doi.org/10.1016/j.aca.2015.06.056
97. Xu Q-S, Liang Y-Z (2001) Monte Carlo cross validation. Chemom Intell Lab Syst 56:1–11. https://doi.org/10.1016/S0169-7439(00)00122-2
98. Esbensen KH, Geladi P (2010) Principles of proper validation: use and abuse of re-sampling for validation. J Chemom 24:168–187. https://doi.org/10.1002/cem.1310
99. International Conference on Harmonization (2005) ICH Q2 (R1): validation of analytical procedures
100. European Medicines Agency (2014) Guideline on the use of Near Infrared Spectroscopy (NIRS) by the pharmaceutical industry and the data requirements for new submissions and variations. EMEA/CHMP/CVMP/QWP/17760/2009 Rev2
101. U.S. Food and Drug Administration (2021) Development and submission of near infrared analytical procedures. FDA-2015-D-0868
102. Wise BM, Roginski RT (2015) Model maintenance: the unrecognized cost in PAT and QbD. Chim Oggi Chemistry Today 33:38–43
103. Andries E, Kalivas JH, Gurung A (2019) Sample and feature augmentation strategies for calibration updating. J Chemom 33:1–20. https://doi.org/10.1002/cem.3080
104. International Conference on Harmonization (2011) ICH Q8/Q9/Q10 (R4) points to consider
105. Igne B, Liu Y, Shi Z et al (2021) Multivariate spectroscopic method lifecycle management as part of the quality management system. J Pharm Sci 110:2925–2933. https://doi.org/10.1016/j.xphs.2021.03.013
106. International Conference on Harmonization (2019) ICH Q12: lifecycle management
107. International Conference on Harmonization (2022) ICH Q14: analytical procedure development
108. European Pharmacopoeia (2023) 5.21. Chemometric methods applied to analytical data. 11.1
109. Algorri M, Abernathy MJ, Cauchon NS et al (2022) Re-envisioning pharmaceutical manufacturing: increasing agility for global patient access. J Pharm Sci 111:593–607. https://doi.org/10.1016/j.xphs.2021.08.032
110. Sohail Arshad M, Zafar S, Yousef B et al (2021) A review of emerging technologies enabling improved solid oral dosage form manufacturing and processing. Adv Drug Deliv Rev 178: 113840. https://doi.org/10.1016/j.addr.2021.113840
111. de Oliveira RR, Avila C, Bourne R et al (2020) Data fusion strategies to combine sensor and multivariate model outputs for multivariate statistical process control. Anal Bioanal Chem 412:2151–2163. https://doi.org/10.1007/s00216-020-02404-2
112. Dhondt J, Bertels J, Kumar A et al (2022) A multivariate formulation and process development platform for direct compression. Int J Pharm 623:121962. https://doi.org/10.1016/j.ijpharm.2022.121962
113. Hemanth Kumar S, Talasila D, Gowrav MP, Gangadharappa HV (2020) Adaptations of pharma 4.0 from industry 4.0. Drug Invent Today 14:405–415
114. Chen Y, Yang O, Sampat C et al (2020) Digital twins in pharmaceutical and biopharmaceutical manufacturing: a literature review. PRO 8:1088. https://doi.org/10.3390/pr8091088
115. Beke ÁK, Gyürkés M, Nagy ZK et al (2021) Digital twin of low dosage continuous powder blending – artificial neural networks and residence time distribution models. Eur J Pharm Biopharm 169:64–77. https://doi.org/10.1016/j.ejpb.2021.09.006

Part 2
Analytical Quality by Design (AQbD)

Chapter 8
Analytical Quality by Design Fundamentals and Compendial and Regulatory Perspectives

Amanda Guiraldelli Mahr, Felipe Rebello Lourenço, Phil Borman, Jane Weitzel, and Jean-Marc Roussel

Abstract Over the last years, the pharmaceutical industry and regulatory agencies have been embracing the Quality by Design (QbD) principles. Significant quality paradigm shifts have been occurring in the pharmaceutical quality environment to enable manufacturing modernization and innovation with the aim to promote continuous quality improvement of medicines. A key component behind this quality culture transformation is the evolution of validation concepts to a lifecycle risk-based approach driven by QbD principles. This chapter will discuss about the evolution of QbD and analytical QbD principles applied in the pharmaceutical environment, including new International Council for Harmonisation of Technical Requirements for Pharmaceuticals (ICH) guidelines (e.g., ICH Q14 and Q2(R2)) and compendial standards that outline quality risk management (QRM) principles. The three-stage analytical procedure lifecycle framework described in the USP General Chapter <1220> will be discussed in this chapter along with the role of the different procedure lifecycle stages and elements (e.g., analytical target profile (ATP), method operable design region (MODR), analytical control strategy (ACS), replication strategy, quality risk management process, and procedure performance program).

A. G. Mahr (✉)
United States Pharmacopeia (USP), Rockville, MD, USA

F. R. Lourenço
United States Pharmacopeia (USP), Rockville, MD, USA

School of Pharmaceutical Sciences, University of São Paulo, São Paulo, Brazil
e-mail: feliperl@usp.br

P. Borman
United States Pharmacopeia (USP), Rockville, MD, USA

USP Measurement and Data Quality Expert Committee, United States Pharmacopeia, Rockville, MD, USA

J. Weitzel · J.-M. Roussel
USP Measurement and Data Quality Expert Committee, United States Pharmacopeia, Rockville, MD, USA

M. C. Breitkreitz, H. C. Goicoechea (eds.), *Introduction to Quality by Design in Pharmaceutical Manufacturing and Analytical Development*, AAPS Introductions in the Pharmaceutical Sciences 10, https://doi.org/10.1007/978-3-031-31505-3_8

8.1 Introduction to Analytical Quality by Design Principles

Over the last years, medicines have evolved into more complex manufacturing of sophisticated drug products, dosage forms, and complex therapies. In parallel, regulatory agencies started formally recognizing the opportunities to improve efficiency and quality of medicines through innovation in pharmaceutical development and manufacturing [1]. Along with that, the manufacturing technologies continue to evolve as the internet of things, artificial intelligence, and advanced computing begin to challenge the traditional approaches for the manufacture of pharmaceuticals [1]. A new industrial revolution, called "Industry 4.0," has resulted from technological advancements during the last decade, where the term "Pharma 4.0" refers to the implementation of these technological advances in the pharma industry and efforts to adapt to the new era of digital transformation [1]. A fundamental piece of this framework is the concept of quality, which has been called "Quality 4.0," where traditional validation practices are undergoing a paradigm shift ("Validation 4.0"), requiring an in-depth understanding of products and processes.

These recent changes in the pharmaceutical industry have led to significant paradigm shifts in the pharmaceutical quality environment [2], moving from the application of compliance-driven and quality by testing approaches toward risk-based and Quality by Design (QbD) approaches. This new quality paradigm started to take shape in the late 1990s when the concepts of process validation started to evolve from a traditional "fixed-point" manufacturing process to a "lifecycle" risk-based approach [3], allowing the identification of risks and design of controls to minimize potential risks within the manufacturing process enabling innovation and proactive continuous improvement of processes and product quality.

Since 2000, regulatory agencies and industry have been embracing the adoption of QbD principles in the pharmaceutical development. Examples are FDA's initiative announced in 2002 to encourage the industry to adopt modern quality management process and risk-based approaches and the 2004 FDA's report underlining these principles entitled "Pharmaceutical cGMPs for the 21st Century – A Risk Based Approach" [4]. The implementation of quality risk management (QRM) was also recognized by the ICH and resulted in the development of a series of ICH guidelines outlining QbD principles [5]. QbD is a concept first developed by the quality pioneer Dr. Joseph M. Juran [6]. In the ICH Q series, the QbD principles were first introduced in 2004 in the ICH Q8 [7], and in 2005, the importance of QRM as enabler of QbD approach was underlined in the ICH Q9 [8]. This ever-changing scenario has led to significant paradigm shifts in the pharmaceutical quality environment to enable manufacturing modernization and innovation, placing more emphasis on product and process understanding based on sound science. More recently, a holistic framework based on product lifecycle management has been proposed in the ICH Q12 [9], also covering analytical condition change management. In July 2021, ICH Q13 [10] on Continuous Manufacturing reached step 2. This guide describes scientific and regulatory considerations for lifecycle management of continuous manufacturing [10] highlighting the development of process analytical technologies (PATs) and real-time release testing (RTRT), which are also described in the 2004 FDA's regulatory framework [11].

Another aspect of this transformation is the recognition by regulators of supply disruptions associated with manufacturing or quality issues as common causes of drug shortages [12], leading to launch of initiatives to support and encourage the adoption of new quality paradigms. As an example, besides FDA's initiatives to encourage QbD adoption, the agency has been developing a framework to rate the quality management maturity (QMM) of pharmaceutical manufacturing sites (e.g., quality metrics) to drive continuous improvement efforts in manufacturing to enable continual improvement and mitigation of quality risks [13, 14]. The concept of analytical procedure validation has also evolved over the last years, where an emphasis has been placed on the importance of incorporating quality risk and knowledge management principles which are the major enablers of QbD. The term analytical quality by design (AQbD) is derived from QbD principles and refers to the application of QbD in the analytical procedure lifecycle management allowing for the quality to be built into the procedure design. AQbD can be described as a systematic approach to procedure development and validation that begins with predefined objectives (analytical target profile) and emphasizes procedure performance understanding and control, based on sound science and QRM [15].

8.1.1 Compendial Perspectives on AQbD

After the ICH Q9 was published, several groups and pharmaceutical associations (e.g., joint working groups of the European Federation of Pharmaceutical Industries and Associations (EFPIA) and the Pharmaceutical Research and Manufacturers of America (PhRMA)) started exploring the application of QbD principles to the development, validation, and verification of analytical procedures [16, 17]. Previously, QbD principles had been explored only in pharmaceutical development and manufacturing, and current practices for procedure validation usually did not consider a comprehensive assessment of sources of variability and the design of analytical control strategies based on knowledge acquired about risks to quality of the reportable value. Under this traditional paradigm, little focus has been given on demonstrating procedure fit for use and minimal understanding of total measurement error, resulting in insufficient connection with the decisions made using the procedure results. The US Pharmacopeia formed in 2013 an expert panel to envision the future of analytical procedure validation. Between 2013 and 2016, four USP stimuli articles were published: PF 39(5) Lifecycle Management of Analytical Procedures [18], PF 42(2) Fitness for Use [19], PF 42(5) Analytical Target Profile [20], and PF 42(5) Analytical Control Strategy [21]. These articles covered key aspects of the procedure lifecycle and built the basis for the new USP General Chapter <1220> Analytical Procedure Lifecycle, which was published in PF46(5) December 1, 2020 (official as of May 1, 2022, in USP-NF). The USP <1220> presents the analytical procedure lifecycle as an alternate and enhanced approach driven by AQbD principles that provides connectivity between the different stages allowing for knowledge management. This approach highlights the importance of sound scientific approaches and quality risk management for the development, control, establishment, and use of analytical procedures [22]. Two other stimuli articles were published by USP later in 2018–2019: PF 44(1) Measurement Uncertainty for the Pharmaceutical Industry [23] and PF 45(6) Distinguishing the Analytical Method from the Analytical Procedure to Support the USP Analytical Procedure Lifecycle Paradigm [24].

The procedure lifecycle concept also underlies the standard ISO/IEC 17025 [25]. The British Pharmacopeia (BP) published a new supplementary chapter on AQbD in BP 2022 [26]. Previously in 2019, BP and MHRA published a consultation that explored the inclusion of AQbD elements in compendial [27]. In 2020, the consultation response was made available [29]. Later, BP published the atorvastatin tablets monograph in BP 2023 including an assay procedure for atorvastatin incorporating the ATP and operating ranges as non-compendial tools. Both BP and USP chapters are informational chapters and aim to provide guidance for the implementation of the enhanced approach to support procedure lifecycle management. A few other USP chapters cover principles and considerations for stages of the analytical procedure lifecycle, including *USP* <1225>Validation of Compendial Procedures [29], *USP* <1226> Verification of Compendial Procedures [30], *USP* <1224> Transfer of Analytical Procedures [31], *USP* <1210> Statistical Tools for Procedure Validation [32], *USP* <1010> Analytical Data—Interpretation and Treatment [33], and *USP* <1039> Chemometrics [34].

8.1.2 Analytical Procedure Lifecycle Approach and AQbD Fundamentals

Fit for use/purpose is the key concept in the lifecycle approach. Decisions are made with the reportable values generated with the analytical procedure. It is imperative that these values are suitable for these decisions and the probability of making a wrong decision is understood. The major driver for adopting the lifecycle approach is to ensure that the reportable value is fit for use. The procedure lifecycle approach is an enhanced approach that holistically incorporates all the events that take place over the procedure lifecycle to demonstrate a procedure is fit for use [22]. The intended purpose of a procedure can be varied such as (a) procedure application (e.g., preclinical safety evaluations and pre-formulation studies, support clinical studies, stability studies, release a lot, etc.) and (b) determining product quality attributes (e.g., identity, purity, assay, etc.). It includes all types of procedures (e.g., quantitative, limit tests, qualitative procedures). It applies to procedures performed online, offline, at-line, and in-line. Currently, there is no clear, commonly used, and internationally agreed-upon process to follow that defines whether a reportable

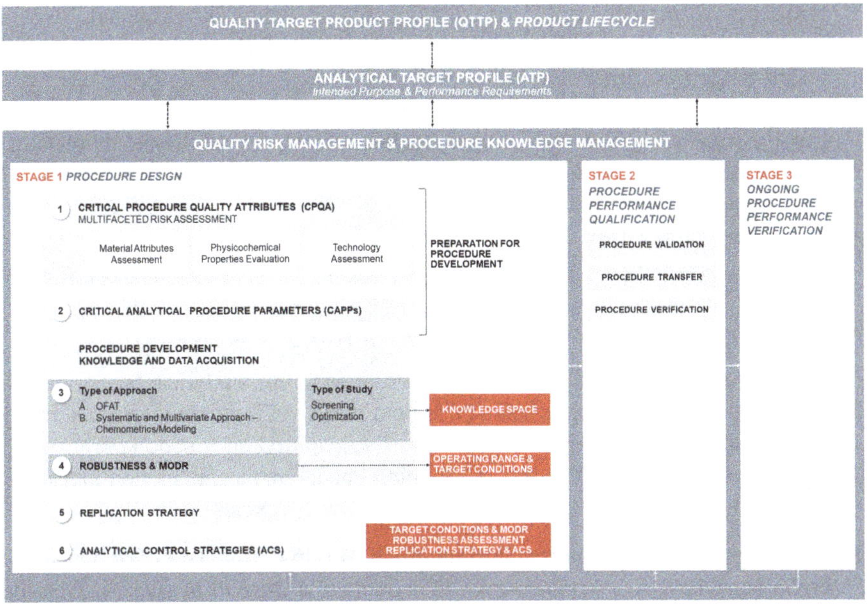

Fig. 8.1 Analytical procedure lifecycle and key elements involved in the framework

value is fit for use [19]. Thus, different approaches can be used and will be discussed in Chap. 9. Under this new paradigm, QRM plays an essential role to allow for the implementation of the procedure lifecycle approach. QRM will be discussed in more details in Sects. 8.3 and 8.5. The analytical procedure lifecycle is comprised of the analytical target profile (ATP) and three stages: procedure design (stage 1), procedure performance qualification (stage 2), and ongoing procedure performance verification (stage 3). This framework is depicted in Fig. 8.1.

Analytical Target Profile (ATP) The procedure lifecycle is initiated by the establishment of the ATP which provides the definition of the procedure scope, specifies the quality requirements of the reportable value, and defines the requirements for procedure performance characteristics that are linked to the intended application and quality attribute to be measured. The ATP should be aligned with the quality target product profile (QTPP) and can be reviewed whenever necessary, based on knowledge acquired and risks identified during other stages of the procedure lifecycle. The ATP will be discussed under Sect. 8.2.

Stage 1 Procedure Design The main focus of stage 1 is the acquisition of understanding about how procedure parameters affect the procedure quality attributes, allowing for the establishment of critical procedure quality attributes (CPQAs – analytical responses that reflect the procedure quality and ensure fitness for use) and critical analytical procedure parameters (CAPPs – procedure parameters whose variability has an impact on CPQAs). This stage encompasses the following activities/steps:

- Initial risk assessment called "preparation for procedure development," which includes knowledge gathering through assessment of material attributes, compound properties, and technology.
- Procedure development where CAPPs and CPQAs are usually assessed through experimentation and/or predictive modeling to acquire additional knowledge about procedure performance and robustness allowing for the establishment of knowledge space.
- Robustness assessment and design of the method operable design region (MODR) (operating range where all study factor combinations provide suitable mean performance and robustness).
- Establishment of replication strategy and analytical control strategy (ACS).

The output of stage 1 includes the final conditions for the analytical procedure and MODR, evidence of procedure robustness, and initial ACS and replication strategy concluding the initial development activities. The acquired information during stage 1 will always be available to support condition change management and investigations about lack of performance along the procedure lifecycle.

Stage 2 Analytical Procedure Performance Qualification (APPQ): Once the preliminary analytical control strategy is defined at the end of stage 1, then the ATP can be redefined if necessary, and the procedure is ready to undergo APPQ. This will be discussed with more details under Sect. 8.6.

Stage 3 Ongoing Procedure Performance Verification (OPPV): This is a continuous exercise to confirm the fitness of analytical procedure. OPPV will be discussed under Sect. 8.7.

8.1.3 Regulatory Perspectives on AQbD

In late 2018, an ICH expert working group (EWG) was formed and finalized a long-anticipated concept paper [35] supporting the creation of a new guideline ICH Q14 [36] on Analytical Procedure Development (analogous to Q8 [7], Q11 [37], and stage 1 of *USP* <1220>) and a revision of ICH Q2(R1) [38] on Validation of Analytical Procedures (similar to stage 2 of *USP* <1220>). The combined topic Q2 (R2)/Q14 represents an opportunity to provide guidance on how to apply enhanced development approaches to analytical procedures (AQbD) and how to use the knowledge obtained to support routine use of procedures. Q2(R2)/Q14 will also have the potential to facilitate the selection or identification of development approaches that will reduce the risk incurred by post-approval changes to analytical procedures discussed in Q12 [9], thus enabling more efficient and science-based change management. The intent of Q14 will be in line with Q8 and Q11, showing minimal (traditional) and "enhanced" approaches for developing analytical procedures while utilizing QRM tools (Q9 [8]) within pharmaceutical quality management systems (Q10 [39]). Q14 will aim to define performance criteria (linked to the control strategy) for analytical procedures and define requirements for analytical development and validation. It will also facilitate a greater understanding of analytical procedures being shared in the regulatory submission and create a basis for more flexible regulatory approaches. Drafts of Q14 and Q2(R2) [40] were issued by the EWG in March 2022 (step 2, public consultation), and there should be no additional expectations/mandated requirements for pharmaceutical analytical scientists [41]. Guiraldelli, A. and Weitzel, J. highlighted recently the similarities and key differences between ICH Q14/Q2(R2), USP <1220>, and the ISO/IEC 17025:2017 [42]. Reference [15] shows a timeline with relevant ICH guidelines, regulatory guides, and USP chapters that are related to QbD and lifecycle approaches applied to pharmaceutical development and analytical procedures.

8.1.3.1 ICH Q14 Analytical Procedure Development

The objective of ICH Q14 is to describe science and risk-based approaches for developing and maintaining analytical procedures suitable for the assessment of the quality of chemical and biological/biotechnological drug substances and drug products [36]. Minimal and enhanced approaches to analytical procedure development are described. The minimal approach reflects the current state and is still deemed acceptable. Some or all of the enhanced approaches can be used to support the development and lifecycle management of analytical procedures if they are considered useful. The ATP concept is introduced as the first enhanced development step and is defined as a prospective summary of the performance characteristics describing the intended purpose and the anticipated performance criteria of an analytical measurement. The ATP is informed by product/process requirements and is not linked to any specific analytical technology.

The ATP therefore enables the analyst to select the best analytical technology [43] to develop the procedure. Other elements included in the enhanced approach include the use of risk assessment [44, 45], prior knowledge [43], and multivariate experiments [46, 47] and the definition of an analytical procedure control strategy [16, 21, 22]. This information can subsequently be used (and shared in the regulatory dossier) to define a lifecycle change management plan with clear definitions and reporting categories of established conditions (ECs), proven acceptable ranges (PARs), or MODR as appropriate [7, 9, 36]. Q14 presents the procedure lifecycle framework including the ATP and the same stages discussed in the previous Sect. 8.1.2. The ATP is informed by product/process requirements and is not linked to any specific analytical technology.

The ATP therefore enables the analyst to select the best analytical technology [43] to develop the procedure. Other elements included in the enhanced approach include the use of risk assessment [44, 45], prior knowledge [43], and multivariate experiments [46, 47] and the definition of an analytical procedure control strategy [16, 21, 22]. This information can subsequently be used (and shared in the regulatory dossier) to define a lifecycle change management plan with clear definitions and reporting categories of established conditions (ECs), proven acceptable ranges (PARs), or MODR, as appropriate [7, 9, 36]. The following points represent other key aspects in Q14.

- ICH Q14 section 5 states the concept of robustness and says that "data from validation studies (e.g., intermediate precision) may be used to complement robustness evaluation." This implies that factors that are typically studied in what have previously been termed as ruggedness studies [48, 49] (e.g., intermediate studies [50] and/or reproducibility) can be used as part of a robustness assessment. Q14 also highlights that robustness evaluation is conducted during development for most procedures rather than as part of the validation exercise.
- The term platform analytical procedure is defined as a multiproduct method suitable to test quality attributes of different products. Such platform procedures may be leveraged to evaluate the attributes of a specific product without conducting additional procedure development as well as to reduce the number of validation tests required.
- A definition is provided for the reportable result that highlights the need to consider how the result is calculated/processed and to ensure that replication has been applied (where applicable). The potency example in Annex A stresses the importance of defining an appropriate replication strategy [24, 51, 52].
- New sections dedicated to the development of multivariate and real-time release testing analytical procedures have been included, building off existing guidance [53, 54].

The most significant section in Q14 for industry is the Chap. 7 entitled "Lifecycle Management and Post-Approval Changes of Analytical Procedures." The mechanisms described in that chapter have the potential to reduce the existing burden felt by industry when post-approval changes to analytical procedures are being made [55]. The tools and enablers discussed in Q12 are applicable to analytical procedures such as ECs, post-approval change management protocols (PACMPs), and the product lifecycle change management (PLCM) document. Q14 presents a decision tree for a risk-based approach for the identification of ECs and reporting categories for associated changes in the enhanced approach. Q14 also mentions that the ATP could also form the basis of a PACMP, which would allow changes between technologies to be reported at a lower reporting category provided that the predefined requirements for a change are met. This is where the real benefit of Q14 could be realized for the pharmaceutical industry.

The use of QRM is advocated to enable evaluation of the impact of changes and reconfirm that the originally agreed reporting category is still appropriate. The outcome of such risk assessments together with an assessment of available prior knowledge can inform the design and extent of the studies required to support the change, including an appropriate bridging strategy to demonstrate that the revised or new procedure is fit for purpose. Two comprehensive examples of how the concepts described in Q14 can be applied are provided in Annex A: (1) measurement of stereoisomers as specific process-related impurities in a small-molecule drug substance and (2) measurement of potency for a monoclonal antibody. These examples include construction of the ATP, selection of the most suitable analytical technology [43], development of the analytical procedure, use of the ATP to aid the design of the validation study, identification of ECs, and how post-approval changes to analytical procedures can be facilitated.

8.2 Analytical Target Profile (ATP)

The ATP is a fundamental component of the lifecycle approach, and its definition is the first step in the AQbD framework [22]. The ATP is the predefined intended purpose of the procedure and establishes the criteria for the analytical procedure performance characteristics that are required to an intended use, which includes the quality attributes to be measured (e.g., accuracy and precision for a quantitative procedure). In other words, the ATP defines the required quality of the reportable value, since the measured value provides the basis for conformity or nonconformity assessment decisions of a tested item or batch against the regulatory, compendial, or manufacturing specification limits [22, 56–58]. Refer to Chap. 9 for a detailed discussion on ATP.

8.3 Quality Risk Management

The QRM is embedded along the entire analytical procedure lifecycle and supports science-based and transparent/scientific rationale for decision-making when integrated into analytical procedure lifecycle management. The QRM for an analytical procedure is a systematic process for the assessment, control, communication, and review of risk to the quality of the reportable value (which is the product of the analytical procedure) across the analytical procedure lifecycle [8, 21, 22]. In the context of the analytical procedures, risk to quality may include analytical factors and their interactions that may lead to lack of robustness and procedure bad performance (factors that are not able to meet established acceptance criteria). The main goals of applying QRM throughout the procedure lifecycle are:

- Assessment of the analytical procedure conditions and identification of sources of bias and variability.
- Acquisition of knowledge for the design of appropriate controls on the analytical procedure parameters and material attributes aiming performance optimization, minimization or even elimination of risks that will ensure the procedure meets the ATP and other defined performance requirements.
- Allow for continual knowledge management and improvement of quality by ensuring that the procedure is fit for use along the procedure lifecycle increasing the confidence on decision-making of pharmaceutical product's conformance.

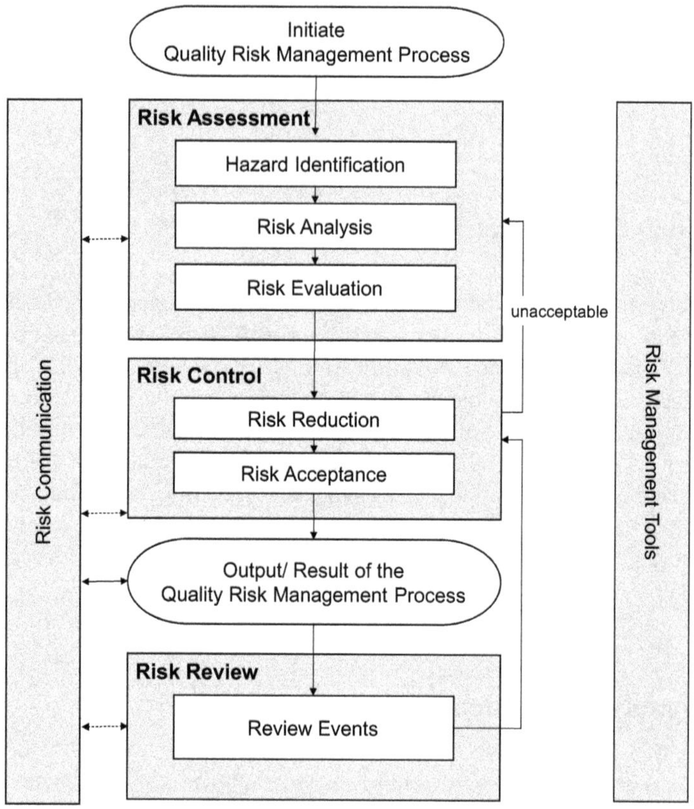

Fig. 8.2 Overview of a typical QRM process. (Reproduced from ICH Q9(R1))

8.3.1 Quality Risk Management Principles and Framework

Quality risk management (QRM) is a process driven by few principles that provides a rationale to understand risk and mitigate it through appropriate and robust controls [59]. The QRM framework presented in ICH Q9 guideline (Fig. 8.2) can be applied to analytical procedures [8, 22, 36].

Risk assessment should be based on scientific knowledge where major sources of variability can be identified and a plan to control the risks can be defined to reduce or even eliminate the risks (e.g., by ensuring the appropriate technology and procedure conditions) [22]. Ultimately, this should be linked to the QTPP and product specifications to ensure product safety, efficacy, and quality. The level of effort, formality, and documentation should be commensurate with the level of risk [22] being QRM a dynamic and iterative process which should be responsive to necessary analytical changes or additional knowledge gained on factors that may impact significantly on performance at any stage of the procedure lifecycle [60].

8.3.2 *Risk Assessment*

This stage includes the acquisition of understanding about how analytical process parameters affect the procedure quality attributes allowing for the identification, analysis, and evaluation of risks.

8.3.2.1 Initiating a QRM Process

Initially, a well-defined problem description or risk question should be established [8] based on the procedure intended use and performance requirements established in the ATP. In parallel, relevant knowledge gathering should be exercised based on risk-based approach. During this initial stage, a multifaceted risk assessment is performed based on scientific principles and experience by gathering prior knowledge on analytical factors that may affect procedure performance allowing for the selection of CPQAs and CAPPs. This may include the following:

- *Material Attribute's Assessment*: This may include the identification of the presence of potential matrix compounds in the matrix through the assessment of sample constitution (e.g., excipients, process impurities, degradation products, APIs, etc.). Knowledge should be gathered by evaluating available information (e.g., compendial monographs, papers, books), knowledge about product development and manufacturing (e.g., drug master file, etc.), and any other existing procedures/resources.
- *Evaluation of Compound's Physicochemical Properties*: A detailed evaluation of physicochemical properties of analytes potentially present in the sample should be performed. Several sources of knowledge may be used (e.g.. chemicalize.org, PubChem, drugbank.ca, in-silico prediction tools/software), and different properties may be assessed (e.g., volatility/vapor pressure; boiling point; hydrophobicity and lipophilicity, solubility, polarity, pKa, stability (thermally labile, redox potential, etc.)). Information about compounds' behavior and stability should also be gathered and may include the assessment of potential degradation reactions which may occur (e.g., through pilot stress testing, predictive tools, and prior knowledge) and evaluation of environment conditions (reagents/solution used for sample preparation, pH of extractor solvent/diluent, temperature, type of analytical technique, etc.).
- *Technology Assessment*: During technology assessment phase, several techniques may be compared in terms of capability of the analytical methods to meet the ATP requirements. Prior knowledge on sample constitution and compound's properties/stability will assist in the selection of the most suitable technique and CPQAs and CAPPs to be further assessed. Brainstorming exercises with subject matter experts are highly recommended for knowledge gathering, especially during the initial risk assessment stage. Examples can be found in reference [43].

The initial risk assessment will provide the necessary prior knowledge for the three steps involved in the risk assessment stage: hazard identification, risk analysis, and risk evaluation [8, 21, 22].

8.3.2.2 Risk Assessment

- *Hazard Identification*: At this stage, the focus will be on the identification of potential factors which can pose a risk to the quality of the procedure and related hazards, including the possible consequences. This stage addresses the question "What might go wrong?" [8].
- *Risk Analysis*: This stage includes the estimation of the risk associated with each analytical factor (and their interaction, if significant). This stage addresses the question "What is the likelihood (probability) it will go wrong?". It is the qualitative or quantitative process of linking the probability of occurrence of harms and their severity. Detectability of risk may also factor into its estimation [8, 21].
- *Risk Evaluation*: This stage includes the comparison of identified and analyzed risk against the previously established given risk criteria. This stage addresses the question "What are the consequences (severity)?". Risk evaluation considers the strength of evidence for all three of the fundamental questions [8].

The result of a risk assessment can be either a quantitative estimate of risk (where a numerical probability is used) or a qualitative description of a range of risk. A "risk score" may be used to further define descriptors in risk ranking [8]. At this point, the risks may be classified into low, medium, or high risk priority based on the ability to meet the ATP and other critical attributes (e.g., using heat maps as illustrated in Fig. 8.3). An example is illustrated in the USP Stimuli Article Analytical Control Strategy [21].

In general, different approaches can be used to assess and manage the risks to quality including formal or informal approaches (using or not risk assessment tools) based on prior knowledge without experimentation and/or based on predictive modeling and experimentation. For analytical factors that may affect significantly the procedure performance (consequently may have a higher associated risk), additional knowledge can be gained through experimentation and/or predictive

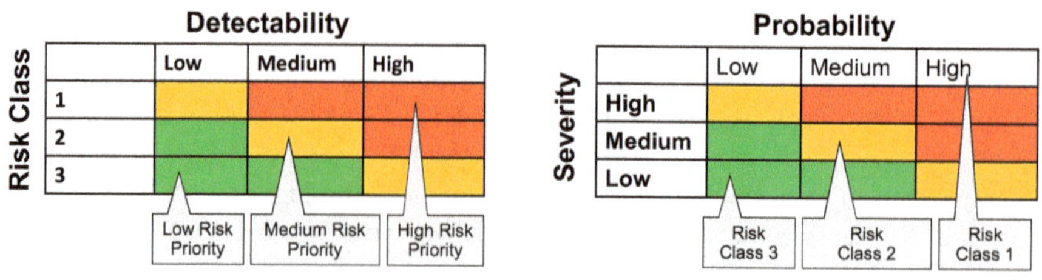

Fig. 8.3 Example of heat maps for risk priority and class definition

modeling. This will be discussed in more details in Sect. 8.4. The level of formality and extent of assessment will be dictated by the risk in question, complexity of sample constitution and technique used, and intended purpose of the study (e.g., screening and optimization studies).

8.3.3 Risk Control

Once the risk associated with each procedure parameter (and/or with their interaction) has been assessed, planning for how to manage those risks occurs [22]. Risk control activities usually involve the establishment of control strategies and measures which may reduce or control the risk below an acceptable level (risk criteria). The establishment of risk controls is a decision-making activity designed based on knowledge of how factors and variables impact procedure performance aiming to reduce and/or accept risks. Below are some examples of control strategies which can help to ensure procedure fitness for use:

- Controllable factors with higher risk: These are usually CAPPs which may have a significant effect on performance (main and/or factor interaction effect); the risk may be mitigated by controlling them within a specific range which ensures procedure fitness for use (e.g., designing the operating range to ensure good performance and the desired level of robustness and minimizing variability) and strategies to reduce the probability of occurrence of harm and increase the detectability (e.g., establishment of system suitability requirements, etc.).
- Uncontrollable factors: Few variables may be difficult or impractical to control, and the risks associated with them will need to be accepted (noise variables).
- Acceptance criteria establishment based on measurement variability understanding (e.g., establishment of suitable performance requirement in the ATP, use of decision rules considering guard bands or indecision zones, tightening of specification, etc.).
- Establishment of routine monitoring program for ongoing procedure performance verification.
- Establishment of purpose-driven acceptance criteria and tests for procedure validation and transfer.

8.3.4 Risk Communication and Review

Risk can only be effectively managed when it is identified, assessed, considered for further mitigation, and communicated [61]. Appropriate systems should be in place to ensure that the output of the QRM process is periodically monitored and reviewed, as appropriate, to assess new information that may impact on the original QRM decision. Examples of such changes include changes to control systems, changes to equipment and analytical conditions, etc. [61].

8.3.5 *Risk Management Tools*

A key early step in the execution of a risk analysis is to determine the appropriate risk assessment tool [61]. Several risk assessment methodologies can be used (alone or in combination) throughout the QRM process, supporting science-based decisions [8, 22, 62]. The following techniques are most commonly used:

8.3.5.1 Basic Tools

- Diagram analysis: Simple techniques that are commonly used to gather and organize data, structure RM processes, and facilitate decision-making. Examples: flowcharts, cause-effect diagrams, check sheets, and process mapping.
- Risk ranking and filtering: Methods to compare and rank risks which typically involves evaluation of multiple diverse quantitative and qualitative factors for each risk and weighting factors and risk scores. Useful for situations when the risks and underlying consequences are diverse and difficult to compare using a single tool. Examples: heat maps.
- Supporting statistical tools: Control charts, DOE, histograms, Pareto charts, and process capability analysis.

8.3.5.2 More Advanced Tools

- Fault tree analysis (FTA): Used to identify all root causes of an assumed failure or problem. Relies heavily on full performance understanding to identify causal factors.
- Failure modes and effects analysis (FMEA): Assesses potential failure modes for processes, and the probable effect on outcomes and/or product performance, allowing for the establishment of risk reduction actions. Highly dependent upon strong understanding of product, process, and/or facility under evaluation. Output is a relative "risk score" for each failure mode [63].
- Other tools: Hazards analysis and critical control points (HACCP) [64], etc.

This is a non-exhaustive list and several other methodologies not listed may be applied. Examples of tools are also listed in several guidance [8, 21, 59, 61, 65]. It is worth mentioning that no single tool is applicable to all cases and specific risks do not always require the same tool. The selection of appropriate tool should be based on the given process and should be commensurate with level of risk. QRM applied to procedure development will be further discussed in Sect. 8.4.

8.4 Analytical Procedure Development and Quality Risk Management

The goal of procedure development is the acquisition of understanding on how analytical parameters impact on critical procedure attributes that express procedure performance. This is usually performed through experimentation and/or predictive modeling, which can ultimately allow for the identification of analytical conditions that optimize performance and minimize or avoid bias, reducing variability to establish operating conditions that ensure procedure fitness for use [22]. Two types of approaches are generally used:

- Enhanced approach: Systematic and multivariate approach to assess the impact of procedure parameters on performance where multiple factors are generally simultaneously varied with the aim of evaluating the main and factor interaction effects, amplifying the extension of knowledge acquired increasing the probability of identifying risks and sources of variability. Experimentation and/or predictive modeling are used for knowledge acquisition with the advantage of providing understanding about performance even if experiments have not been run. This approach emphasizes the importance of QRM and establishment of a lifecycle change management plan, allowing for a more comprehensive risk and knowledge management facilitating the establishment of optimized and robust conditions (instead of fixed-set points) and suitable analytical control strategies (ACS).
- Minimal approach (also known as traditional approach) [36]: Typically includes procedure development using one-factor-at-a-time (OFAT) approach (univariate experiments), where procedure performance characteristics are evaluate based on the study of parameters changed one at a time. In general, it results in the design of procedure with fixed-set point conditions.
- Minimal versus enhanced approach: Effects of factor interaction and conditions not experimentally tested cannot be assessed when applying OFAT approach which is only able to provide understanding of main effects and access information if experimentally tested. Consequently, there is a reduced probability of designing appropriate controls to ensure fitness for use and optimize procedure performance. Usually, under the procedure lifecycle approach, the enhanced approach is used.

The material complexity and procedure intended used are factors that dictate the level of development needed in terms of the experimentation extent and any modeling that may be used to optimize performance [22]. Procedure development generally includes one study or a sequence of studies with different intended purposes such as systematic screening and/or optimization studies. Ultimately, the main outputs of the procedure development stage are the establishment of operating conditions (where the desired level of robustness is built during design) and ACS, providing knowledge at hand for proper lifecycle management.

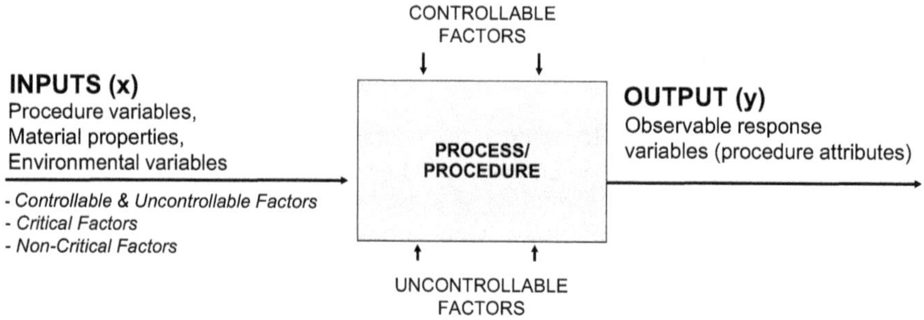

Fig. 8.4 General model for an analytical procedure or process

8.4.1 Intended Purpose of the Study and Quality Risk Management

In general, experiments can be used to study the performance of processes and procedures (Fig. 8.4).

The objective of the experiment may be to determine [66]:

- Which variables (x) are most influential on responses (y).
- Where to set the influential variable (x) value so that the response (y) is almost always near to the expected procedure performance.
- Where to set the influential variable (x) value so that the variability in the output (y) is small or less than the established requirement.
- Where to set the influential variable (x) value so that the effects of the uncontrollable variables on output (y) are minimized.

At this stage, critical procedure quality attributes (CPQAs, outputs) and analytical process parameters (inputs) should be selected for experimentation. The CPQAs may include (a) primary analytical responses which quantitatively or qualitatively express the desired procedure performance acting as surrogate (e.g., chromatographic resolution between target compounds expressing selectivity, retention factor k, signal-to-noise ratio expressing sensitivity, etc.) and/or (b) reportable value and measurement variability (e.g., content of API in drug product and its measurement uncertainty). The analytical procedure parameters may include different types of factors [22]:

1. Factors related to sample preparation or material attributes (e.g., reagents, solvents, etc.)
2. Analytical procedure factors (e.g., equipment/instrument settings).
3. Environmental factors (e.g., light, moisture, or temperature).

These procedure parameters can be further classified into "critical" and "non-critical" parameters and may be controllable or uncontrollable factors. Not all these parameters need to be studied during experimentation if they are classified as low risk priority (non-critical). Critical analytical procedure parameters (CAPPs) are the ones that have a significant impact on CQPAs, and these should be selected based on prior knowledge about critical conditions that may significantly affect the procedure attributes for further assessment through experimentation and/or modeling. This results in additional knowledge for the establishment of controls (e.g., establishment of specific ranges) to eliminate/reduce risks to quality. The selection of CAPPs should also be based on the type and intended purpose of the study (e.g., screening and/or optimization).

8.4.2 Screening Studies

The intended purpose of screening studies is to determine which factors affect the analytical procedure performance and identify CAPPs and their workable ranges, providing additional knowledge for comprehensive risk assessment. To accomplish this, an experiment protocol can be designed to enable the estimation of the magnitude and direction of the factor effects, where, in general, wider ranges of each factor are studied to allow for risk identification. The primary goal is to learn about (1) main factor effects (how much the analytical response variables change when each factor is changed), (2) factor interaction effects if they exist (how much the analytical response change when changing the factors together. Is there a combined effect of factors on the response?), and (3) factors and their ranges with best performance with a primary focus on selectivity and sensitivity.

Typically, chemometric methodologies are used, being DoE for screening widely used, such as factorial design, fractional factorial design, Plackett-Burman, mixture designs, etc. Optimal designs may also be useful at this stage especially for few analytical techniques where factor interaction effects may be larger than main factor effects (e.g., chromatography). For instance, the development of chromatographic procedures may consider categorical variables with several levels (e.g., screening of LC columns and types of solvents used in the mobile phase), and very often experimental constraints must be applied due to non-compatibility of levels between factors (e.g., non-compatibility of few LC column levels with pH of mobile phase level). In these cases, the use of optimal design represents a strategy to reduce the number of experiments and prevent special factor combination, allowing for testing irregular regions. Additionally, it allows for fitting special regression models (deletion of terms or inclusion of high-order terms).

Predictive modeling may also be deemed useful at the screening stage for additional knowledge generation. Several QRM tools may be used at this stage to facilitate risk identification, such as cause-effect diagram, heat maps, process maps, Pareto chart, mean effect plots, histograms, interaction plots, etc. Other tools such as acceptance performance region graphics, contour plots, and response surface (if appropriate design is used) are great tools that facilitate the comparison of risk against the established risk criteria (risk analysis and evaluation). During screening studies, risk control may be established by selecting best performance conditions (usually with focus on selectivity and efficiency) and workable regions that have the ability to meet established risk criteria. The identified CAPPs (factors and ranges) during screening can be further optimized later using appropriate methodologies for optimization.

8.4.3 Optimization Studies

The intended purpose of optimization studies is to determine the optimum region of the analytical factor space in which performance requirements are satisfied and performance is optimized. At this stage, the goals are (1) identification of factors and their interactions which may have significant impact not solely on selectivity but also on accuracy and precision and (2) identification of sources of variability and parameter ranges which minimize variability to a level below the risk criteria allowing the procedure to meet specification and optimize performance. Typically, optimization experiments based on chemometric methodologies are used, such as DoE for optimization (e.g., central composite design (CCD), Box-Behnken, Doehlert, mixture designs, optimal designs, etc.). Predictive modeling is commonly used during optimization studies as a strategy to gain additional knowledge about procedure performance using empirical, mechanistic, or hybrid models. Empirical models or "statistical models" are the ones generated based on observations acquired through experimentation. These are probabilistic models which best describe the relationship between dependent and independent variables (e.g., models derived from DoE or other chemometric methods). There are certain situations where the scientific phenomena are well understood that useful results, including mathematical models, can be developed directly by applying these principles [67]. These are classified as mechanistic models which can predict beyond the calibration space. Examples of mechanistic models are the ones built based on fundamentals of chromatographic separation mechanism used to predict retention time which are available in few software packages used to assist procedure development (e.g., models based on solvophobic theory [68–70] or linear free energy relationships (LFERs) [71, 72]). Although some phenomena related to analytical techniques are well understood, it may be needed to use hybrid models combining knowledge from fundamentals and experimentation (i.e., combination of mechanistic and empirical models to describe a system of study).

Chemometric methods and their particularities are out of scope of this chapter; however, it is worth mentioning that chemometrics is commonly used during procedure design for knowledge gathering and application, supporting risk assessment in a comprehensive way. As per USP General Chapter <1039> *Chemometrics* [34], chemometrics can be defined as a chemical discipline that uses mathematical tools and statistical methods to extract information from chemical data, which can be considered as a subset of the machine learning (ML) domain, which is a subset of artificial intelligence (AI) [65]. In general, the goals of applying chemometrics are (1) to design or select optimal measurement procedures and experiments and (2) to provide the maximum amount of relevant chemical information by analyzing chemical data (usually with several variables and large dataset) [34, 53]. Chemometrics can be performed by both linear and non-linear methods and can be applied for (1) quantitative properties' quantification (e.g., impurity profiling) and (2) sample classification (e.g., adulteration of samples). Linear methods include principal component analysis (PCA), hierarchical cluster analysis (HCA), principal component regression (PCR), partial least squares regression (PLS), soft independent modeling of class analogy (SIMCA), linear discriminant analysis (LDA), partial least squares discriminant analysis (PLS-DA), etc. [73]. Non-linear methods include artificial neural networks (ANN), support vector machine (SVM), and decision tree methods (random forest, boosting method, etc.) [73]. Chemometrics is divided into two types, supervised (predict the value of an output variable based on several input variables) and unsupervised (describe the associations and patterns among a set of input variables without an output variable), and includes the fields of predictive modeling (using different regression analysis techniques for multivariate analysis) and experimental designs (i.e., DoE). Chemometrics can be used in different applications where procedures are used to monitor pharmaceuticals attributes to ensure their quality (e.g., establishment of CPQAs and specifications for very complex materials such as herbal medicines and biotherapeutics), identification of target compounds to be monitored using univariate analytical procedures (e.g., sample adulteration analysis and authentication); development of multivariate analytical procedures (e.g., process analytical technologies – PATs [53]). This topic was also introduced in ICH Q14/Q2(R2).

Procedure development can also be assisted by the use of QbD software based on DoE or mechanistic/hybrid models (e.g., ACD/AutoChrom, ChromSword, Drylab, Fusion, Design-Expert, and any other statistical software such as Statistica, Minitab, R program, etc.).

Screening and Optimization Study Using DOE: Application of QRM Tools

A procedure for the quantification of organic impurities in a drug substance was developed by reversed-phase liquid chromatography (RPLC-UV) using DoE (optimal design). CPQAs and CAPPs were selected using a cause-effect diagram to assist the initial risk assessment based on prior knowledge for the initial screening experiments. Pareto chart was used to help identify main factor and factor interactions with larger effect on performance. An example of how to conduct risk assessment using cause-effect diagram, Pareto chart, and acceptance performance region graphics is shown in Fig. 8.5.

The use of ion-pair reagents 3 and 4 showed large and positive effect leading to an increased retention factor of target impurity which may allow for suitable precision and accuracy in quantification. Knowledge from Pareto charts and acceptance region graphics based on predictive modeling were used for the selection of best performance conditions for optimization. Figure 8.5c shows the acceptance performance region graphic built based on prediction models for all CPQAs derived from DoE for optimization. Conditions within the white area represent the region where all study factor combinations meet the established risk criteria.

(continued)

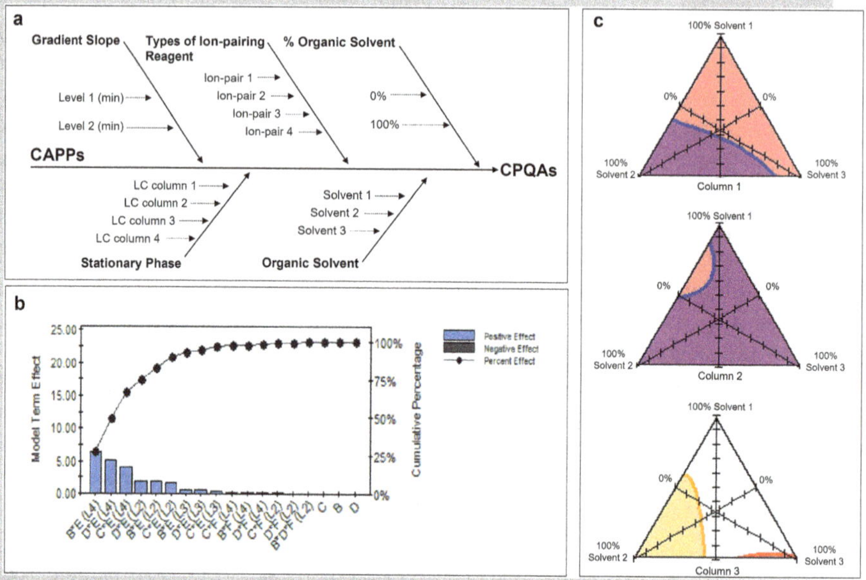

Fig. 8.5 Risk assessment tools: (**a**) Cause-effect diagram for CPQAs and CAPPs selection for screening, considering the CPQAs y1 – number of peaks with Rs ≥ 1.5, y2 – tailing factor of the most basic impurity and y3 – retention factor of the most polar compound; (**b**) Pareto Chart for retention factor (k) of the most polar impurity, where A, B, C, D, E-L2, E-L3, E-L4, F-L2, F-L3 and L-F4 represent gradient time, solvent 1, solvent 2, solvent 3, ion-pair 2, ion-pair 3, ion-pair 4, LC Column 1, 2 and 3, respectively; (**c**) Overlay of analytical responses using ion-pair 4, considering the following responses: blue – number of peaks with Rs ≥ 1.5, orange – tailing factor between 0.8–1.1 and red – critical pair Rs not less than 1.8

During optimization studies, robustness assessment plays a very important role to allow for comprehensive risk assessment, since it allows for the estimation of measurement variability associated with monitored responses (for risk identification) and probability assessment of measurements to fall within the specification limit (risk analysis and evaluation). Robustness assessment will be further discussed under Sect. 8.4.3 and Chap. 9. In terms of QRM tools, the same tools described for screenings can be applied to optimization. "box[title] ends".

8.5 Robustness Assessment and Method Operable Design Region (MODR)

Within the AQbD framework, robustness assessment should be initiated at stage 1 (procedure design) with the aim of providing acquisition of a high degree of understating about the sources of variability that can affect the procedure bias and precision. This knowledge allows for the establishment of proper risk control strategies to reduce variability associated with the procedure and ultimately supports the design of an operating range called method operable design region (MODR). The MODR is considered as a multidimensional combination and interaction of procedure parameters where all study factor combinations have been demonstrated to provide acceptable mean performance and robustness. Refer to Chap. 9 for a discussion on the initial robustness assessment conducted at stage 1, design of the MODR, and establishment of initial analytical control strategy (ACS).

8.6 Replication Strategy and Analytical Control Strategies

During the development of an analytical procedure, an associated analytical procedure control strategy (APCS) can be identified that ensures the analytical procedure performance and the quality of the measured result [22, 36]. The APCS is a set of controls which are needed to ensure the analytical procedure performs as expected and plays a key role in ensuring that the ATP is met throughout the lifecycle. The APCS includes analytical procedure parameters needing control and the system suitability test (SST) which are part of the analytical procedure description. A precision criterion is typically included in the ATP and the SST. An understanding of the main sources of variability within an analytical procedure can be used to develop a replication strategy [51] where the factors of the procedure that contribute to a large part of the overall variation can be replicated in order to improve the precision [52]. While replication of factors can reduce the uncertainty of an analytical procedure, there are limits to how much the uncertainty can be improved, and replication should not be used as a way of masking a poorly developed analytical procedure. Where replication of factor(s) exists in an analytical procedure, the SST should include criteria for acceptable variability among individual measurements to avoid the unscientific practice of comparing individual measurements against specification limits.

8.6.1 Statistical Considerations and Uncertainty

A general equation (Eq. 8.1) which combines all the factors which can be replicated within a procedure is used to generate the standard deviation of measurements from a procedure. Eq. 8.1 uses a sum of variabilities (variance components symbolized s_{Factor}) and the number of replicates (or unique levels, e.g., of intermediate precision [52] factors such as analyst or equipment) of the individual factors (n_{Factor}).

$$s_{Total} = \sqrt{\sum s_i^2/n_i} \tag{8.1}$$

Factors are often hierarchal in nature, such as replicate injections of multiple sample preparations in HPLC. In these cases, the degrees of replication are compounded leading to an adaptation of Eq. 8.1 for the estimation of standard deviation of measurements from a hierarchal (nested) design:

$$s_{Total} = \sqrt{\sum s_i^2/n_i + \sum s_{j(i)}^2/n_i n_j} \tag{8.2}$$

Interpretation of the impact of within procedure replication is incomplete without consideration of two additional points. The first is the impact of factors which impact variability over the long-term use of a procedure. These "between-procedure" (or between-run) factors are defined by ICHQ2(R1) [38, 74] as intermediate precision [50] (within laboratories) or reproducibility (within and between laboratories) factors and include laboratories, analysts, instruments, and reagent lots. In some procedures, these are negligible and can be ignored. Most notably, the use of a standard can reduce the influence of these factors on variability. If non-negligible, the variability due to intermediate precision factors should be included in the calculation of s_{Total}. These intermediate precision factors may likewise be replicated to manage the total long-term variability of the reportable value. In this case, the reportable value could be a summary measure from replicate runs.

Another consideration is the translation of total variability to uncertainty (U) associated with a reportable value generated by an analytical procedure. This can be expressed simply in Eq. 8.3 as:

$$U = k \cdot s_{Total} \tag{8.3}$$

Here, k is a statistical constant which is associated with "confidence." Thus, $k = 2$ is sometimes used to express the uncertainty associated with 95% confidence. The need to express total variability as uncertainty is required to evaluate a procedure's capability against the target measurement uncertainty (TMU) which can be specified in the analytical target profile (ATP) [20, 43, 75]. USP <1220> suggests that a replication design can be used which reduces uncertainty to the point of satisfying the ATP and further suggests that the replication design can be adapted to ensure conformance to the ATP after "qualification."

An extension of the formula in Eq. 8.3 where the overall variability has been divided into inter-run and intra-run components is provided in Eq. 8.4.

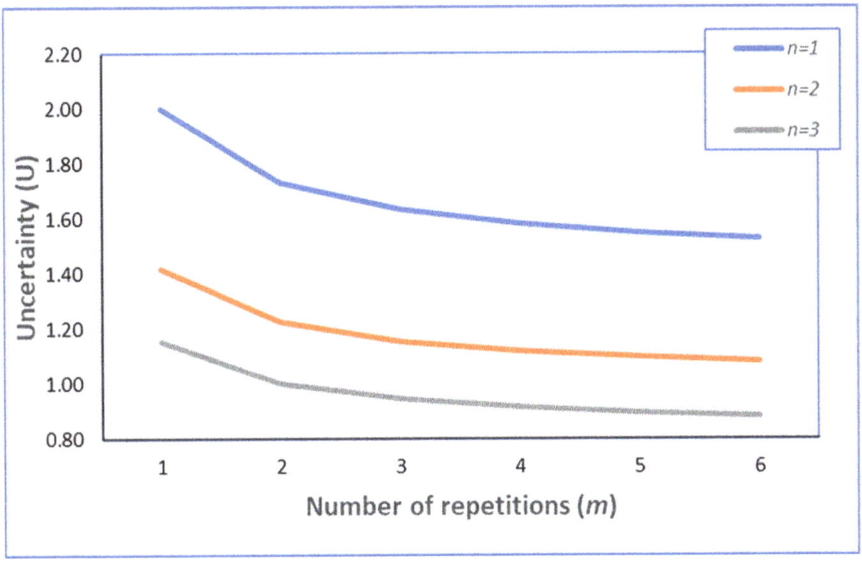

Fig. 8.6 Predicted uncertainty (U) of a reportable value for Var(inter-run) = Var(intra-run) = 0.50 and k = 2 (i.e., 95% probability)

$$U = k \cdot \sqrt{\frac{s_{inter}^2}{n} + \frac{s_{intra}^2}{nm}} \tag{8.4}$$

n is the number of independent runs, and m is the number of independent replicates performed on a test article within each run. The uncertainty is reduced with increases in m and/or n. Inter-run replication is most effective in reducing the overall variability due to its action on both components of variability (i.e., n is in the denominator of both terms). When using replication, the overall uncertainty (U) can be reduced through strategic selection of n and m. Note that the ideal strategy is to invest in the reduction of variability due to the inter-run component. The cost of reducing the impacts of Var(inter-run)and Var(intra-run) should also be considered. Figure 8.6 illustrates the impacts of the number of runs (n) and number of replicates within runs (m) on uncertainty (U).

As previously mentioned, uncertainty is reduced more effectively with an increased number of inter-run replicates, while there are diminishing returns with more than three intra-run replicates. Borman et al. [51] describe how the uncertainty of common analytical procedures such as HPLC and bioassay can be reduced via efficient replication.

8.7 Validation in the Analytical Procedure Lifecycle Framework

Under the procedure lifecycle approach, stage 1 "procedure design" is followed by stage 2, referred to as "analytical procedure performance qualification" (APPQ) [22]. APPQ is a term inspired by FDA's guidance for industry entitled "Process Validation: General Principles and Practices" [76] where the term "process validation" was revised to "the collection and evaluation of data, from the process design stage through commercial production, which establishes scientific evidence that a process is capable of consistently delivering quality product" recognizing that process validation involves a series of activities taking place over the product lifecycle. In the FDA guidance, the new term "process performance qualification" is used to refer to activities that confirm that the commercial manufacturing process design and performance are as expected. In the analytical procedure context, APPQ refers to all activities performed in stage 2, conducted to confirm (or qualify) that the procedure is fit for its intended purpose and meets the requirements of the ATP and consequently demonstrate that the procedure is suitable for routine use in the laboratory [22, 77]. APPQ may include "traditional" procedure validation (e.g., as described in USP <1225> and Q2(R1)), transfer, and verification, described in other guidelines [22]. To sum up, under the new paradigm for procedure validation (procedure lifecycle), "analytical procedure validation" refers to a broader concept which encompasses all activities that confirm that a procedure is suitable for use and that take place over the entire lifecycle of the procedure and not just activities restricted to stage 2 and "traditional validation." In this context, APPQ is part of "analytical procedure validation" (Fig. 8.7).

This more holistic concept was already introduced in the USP <1220>, which expands the well-known activities to integrate them into the procedure lifecycle and include them in a process that incorporates the ATP, while other existing guidance, such as USP <1225> and ICH Q2(R1), still describe the "traditional" procedure validation concept only, where "traditional" validation activities expected for a submission application are described. ICH Q2(R1) is currently under revision to include risk-based approach principles which has resulted in the issue of ICH Q2(R2) draft

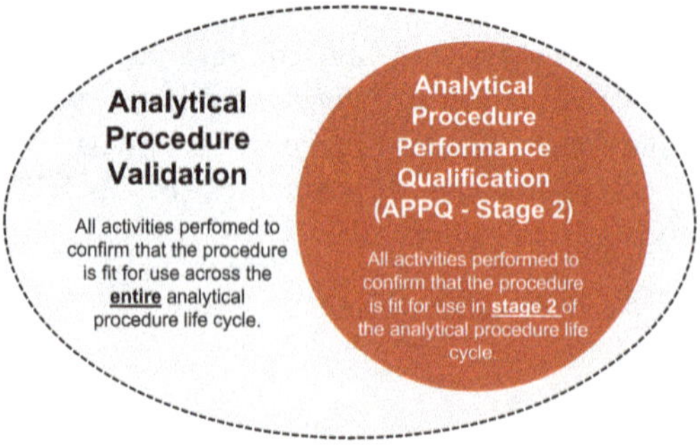

Fig. 8.7 Concept of analytical procedure validation under the new paradigm for validation "procedure lifecycle." APPQ may include "traditional validation," transfer, and verification

guideline (step 2, public consultation). Although ICH Q2(R2) attempts to align with risk and knowledge management principles and provides a clearer link between the attribute measurements and the product control strategy, it still does not recognize the term "procedure qualification" and the holistic "procedure validation" concept.

8.7.1 *Stage 2: Analytical Procedure Performance Qualification (APPQ)*

All activities covered in stage 2 are governed by the same documentation including (a) protocol and study design and (b) results and documentation. The extent of the APPQ study design should be based on the ATP and on an assessment of the risks to the procedure performance [22]. This depends on several factors including the type of activity and the type of technology of the procedure being qualified [22] Typical types of activities include:

- "Traditional" validation: Validation of an analytical procedure is the process by which it is established, by laboratory studies, that the performance characteristics of the procedure meet the requirements for the intended analytical applications [29].
- Procedure verification: This is a "documented process that qualifies a laboratory (the receiving unit) to use an analytical test procedure that originated in another laboratory (the transferring unit), thus ensuring that the receiving unit has the procedural knowledge and ability to perform the transferred analytical procedure as intended" [31].
- Procedure transfer: This is a "documented assessment of whether the compendial procedure can be used for its intended purpose, under the actual conditions of use for a specified drug substance and/or drug product matrix" [30].

8.7.2 *Protocol and Study Design*

The protocol for a qualification study generally includes but should not be limited to:

- The ATP and the acceptance criteria needed to meet the ATP.
- Justification for using pre-existing knowledge (e.g., from development, routine use, or validation) in lieu of qualification activities.
- Justification for abbreviated qualification activities if an established platform analytical procedure is being leveraged [32].
- Description of the qualification experiments that will be performed including:
  - Procedure-specific performance characteristics and acceptance criteria.
  - Statistical approach to be used to analyze the data.

- Appropriate guidance for qualification should also be referred in the protocol (e.g., ICH Q2(R1)/Q2(R2) [38, 40], USP <1225> [24], FDA guidance [78], AOAC guidance [79], ISO 5725 [81–85]).

ICH QR(R2) states which of the following four procedure performance characteristics need to be demonstrated for different types of qualitative and quantitative measurements: specificity/selectivity, working range (suitability of calibration model and lower range limit verification – quantitation limit and detection limit), accuracy, and precision. It is important not to use these characteristics in a check box manner but to consider the intended use of the procedure. These four performance characteristics are also typically provided in the ATP along with performance criteria as in the two examples in the draft ICHQ14 (found in Annex A).

Although a detailed discussion on how to assess and report the different performance characteristics is out of scope of this chapter, it is worth mentioning that USP <1225>/ICH Q2(R2) and ISO guide 5725 for validation describe "accuracy" in different ways. Table 8.1 lists the concepts of precision and accuracy described in these guides and highlights the differences in definition of accuracy.

ICH Q2(R2) also included an alternative combined approach to validate both precision and accuracy (bias), which is in line with USP <1210> *Statistical Tool For Procedure Validation* [32]. In USP <1210>, examples for estimating prediction intervals and tolerance intervals are described to validate both precision and accuracy [32]. Q2(R2) and Q14 also include guidance on the development and validation of multivariate analytical procedures, building off existing ICH guides (e.g., Q13) supporting the development of PATs and RTRT and any other multivariate procedure.

Table 8.1 Precision and accuracy concepts described in USP <1225>, ICH Q2(R2), and ISO guide 5725

	Accuracy	Trueness (bias)	Precision
USP <1225> and ICH Q2(R1) Q2(R2)	The accuracy of an analytical procedure is the closeness of test results obtained by that procedure to the true value	Referred to as "accuracy"	The precision is the degree of agreement among individual test results when the procedure is applied repeatedly to multiple samplings of a homogeneous sample
ISO 5725	The general term accuracy refers to both trueness and precision	Trueness refers to the closeness of agreement between the arithmetic mean of many test results and the true or accepted reference value	Precision refers to the closeness of agreement between test results

8.7.3 Results and Documentation

After the qualification study is complete and the results meet the predefined acceptance criteria and are deemed acceptable, the study is documented in a report [22]. Additionally, the ACS should be qualified, where a set of controls should be evaluated, including any system suitability criteria, defined in stage 1. Additional controls or modifications to any system suitability criteria may be included to reduce sources of variability that are identified in the routine operating environment [22]. The replication levels may also be revised based on the precision of reportable value assessed during qualification. If the qualification study did not meet predefined acceptance criteria, additional procedure development activities may be required before performing qualification experiments again (stage 2) [22]. After the completion of stage 2 activities, the procedure is released for its intended purpose (usually routine testing) [22].

8.8 Ongoing Procedure Performance Verification and Analytical Procedure Lifecycle Management

Ongoing procedure performance verification (OPPV) is the culmination of all the design, development, and qualification of the analytical procedure and involves monitoring the analytical procedure during use and confirming that the ATP criteria are still being met [22]. This stage includes both (a) routine monitoring of data linked to the performance of the analytical procedure and (b) evaluation of the procedure's performance after changes to determine if the analytical procedure continues to be fit for purpose [22]. The ATP and TMU continue to be the criteria used to evaluate the performance of the analytical procedure during routine use. For example, if the trending shows the standard deviation of the control sample varies over the year, the impact of that change in variability can be evaluated against the ATP and the change in probability of failure used in assessing the impact of the variation in the standard deviation for the control sample. This ensures the pharmaceutical product is fit for its intended purpose.

Figure 8.8 illustrates the link with fit for purpose. In the figure, the product QTPP establishes the use of the reportable value generated by the analytical procedure. The downward pointing arrows on the left side of the pyramid illustrate how fit for purpose links all levels, defines the ATP, and drives analytical procedure design, AQbD, and development and qualification. The upward pointing arrows on the right side of the pyramid illustrate that the analytical procedure performance on an ongoing basis and after change is driven by the use of the reportable value as required by the product QTPP.

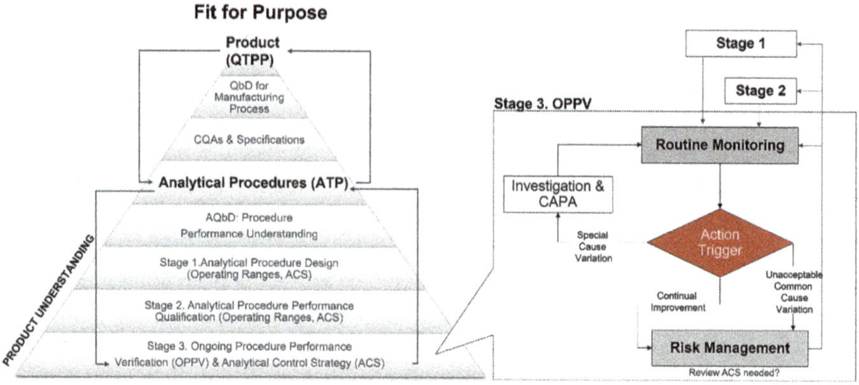

Fig. 8.8 Link between product lifecycle and analytical procedures to ensure fit for purpose and product quality, highlighting procedure lifecycle stage 3

The extent of monitoring is based on risk to both the respective product quality attribute and procedure performance. This ensures the monitoring is adequate, without being excessive and wasting resources. During routine operation, a large number of analyses may be performed providing more information and knowledge on the procedure performance. The monitoring provides an excellent opportunity to use this data obtained during routine use to further refine the ACS, improving its effectiveness and efficiency.

8.8.1 Designing a Routine Monitoring Program

The routine monitoring program should include the development of a sampling plan to collect data at a determined frequency, establishment of strategies for data visualization and analysis, and a defined reaction plan to guide the actions. For a proper verification of performance, relevant and critical attributes must be selected along with suitable acceptance criteria. For instance, the replication strategy can provide the acceptance criteria for the analytical controls, such as control charts, especially if multiple replicates are used. The acceptable range for the concentration of the individual replicate values must be defined and continually confirmed. The FDA Guidance "Investigating Out-of-Specification (OOS) Test Results for Pharmaceutical Production" [86] states: "Acceptance limits for variability among the replicates should also be specified in the method." A range control chart is valuable for this task. The qualification concept under the analytical procedure lifecycle framework confirms the ACS and provides more data to further define controls.

8.8.2 Identification of Analytical Procedure Aspects That Need Further Investigation During Routine Use

Some aspects of the control strategy may need to be further modified after the analytical procedure is placed into routine use. The advantage of the lifecycle approach is that these aspects are identified and a system is put in place to prepare for and control these aspects. For example, in an HPLC procedure, only one column may have been used during development and qualification because the column is unique and custom-made. When a new column is used, the control strategy would trigger an action to review the performance of the new column and ensure it meets requirements.

8.8.3 Tools Used by the Control Strategy

ACS uses tools for controlling the procedure and for trending its performance. Tools used for controlling variables specify their state or limit their range. These tools are:

- Environmental procedure controls: The analytical procedure may require defined environmental conditions, such as incubator temperature or room temperature. The procedure lists these environmental conditions and provides their criteria so that the laboratory can provide and confirm the required conditions.
- Training: Some steps in an analytical procedure are sensitive to their execution. The procedure needs to identify this, and the training program needs to deliver this training and demonstrate the analyst is competent.
- Equipment: Equipment includes measuring instruments, software, measurement standards, reference materials, reference data, reagents, consumables, or auxiliary apparatus that are required to perform the analytical procedure. Any specific requirements for the equipment need to be defined and met. For example, the metrological calibration of reference standards or instruments needs to be done according to a procedure, by a qualified operator, on a defined schedule.
- Documents: A key tool for the ACS is the documents which include the instructions and requirements for the analytical procedure. These can be electronic or paper based.
- Records: When an analytical procedure is executed, the record system captures the history of the performance of the procedure. The development and qualification processes define the information that needs to be captured. The system to capture the information needs to allow the data and information to be tracked, especially so that it can be used for trending.

8.8.4 Tools Used for Trending

During routine use, variation will occur and can be of different types:

- Acceptable common cause. The procedure is considered fit for purpose when all observations are within the predefined upper and lower limits. The common cause variation is acceptable and no special cause variation exists.
- Special cause variation (e.g., new, unexpected variation) can be due to signal shifts, drift, and observations which might fall well above or below the performance requirement.
- Unacceptable common cause variation occurs when some observations fall outside the performance requirement (e.g., a noise variable occurs and was not identified during method development).

The variation should be from acceptable common cause. The procedure should be under statistical process control (SPC) which uses statistical-based techniques to monitor and control a process to advance the quality or uniformity of the output of a process. The ACS is designed to monitor variation and react as required. When special cause variation is detected, an investigation can be conducted to determine the root cause and appropriate corrective actions.

Relevant and critical quality attributes should be selected for procedure performance verification. This can be done by using the established analytical control strategies (ACS) identified during stage 1 (e.g., SST attributes such as system precision, signal-to-noise ratio, and peak symmetry; other attributes or performance characteristics necessary to compare with ATP requirements). Operational attributes not directly linked to the ATP may not be well suited for the performance monitoring program. The extent of monitoring is based on the risk associated with the respective quality attribute and with the analytical procedure. The trend reviews are defined and periodic. If there is an indication that the procedure is not in control, an investigation should be conducted and corrective actions taken. Tools for data trending and monitoring are based on SPC and include investigation of failures (e.g., out-of-specification or out-of-trend results, SST failures), stability trends, SSTs, analysis of control samples, and control charts.

- Investigation of Failures: An analytical procedure control failing to meet its limits, a product being OOS, and an SST failure must be investigated with a cause analysis. Such failures are indicators of a possible common cause variation that was not recognized before or of a special cause. These causes need to be identified and appropriate controls put in place to either eliminate or control them.
- Stability Trends: When the procedure is used for stability programs, the stability studies can provide useful information on the performance of an analytical procedure. Using the data from both release and stability testing increases the opportunity to identify special causes and reduce or eliminate sources of variability.

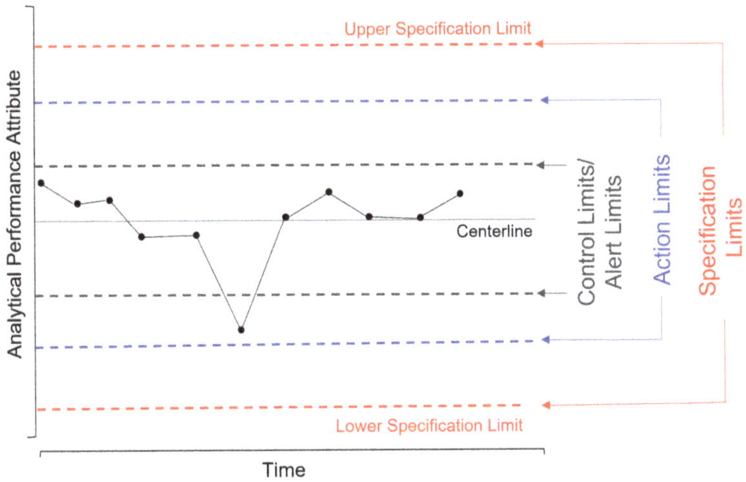

Fig. 8.9 Example of control chart including control, action, and specification limits

- SSTs: A system suitability test (SST) is designed to verify selected analytical procedure attributes. The acceptance criteria are based on the analytical procedure performance. Validity of the results and reportable values depends upon the SST meeting criteria. For multivariate methods, software tools may be needed to verify the SST.
- Control Samples: A control sample is included with the analyses and mimics the behavior of the actual laboratory sample as much as possible. There should be sufficient control sample to be used in many analyses over time.
- Control Charts: Although various trending methods exist, control charts are one of the most simple and effective graphical tools for such analysis. "A control chart is a time sequenced chart with upper and lower control limits on which values of some statistical measure for a series of samples or subgroups are plotted" (Fig. 8.9).

The chart may show a central line to help detect a trend of plotted values toward either control limit. A control chart records data so that one can see an unusual event, such as a value outside the assigned limits. Control charts help distinguish between acceptable common cause variation and special cause variation. There are many types of control charts (e.g., individual chart, X-car chart, range chart, *s* chart, etc.). More information on control charts can be found in the USP General Chapter <1010> [33] and other references [87, 88]. It is important to select the correct chart for the type of data, e.g., a Shewhart individual control chart tracks individual values and not averages.

8.8.5 *Procedure Performance Monitoring After Analytical Condition Changes*

Changes to analytical procedures may be needed over the lifecycle and can be prompted by several events including needed changes identified through a routine monitoring program, the adoption of new technology, or changes to the ATP [22]. Changes may be necessary due to changes occurring in stages of the drug product lifecycle (e.g., change in formulation, change in the API synthetic route, change in manufacturing process and increasing regulatory requirements). Necessary analytical condition changes should be risk-assessed for their impact. In addition, appropriate change management approaches and documentation should be used when making changes to a procedure. The ACS can identify when and what analytical procedure changes are required and can provide data to support the change. After the change is made, it can provide data confirming the change is having the impact as expected.

References

1. Arden NS, Fisher AC, Tyner K, et al (2021) Industry 4.0 for pharmaceutical manufacturing: Preparing for the smart factories of the future. Int J Pharm 602
2. Weitzel J, Pappa H, Banik GM et al (2021) Understanding quality paradigm shifts in the evolving pharmaceutical landscape : perspectives from the USP quality advisory group. AAPS J 23:112. https://doi.org/10.1208/s12248-021-00634-5
3. Castillo FC, Cooney B, Levine HL (2016) Biopharmaceutical manufacturing process validation and quality risk management. Pharm Eng
4. FDA (2004) Pharmaceutical CGPMs for the 21st Century – a risk based approach. https://www.fda.gov/media/77391/download
5. Nethercote P, Ermer J (2014) Chapter 1. Analytical validation within the pharmaceutical lifecycle. In: Method validation in pharmaceutical analysis: a guide to best practice. Wiley, pp 1–10
6. Juran J (1992) Juran on quality by design: the new steps for planning quality into goods and services. The Free Press, New York
7. ICH (2009) ICH Guideline Q8(R2) Pharmaceutical Development
8. ICH (2005) ICH Guideline Q9 Quality Risk Management
9. ICH (2019) ICH Guideline Q12 Technical and regulatory considerations for pharmaceutical product lifecycle management
10. ICH (2021) ICH Draft Guideline Q13 Continuous Manufacturing of Drug Substance and Drug Products
11. (2004) The PAT group. Innovation and Continuous Improvement in Pharmaceutical Manufacturing
12. Woodcock J (2019) To help reduce drug shortages, we need manufacturers to sell quality – not just medicine. In: FDA. https://www.fda.gov/news-events/fda-voices/help-reduce-drug-shortages-we-need-manufacturers-sell-quality-not-just-medicine. Accessed 13 July 2022
13. FDA (2022) Quality Management Maturity: Essential for Stable U.S. Supply Chains of Quality Pharmaceuticals. https://www.fda.gov/media/157432/download. Accessed 13 July 2022
14. FDA (2022) Quality Metrics Reporting Program Fed Regist 87:

15. Guiraldelli A, Weitzel J (2022) Evolution of analytical procedure validation concepts: part I – analytical procedure life cycle and compendial approaches. Pharmaceutical Technology
16. Borman P, Chatfield M, Nethercote P et al (2007) The application of quality by design to analytical methods. Pharm Technol 31:142–152
17. Schweitzer M, Pohl M, Hanna-Brown M et al (2010) Implications and opportunities of applying QbD principles to analytical measurements. Pharm Technol 34:52–59
18. Martin GP, Barnett KL, Burgess C et al (2013) Lifecycle Management of Analytical Procedures: method development, procedure performance qualification, and procedure performance verification. USP Stimuli Article. Pharmacopeial Forum 39(5)
19. Burgess C, Curry P, Leblond DJ et al (2016) Fitness for use: decision rules and target measurement uncertainty. USP Stimuli Article. Pharmacopeial Forum 42(2)
20. Barnett KL, Mcgregor PL, Martin GP et al (2016) Analytical target profile: structure and application throughout the analytical lifecycle. USP Stimuli Article. Pharmacopeial Forum 42(5)
21. Kovacs E, Ermer J, McGregor PL et al (2016) Analytical control strategy. USP Stimuli Article Pharmacopeial Forum 42(5)
22. USP (2022) USP General Chapter <1220> Analytical Procedure Lifecycle
23. Weitzel J, Meija J, LeBlond D, Walfish S (2018) Measurement uncertainty for the pharmaceutical industry. USP Stimuli Article. Pharmacopeial Forum 44(1)
24. Schofield T, van den Heuvel E, Weitzel J et al (2019) Distinguishing the analytical method from the analytical procedure to support the USP analytical procedure life cycle paradigm. USP Stimuli Article. Pharmacopeial Forum 45(6)
25. Feinberg M, Boulanger B, Dewé W, Hubert P (2004) New advances in method validation and measurement uncertainty aimed at improving the quality of chemical data. Anal Bioanal Chem 380:502–514. https://doi.org/10.1007/s00216-004-2791-y
26. BP (2022) BP Supplementary chapter on the use of Analytical Quality by Design concepts for analytical procedures
27. MHRA (2019) Technical review of MHRA analytical quality by design project
28. MHRA (2020) MHRA response and strategy for the application of analytical quality by design concepts to Pharmacopoeial standards for medicines
29. USP General Chapter <1225> Validation of Compendial Procedures. https://doi.org/10.31003/USPNF_M99945_04_01
30. USP General Chapter <1226> Verification of Compendial Procedures. https://doi.org/10.31003/USPNF_M870_03_01
31. USP General Chapter <1224> Transfer of Analytical Procedures
32. USP General Chapter <1210> Statistical Tools for procedure Validation. https://doi.org/10.31003/USPNF_M8646_06_01
33. USP General Chapter <1010> Analytical Data – Interpretation and Treatment. https://doi.org/10.31003/USPNF_M99740_05_01
34. USP General Chapter <1039> Chemometrics. USP-NF. https://doi.org/10.31003/USPNF_M2345_02_01
35. ICH (2018) Final Concept Paper: ICH Q14: Analytical Procedure Development and Revision of Q2(R1) Analytical Validation
36. ICH (2022) ICH Draft Guideline Q14 Analytical Procedure Development
37. ICH (2012) ICH guideline Q11 Development and Manufacture of Drug Substances (Chemical Entities and Biotechnological/Biological Entities)
38. ICH (1994) ICH guideline Q2(R1) Validation of Analytical Procedures
39. ICH (2008) ICH guideline Q10 Pharmaceutical Quality System
40. ICH (2022) ICH Draft Guideline Q2(R2) Validation of Analytical Procedures
41. Teasdale A, Borman PJ, Mullen AK (2022) Regulatory highlights. Org Process Res Dev 26: 1029–1037. https://doi.org/10.1021/acs.oprd.2c00054

42. Guiraldelli A, Weitzel J (2022) Evolution of analytical procedure validation concepts: part II – incorporation of science and risk-based principles in ICH Q14 and Q2(R2) guidelines. Pharm Technol
43. Borman P, Campa C, Delpierre G et al (2022) Selection of analytical technology and development of analytical procedures using the analytical target profile. Anal Chem 94:559–570. https://doi.org/10.1021/acs.analchem.1c03854
44. Almeida J, Bezerra M, Markl D et al (2020) Development and validation of an in-line API quantification method using AQbD principles based on UV-vis spectroscopy to monitor and optimise continuous hot melt extrusion process. Pharmaceutics 12. https://doi.org/10.3390/pharmaceutics12020150
45. Borman P, Chatfield M, Jackson P et al (2010) Reduced-method robustness testing of analytical methods driven by a risk-based approach. Pharm Technol 34:72–86
46. Hindocha D, Smith NW, Borman PJ (2001) Validation and robustness testing of a CEC method used to determine impurities in a pharmaceutically active compound. Chromatographia 53: S276–S283
47. Champarnaud E, Laures AMF, Borman PJ et al (2009) Trace level impurity method development with high-field asymmetric waveform ion mobility spectrometry: systematic study of factors affecting the performance. Rapid Commun Mass Spectrom 23:181–193. https://doi.org/10.1002/rcm.3844
48. Borman PJ, Chatfield MJ, Damjanov I, Jackson P (2011) Method ruggedness studies incorporating a risk based approach: a tutorial. Anal Chim Acta 703:101–113
49. Borman PJ, Chatfield MJ, Crowley EL et al (2008) Development, validation and transfer into a factory environment of a liquid chromatography tandem mass spectrometry assay for the highly neurotoxic impurity FMTP (4-(4-fluorophenyl)-1-methyl-1,2,3,6-tetrahydropyridine) in paroxetine active pharmaceutical ingredient (API). J Pharm Biomed Anal 48:1082–1089. https://doi.org/10.1016/j.jpba.2008.08.026
50. Borman P, Chatfield M, Asahara H et al (2019) Risk-based intermediate precision studies for analytical procedure validation. Pharm Technol 2:12–22
51. Borman P, Schofield T, Lansky D (2021) Reducing uncertainty of an analytical method through efficient use of replication. Pharm Technol 45:48–56
52. Ermer J, Agut C (2014) Precision of the reportable result. Simultaneous optimisation of number of preparations and injections for sample and reference standard in quantitative liquid chromatography. J Chromatogr A 1353:71–77. https://doi.org/10.1016/j.chroma.2014.03.043
53. Ferreira AP, Tobyn M (2015) Multivariate analysis in the pharmaceutical industry: Enabling process understanding and improvement in the PAT and QbD era. Pharm Dev Technol 20:513–527
54. Brereton RG (2000) Introduction to multivariate calibration in analytical chemistry. Analyst 125:2125–2154
55. Popkin ME, Borman PJ, Omer BA et al (2019) Enhanced approaches to the identification, evaluation, and control of impurities. J Pharm Innov 14:176–184. https://doi.org/10.1007/s12247-018-9363-8
56. Separovic L, Simabukuro RS, Couto AR et al (2021) Measurement uncertainty and conformity assessment applied to drug and medicine analyses – a review. Crit Rev Anal Chem. https://doi.org/10.1080/10408347.2021.1940086
57. Separovic L, Lourenço FR (2020) Frequentist approach for estimation of false decision risks in conformity assessment based on measurement uncertainty of liquid chromatography analytical procedures. J Pharm Biomed Anal
58. Williams A, Magnusson B (eds) (2021) Eurachem/CITAC Guide: Use of uncertainty information in compliance assessment, 1st edn
59. WHO (2010) WHO guidelines on quality risk management. Annex 2
60. Gupta NV, Reddy V, Gupta V et al (2014) Quality risk Management in pharmaceutical industry: a review. Int J PharmTech Res 6:908–914

61. Frank T, Brooks S, Creekmore R et al (2011) Quality risk management principles and industry case studies. Pharm Technol 35
62. WHO (2010) WHO guideline on Quality Risk Management
63. Stamatis DH (2003) Failure mode and effect analysis: FMEA from theory to execution, 2nd edn. Willian A. Tony
64. Chaudhary M, Priya (2021) Hazard analysis and critical control points as a quality risk management tool in the pharmaceutical industry: a systematic review. J Drug Deliv Ther 11: 167–175. https://doi.org/10.22270/jddt.v11i5-s.5094
65. Frank T, Brooks S, Creekmore R, et al Quality risk management principles and industry case studies
66. Montgomery D (2013) Designs and analysis of experiments, 8th edn. Wiley, New York
67. Rajamanickam V, Babel H, Montano-Herrera L et al (2021) About model validation in bioprocessing. PRO 9
68. Horvath C, Melander W, Molnär I (1976) Solvophobic interactions in liquid chromatography with nonpolar stationary phases
69. Snyder LR, Dolan JW, Carr PW (2004) The hydrophobic-subtraction model of reversed-phase column selectivity. J Chromatogr A 1060:77–116
70. Molnár I (2005) Searching for robust HPLC methods – Csaba Horváth and the Solvophobic theory. Chromatographia 62. https://doi.org/10.1365/s10337-005-0645-1
71. Vitha M, Carr PW (2006) The chemical interpretation and practice of linear solvation energy relationships in chromatography. J Chromatogr A 1126:143–194
72. Kaliszan R, van Straten MA, Markuszewski M, et al (1999) Molecular mechanism of retention in reversed-phase high-performance liquid chromatography and classification of modern stationary phases by using quantitative structure-retention relationships
73. de Carvalho Rocha WF, do Prado CB, Blonder N (2020) Comparison of chemometric problems in food analysis using non-linear methods. Molecules 25
74. Borman P, Elder D (2018) Q2(R1) validation of analytical procedures. In: Teasdale A, Elder D, Nims RW (eds) ICH quality guidelines: an implementation guide, 1st edn. Wiley, pp 127–166
75. Jackson P, Borman P, Campa C et al (2019) Using the analytical target profile to drive the analytical method lifecycle. Anal Chem. https://doi.org/10.1021/acs.analchem.8b04596
76. FDA (2011) Guidance for industry. Process validation: general principles and practices
77. Nethercote P, Borman P, Bennett T, et al (2010) QbD for better method validation and transfer. Pharma Manufacturing
78. FDA (2015) Guidance for industry. Analytical procedures and methods validation for drugs and biologics
79. AOAC (2005) Appendix D: guidelines for collaborative study procedures to validate characteristics of a method of analysis
80. International Standard (1994) ISO 5725-1. Accuracy (trueness and precision) of measurement methods and results. Part I – General principles and definitions
81. International Standard (1994) ISO 5725-2. Accuracy (trueness and precision) of measurement methods and results. Part 2 – Basic method for the determination of repeatability and reproducibility of a standard measurement method
82. International Standard (1994) ISO 5725-1. Accuracy (trueness and precision) of measurement methods and results. Part 3 – Intermediate measures of the precision of a standard measurement method
83. International Standard (1994) ISO 5725-4 Accuracy (trueness and precision) of measurement methods and results – Part 4: Basic methods for the determination of the trueness of a standard measurement method
84. International Standard (1994) ISO 5725-5. Accuracy (trueness and precision) of measurement methods and results – Part 5: Alternative methods for the determination of the precision of a standard measurement method

85. International Standard (1994) ISO 5725-6. Accuracy (trueness and precision) of measurement methods and results – Part 6: Use in practice of accuracy values
86. FDA (2022) Guidance for Industry Investigating Out-of-Specification (OOS) Test Results for Pharmaceutical Production
87. (2016) Standard Practice for Use of Control Charts in Statistical Process Control (ASTM E2587–16). West Conshohocken, PA
88. Wheeler D, Chambers D (2012) Understanding statistical process control, 3rd edn. SPC Press, Knoxville

Chapter 9
Analytical Target Profile (ATP) and Method Operable Design Region (MODR)

Amanda Guiraldelli Mahr, Felipe Rebello Lourenço, Phil Borman, Jane Weitzel, and Jean-Marc Roussel

Abstract A fundamental piece of the analytical procedure lifecycle (APLC) approach involves the acquisition of understanding about total procedure variability (bias and precision) and its association with risk of false conformity/nonconformity decisions about the quality of pharmaceutical products. Concepts such as measurement uncertainty (MU), total analytical error (TAE), and the maximum allowable error or MU (i.e., target measurement uncertainty [TMU]) have been introduced in the APLC framework to draw attention to fitness for use (USP, USP general chapter <1220> analytical procedure lifecycle, 2022). These parameters are commonly used to define the ATP performance requirements for quantitative and semi-quantitative procedures. The ATP element presents a key role in assuring procedure fitness for use, through the establishment of performance requirements, which serve as a basis for the establishment of suitable analytical control strategy (ACS), to

A. G. Mahr (✉)
United States Pharmacopeia (USP), Rockville, MD, USA

F. R. Lourenço
United States Pharmacopeia (USP), Rockville, MD, USA

School of Pharmaceutical Sciences, University of São Paulo, São Paulo, Brazil
e-mail: feliperl@usp.br

P. Borman
United States Pharmacopeia (USP), Rockville, MD, USA

USP Measurement and Data Quality Expert Committee, United States Pharmacopeia, Rockville, USA
e-mail: phil.j.borman@gsk.com

J. Weitzel · J.-M. Roussel
USP Measurement and Data Quality Expert Committee, United States Pharmacopeia, Rockville, USA

M. C. Breitkreitz, H. C. Goicoechea (eds.), *Introduction to Quality by Design in Pharmaceutical Manufacturing and Analytical Development*, AAPS Introductions in the Pharmaceutical Sciences 10, https://doi.org/10.1007/978-3-031-31505-3_9

derive risk criteria for the MODR design (where robustness assessment play an important role), procedure qualification, and performance monitoring program.

9.1 Analytical Target Profile

The ATP is a fundamental component of the lifecycle approach, and its definition is the first step in the AQbD framework [1, 2]. The ATP is the predefined intended purpose of the procedure and establishes the criteria for the analytical procedure performance characteristics that are required to an intended use, which includes the quality attributes to be measured (e.g., accuracy and precision for a quantitative procedure). In other words, the ATP defines the required quality of the reportable value, since the measured value provides the basis for conformity or nonconformity assessment decisions of a tested item or batch against the regulatory, compendial, or manufacturing specification limits [1–5]. Thus, the acceptable level of risk of false conformity/nonconformity decisions should be considered when defining the ATP criteria which are a fundamental piece of ensuring procedure fitness for use [5–7].

Regardless of the procedure, repeated measurements on the same sample will generally produce different results if the system used is sufficiently sensitive. Thus, all measurements have many potential sources of variation, and a close examination of the performance characteristics reveals that they focus on two primary aspects of the measurement: systematic error (bias) and random error (precision) [2, 8]. Figure 9.1 illustrates the bias (difference between the mean replicate value and the assigned reference value) and the precision (dispersion of the obtained values, expressed as a standard deviation).

Measurement accuracy depends on the precision and bias of the procedure, and there are several factors which significantly influence the uncertainty of a measurement. Both precision and bias are unknown, like the true value of the sample, but estimates can be obtained (e.g., from measurements made during validation). The estimation of bias and precision is not the only objective of procedure validation and alone may not ensure fitness for use. The objective of validation is to evaluate the

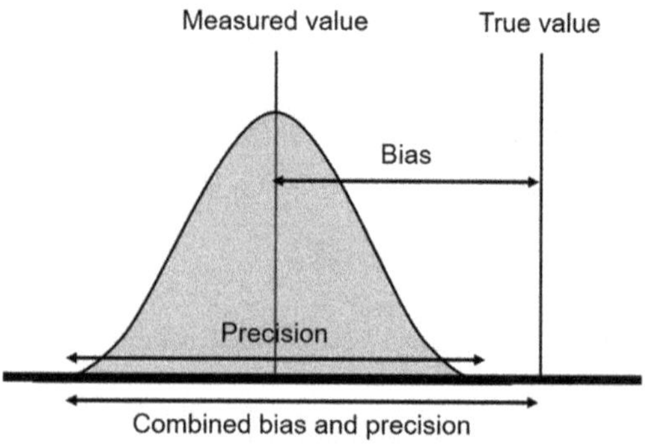

Fig. 9.1 Types of errors in a measurement: random error (precision) and systematic error (bias)

ability of the procedure to satisfy its purpose by ensuring that most of measurements that the procedure will provide routinely in the future are accurate enough and will fall inside the acceptability limits. In other words, the expectation is to evaluate the risk or confidence that measurements are close enough (are accurate) to the unknown true value for the sample submitted to an analysis. The evaluation of combined impact of bias and precision is often useful to assess whether an analytical procedure is fit for use [9, 10]. The degree to which bias impacts the usefulness of an analytical procedure depends in part on precision (e.g., a procedure with a relatively high precision can accommodate a relatively greater value of bias than a procedure with a lower precision) [9]. Hubert, P. et al. [10, 11] published a clear example using total analytical error estimation for establishing acceptability regions (characterized by the true bias and precision), with a predefined probability which is illustrated in Fig. 9.2. On the left graphics of Fig. 9.2, the acceptability regions are represented (as a function of their theoretical true bias and true precision) by the triangles for analytical procedures that give at least 66, 80, and 95% of their measurements falling within acceptability limits previously established of [15%, +15%] [11]. Four procedures are inserted into Fig. 9.2 for illustration purposes as a function of their

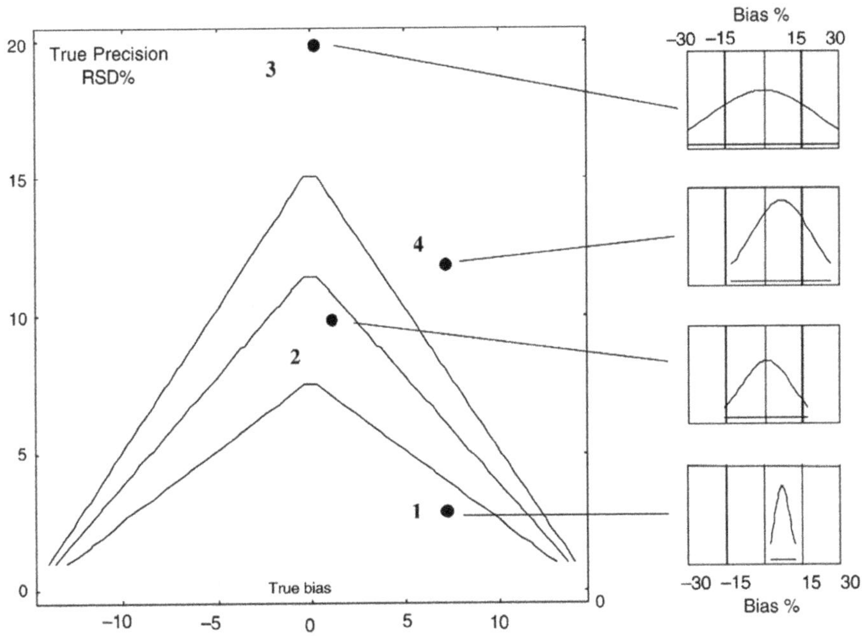

Fig. 9.2 The interiors of the triangles of the left graphic represent the acceptability regions for analytical procedures that give at least 66, 80, and 95% of their measurements within acceptability limits, respectively, as a function of their theoretical true bias (%) and true precision (RSD %). The performance of an analytical procedure is represented by a point in this graphic. The expected distributions of future measurements are represented in the small graphics on the right. The acceptability limit is [−15%, +15%]. (Reproduced from Feinberg et al. 2004)

performances (illustrated as a black point). Procedures 1 and 2 are located inside the acceptability regions that guarantee that at least 95 and 80%, respectively, of their results will be within the acceptability limits. On the other hand, for the same risk of the measurements outside of the acceptance limits, procedures 3 and 4 are not considered as valid [11]. The right graphics illustrate the expected distributions of future measurements for the procedures (corresponding to the four procedures), resulting from the estimation of "β-expectation tolerance interval," which defines where the expected proportion of future results will fall in [11].

The individual assessment of bias and precision may fail to flag that a procedure is not fit for use if non-suitable criteria are established. The quality of an analytical test result may be measured by different approaches, such as by the estimation of total analytical error (TAE) and measurement uncertainty (MU). The final objective is to evaluate if the analytical method will provide reliable or accurate results and then ensure procedure fitness for use [11]. TAE and MU represent the overall error distribution of the test results and are obtained from the combination of random and systematic errors (i.e., precision and bias, respectively) [3, 12, 13]. TAE defines a region around the reference value where the measured analytical results can be found with a defined probability [8]. MU defines a region around the measured analytical results where the conventional true or reference value can be found with a defined probability [11]. TAE and MU can be used to estimate measurement accuracy consequently being suitable approaches to evaluate procedure fitness for use. MU is also usually used to define if a result is compliant with specifications and it can be used during routine applications [3]. Typically, MU assessment involves four steps: (A) measurand definition (e.g., the content of API in the pharmaceutical dosage form, defined as function of input quantities); (B) identification of the sources of uncertainty (identification of the most relevant sources of uncertainties that affect the output quantity); (C) estimation of uncertainty of input sources (estimation of systematic and random errors associated with the most relevant sources of uncertainty and convert to standard uncertainty values); and (D) combine all uncertainties using an appropriate approach, and expand the combined MU by multiplying by a coverage factor k that expands it to a coverage interval with a chosen level of confidence [3, 12, 14].

The most relevant sources of uncertainty in pharmaceutical analysis include the uncertainty from sampling, instrumental effects (e.g., weighting and volume apparatus errors, instrumental repeatability), purity of reagents and reference standards (e.g., humidity, purity, or potency of a certified reference standard chemical substance), matrix interferences (e.g., interferences of impurities and/or excipients), environmental effects (temperatures, humidity, microbiological quality of the environment), and analytical/chromatographic conditions (e.g., mobile phase composition, type of stationary phase, flow rate), among other factors [3, 12, 14, 15]. The standard uncertainty quantified for input sources may be combined using (1) bottom-up approach, in which systematic and random errors from each step of the analytical procedures are combined to obtain the overall uncertainty (e.g., law of uncertainty propagation and Monte Carlo simulation), or (2) top-down approach, in which bias and precision of the analytical procedure are estimated and combined to obtain the overall uncertainty [3, 12, 14].

The bottom-up approach is more laborious and requires more information for the calculation of overall MU; however, it allows one to identify the most significant sources of uncertainty. On the other hand, the top-down approach is less laborious and requires less information for the calculation of overall uncertainty; however, it does not allow one to quantify the individual contribution of each source of uncertainty. When relevant, the uncertainty arising from sampling should also be considered to estimate the overall uncertainty. Usually, the analytical procedure validation data (e.g., accuracy (bias) and precision) can be used to estimate the overall uncertainty using the top-down approach [3, 12, 14]. An example is the estimation of TAE using ranges or intervals of values rather than a single value providing the maximum total error that can be obtained for any observed analytical results. Prediction and tolerance intervals were proposed by several authors [9, 11, 16–20] for TAE estimation. It is worth mentioning that TAE is currently not a concept recognized in the International Vocabulary of Metrology (VIM) [21] and in the "Guide to the expression of uncertainty in measurement (GUM)" [22, 23], as opposed to MU [60, 61] [8].

Since the ATP defines requirements for a reportable result to be fit for use, it generally includes the maximum admissible total analytical error (or target measurement uncertainty – TMU [7, 24, 25]) or, alternatively, the maximum admissible values of precision and accuracy that ensure a reduced risk of false conformity/nonconformity decisions. Usually, the TMU should be set for measurement results close to the regulatory specification limit and a scientific rationale provided. Additionally, the definition of suitable decision rules (DR) helps ensure fitness for use since it provides understanding of maximum variability associated with reportable results and how to accept or reject a product according to its specification and result of measurement. When the analytical result value is close to the regulatory specification limit and/or the measurement uncertainty is large, there may be a significant risk of false conformity/nonconformity decision. Usually, four situations should be considered for each of the four possible outcomes (Fig. 9.3), where the DR defines the compliance judgment: (A) when measured value and its measurement uncertainty interval are outside the specification limits (Fig. 9.3a); (B) when measured value is outside the specification limits, but the measurement uncertainty interval is within the specification limits (Fig. 9.3b); (C) when measured value is within the specification limits, but the uncertainty interval exceeds the specification limits (Fig. 9.3c); and (D) when measured value and its measurement interval are within limits (Fig. 9.3d). The conformity and nonconformity decisions are evident in the situations presented in Fig. 9.3a, d, respectively. However, in the situations presented in Fig. 9.3b, c, there may be a significant risk of false rejection (producer's risk) or false acceptance (consumer's risk), respectively [2, 3, 5].

The concept of decision rules is key to the assessment of compliance. These rules give a prescription for the acceptance or rejection of a product based on the measured quantity value, its uncertainty, and the specification limit or limits, taking into account the acceptable level of the probability of making a wrong decision [25]. The estimation of consumer's and/or producer's risk on a routine basis can be laborious; thus, decision rule using guard band based on TAE or MU may be an

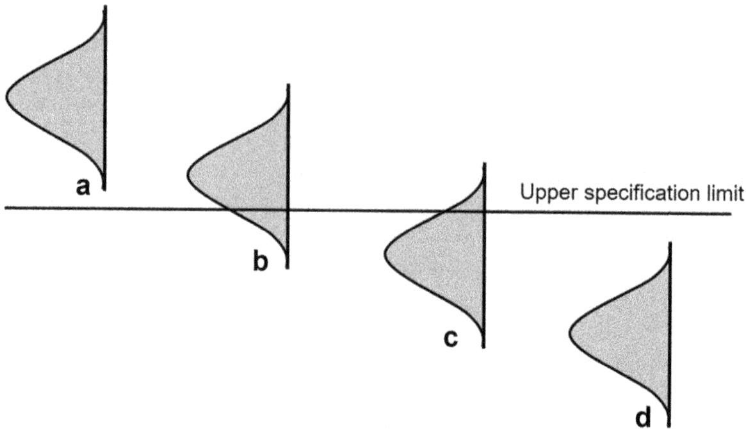

Fig. 9.3 The four situations that should be considered: when measured value and its measurement uncertainty interval are outside the specification limits (**a**); when measured value is outside the specification limits, but the measurement uncertainty interval is within the specification limits (**b**); when measured value is within the specification limits, but the uncertainty interval exceeds the specification limits (**c**); and when measured value and its measurement interval are within the specification limits (**d**)

advantageous alternative to ensure a reduced risk of false decisions. In order to ensure a reduced consumer's risk, an acceptance zone can be defined by adding and/or subtracting a guard band from lower and/or upper specification limits (Fig. 9.4a). On the other hand, a rejection zone can be defined by subtracting and/or adding a guard band to the lower and/or upper specification limits, in order to ensure a reduced producer's risk. When the measured value falls into the acceptance zone, the item or batch is declared conforming, while when the measured value falls into the rejection zone, the item or batch is declared nonconforming. Usually, the guard band is calculated by multiplying the standard uncertainty by an appropriate coverage factor (typically 1.64) that ensures a risk of false decisions of less than 5% [3, 5]. A multistage testing approach can also be used with an indecision zone for the initial stage of testing (Fig. 9.4b). When the reportable value is within the indecision zone, a predetermined supplementary sampling and testing plan and decision rule are used to determine conformance.

Since the focus of the ATP is on the intended use of the reportable value, any method that meets the ATP requirements can be used. The ATP criteria drive the choice of analytical technology (e.g., chromatographic methods and bioassays, among other techniques). Several available analytical techniques may meet the performance requirements defined in ATP criteria. Few examples of ATPs are available in the references [2, 24, 26]. The ATP defines the performance characteristics for the procedure because any combination of precision and bias that meets the ATP requirements is acceptable. This also prevents unnecessary work in developing a procedure. Once the procedure meets the ATP requirements, development has achieved its goal. This is efficient, saving time and precious resources.

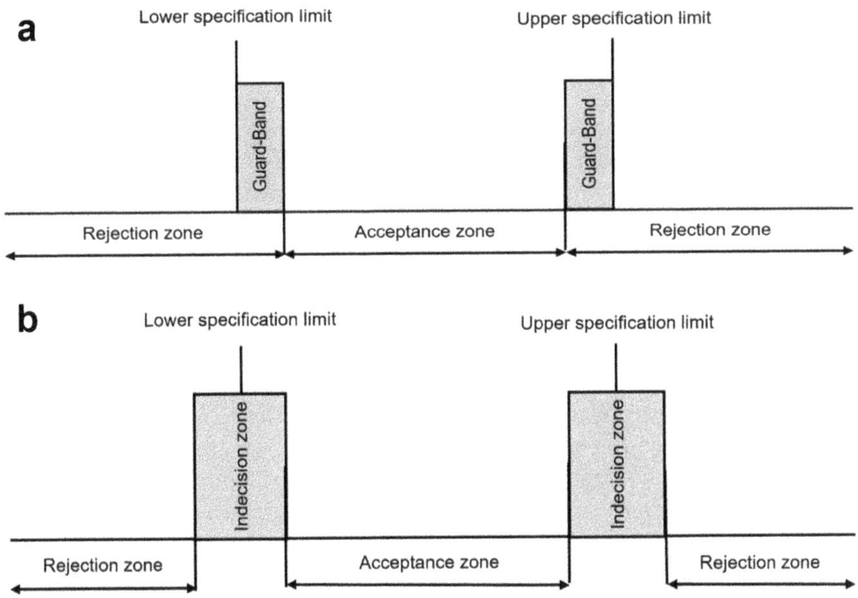

Fig. 9.4 Decision rules using (**a**) guard bands based on MU and (**b**) using indecision zones based on MU

ATP Establishment

An example is taken from reference [27] which applies the ATP to a procedure which quantifies the number of colony-forming units in a probiotic supplement. The paper provides detailed information on how the ATP and decision rule were created and used to demonstrate the procedure is fit for its intended use.

The ATP is as follows: The procedure must be able to enumerate the *Lactobacillus* spp. culturable cell count in CFU/g of powder with cryoprotectant, formulated to 11.462 Log_{10} CFU/g, so the reportable values fall below a $TMU = 0.305$ Log_{10} CFU/g (i.e., the *TMU* associated with the reportable value is <0.305) and the probability of being wrong is ≤5%. The plating range used by the laboratory will cover 9.462–13.462 Log_{10} CFU/g, two Log_{10} above and below the internal release specification.

The decision rule is as follows: The laboratory sample, taken from the batch of *Lactobacillus* spp. probiotic powder (culturable cells, freeze-dried), will be considered compliant with the specification of 10.962 Log10 CFU/g if the reportable value is ≥10.962 Log10 CFU/g, the MU is <0.305 Log10 CFU/g, and the probability of being wrong is ≤5%. Otherwise, it will be considered non-compliant.

MU estimated from the procedure qualification is 0.10 Log_{10} CFU/g. This is less than the *TMU* of 0.305 Log_{10} CFU/g, demonstrating the procedure is fit for its intended use.

9.2 Robustness Assessment and Method Operable Design Region (MODR)

Within the AQbD framework, robustness assessment should be initiated at stage 1 (procedure design) with the aim of providing acquisition of a high degree of understating about the sources of variability that can affect the procedure bias and precision. This knowledge allows for the establishment of proper risk control strategies to reduce variability associated with the procedure. Robustness of an analytical procedure is a measure of its capacity to meet the expected performance requirements during normal use and is usually tested by deliberate variations of analytical procedure parameters [2, 26]. The terms procedure "ruggedness" and "robustness" are often used interchangeably in the literature [28]; however, these terms refer to separate measurable characteristics [29]. The term "procedure robustness" refers to the influences of deliberate variations in procedure-related parameters [28], while "procedure ruggedness" refers to the influence of random variation (precision) which can be assessed by the analysis of the same sample under a variety of normal test conditions. In the USP <1225> [25], procedure ruggedness is also described as intermediate precision referring to the degree of variation of test results obtained by the analysis of the same samples under a variety of normal test conditions, such as different analysts, instruments, elapsed assay times, and days. No specific mention is made in ICH Q2(R1) or Q2(R2) to procedure ruggedness, which is alluded to in the definitions of precision [30, 31].

Despite the current USP <1225> [25] stating that intermediate precision is also known as ruggedness, the ruggedness concept in this chapter extends to include factor conditions explored during repeatability (same analyst/equipment over a short period of time), intermediate precision (within-laboratory variability), and reproducibility (within–/between-laboratory variability) studies, as per previous publication [28]. Ruggedness factors are in general assessed at stage 2 (procedure qualification) during repeatability, intermediate precision, and reproducibility studies which are usually part of the overall procedure precision evaluation [32]. Robustness assessment is then treated in this chapter as a broader concept that encompasses the measurement bias and precision assessment at different stages of the procedure lifecycle. This section will focus on discussing the initial robustness assessment at stage 1 to allow for the delimitation of MODR and initial analytical control strategy (ACS).

9.2.1 Concept of Knowledge Space and Method Operable Design Region (MODR)

After acquisition of knowledge (e.g., through experimentation and/or predictive modeling), the knowledge space can be accessed. The knowledge space is a region where knowledge is available providing understanding on how the analytical

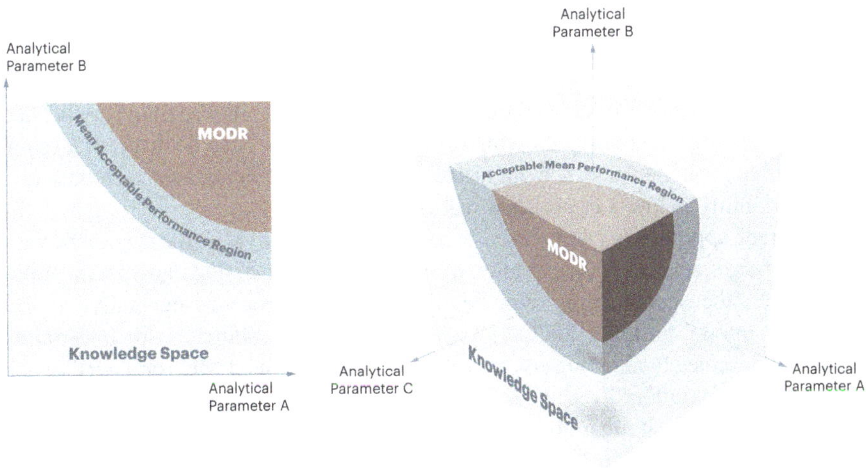

Fig. 9.5 Representation of the knowledge space and MODR accessed during procedure design

responses (and performance) behave when the analytical parameter values are changed. Depending on the approach used to acquire knowledge, for instance, using empirical models (e.g., DOE-derived models), knowledge may be only accessed within the experimental domain. The use of mechanistic/hybrid models may provide the flexibility to access knowledge within a wider region. Within the knowledge space, we may identify a region where the analytical parameters and their ranges will not meet the specifications and can be called as a "non-acceptable performance region," whereas the region with mean responses within specification (risk criteria) can be called as "acceptable mean performance region" (Fig. 9.5).

As per USP General Chapter <1220> and ICH Q14, the MODR is a multivariate space of analytical procedure parameters that ensure the ATP is fulfilled and therefore provide assurance of the quality of the measured value [2, 26]. Thus, the MODR is considered as a multidimensional combination and interaction of procedure parameters where all study factor combinations have been demonstrated to provide acceptable mean performance and robustness (Fig. 9.5). The MODR concept is equivalent to "design space" concept described in the ICH Q8 [33].

Based on the concept described in ICH Q14, the MODR should provide assurance of quality, implying that the MODR is a measure of its robustness [78]. Thus, the design of MODR should consider the evaluation of measurement uncertainty to assess whether the analytical procedure will provide assurance of quality. If a procedure is robust "on the average" but its variation is too large, then it may fail its robustness criteria [34]. At stage 1, robustness assessment is usually performed during optimization studies after analytical factor screening. The goal of robustness assessment is the estimation of analytical measurement variability related to CPQAs and determination of the relative amount of each source of error in order to improve the analytical performance of a procedure by reducing the variability linked to analytical parameters [11]. Simply put, if we can quantify the uncertainty components of a given response (associated with the input factor x's), the combined measurement uncertainty of the analytical response y can be estimated; consequently, a plan can be established to reduce bias and overall variability associated with the analytical response, allowing for the design of the MODR that provides not just suitable mean performance but also a desired level of robustness with measurement variability below a established risk criteria.

Different approaches can be used to estimate total analytical measurement variability (measurement accuracy) and the probability of meeting the specifications (how often the specifications will be met). These approaches generally involve prediction model building for CPQAs, followed by the estimation of uncertainty of model parameters allowing for the estimation of combined MU (or TAE) associated with the monitored responses. Few examples are propagation of distribution using Monte Carlo simulations [22, 35], prediction or tolerance intervals [9, 11, 16–20], Bayesian modeling [34], bootstrapping methods [36], etc. Estimation of prediction and tolerance intervals is a useful strategy to estimate TAE and assess robustness during stage 1. An adequate model for TAE can be defined by a statistical model modeling the experiments and then deriving statistical prediction and/or tolerance intervals for these models. It is worth mentioning that a prediction interval is different from a confidence interval. A confidence interval (CI) quantifies the uncertainty on an estimated population variable (e.g., the mean or standard deviation), giving an indication of the degree of uncertainty of an estimate. The CI helps to decide how precise a sample estimate is by specifying a range of values (defined by lower and upper limits) likely to contain the unknown population value, whereas prediction and tolerance intervals quantify the uncertainty for future observations estimated from the population using the analytical procedure and indicate where you can expect to see the next data point sampled.

The prediction interval (PI) bounds the average of a future sample of a given size and must account for both the uncertainty in estimating the population mean and the random variation of the individual values. Thus, a prediction interval is always wider than a confidence interval. The tolerance interval (TI) contains at least a given proportion of all individual outcomes and may be wider than PI because it attempts to account for the variability observed in the experiment and the unobserved variation from all individual outcomes drawn from the same population. In general, for robustness study (and procedure qualification), the question to be answered to allow for proper risk assessment is how to specify the range in which the response values of future analysis using the procedure are likely to be found. Wider prediction and tolerance intervals can answer this question [20], whereas simple confidence intervals are not the ideal approach to quantify the uncertainty for future observations and consequently may not be suitable to help design the MODR [37]. Different modeling approaches can be used, being DoE a very frequent tool used to generate empirical models. From the DoE results, the effect of each studied procedure parameter, or parameter interaction, can be estimated (if an appropriate design is selected), and its impact on the mean effect (b_o) can be compared to a given acceptance criterion. Considering that the robustness of an analytical method addresses the effect of the possible variations in its operating parameters during day-to-day work in the laboratory, it becomes obvious that assessing individually the effect of the parameters and interactions does not fully answer to the objective of a robustness study and prediction intervals calculated from the DoE experimental results and the other abovementioned approaches can help to estimate this joint effect of the parameters. Mechanistic or hybrid models can also be used.

9.2.2 Risk Assessment and Control

With the MU or TAE at hand, risk assessment is performed again allowing for the identification of sources of variability, design of procedure parameters (e.g., MODR), and definition of ACS. This includes first risk identification, where analytical parameters and their ranges that have a significant impact on performance are flagged. Risk analysis and evaluation are then conducted where different approaches can be used to compare the risk to a given risk criteria for evaluating how often the specifications will be met. Examples are establishment of a maximum allowable MU (e.g., target measurement uncertainty – TMU) and use of appropriate decision rules [24], process capability assessment, and other performance evaluation tools (e.g., prediction or tolerance intervals). At this stage, risk evaluation is performed where the identified risks are compared against the risk criteria allowing for the access of knowledge space.

9.2.2.1 Risk Assessment Based on Target Measurement Uncertainty (TMU)

The TMU is defined as the maximum acceptable uncertainty for the reportable value in order to meet performance criteria and accomplish the fitness-for-purpose requisite [24]. This can also be used to establish the risk criteria during procedure design and MODR delimitation and qualification (usually considering conditions with MU $\leq$ TMU); however, it is more commonly used to define criteria for the ATP. More guidance on how to calculate the TMU can be found in the Eurachem Guide "Setting and Using Target Uncertainty in Chemical Measurement" [7] and USP Stimuli Article on ATP [24].

9.2.2.2 Risk Assessment Based on Process Capability Assessment

Process capability assessment is commonly used to evaluate the ability of a process to meet specifications. Process capability refers to the process (or procedure) performance when it is operating under statistical control, where the process is said to be capable if almost all the measurements fall inside the specification limits [38]. This can be represented by Fig. 9.6a where it is assumed that the population of data values

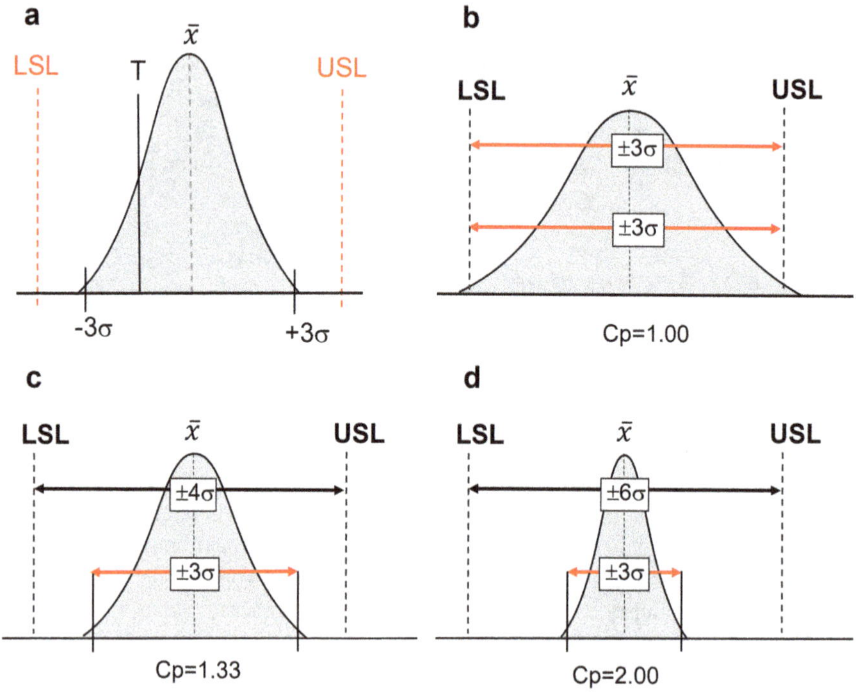

Fig. 9.6 Process capability assessment: (**a**) process variability compared to specification, (**b**) risk level for $Cp = 1.00$, (**c**) risk level for $Cp = 1.33$, and (**d**) risk level for $Cp = 2.00$

is normally distributed, with $\bar{x}$ and σ as the population mean and standard deviation, respectively. Here, it is also assumed that the process has a two-sided specification, where USL and LSL are the upper and lower specification limits, respectively, and T is the target value. In this example, the process spread is within the specification limits, and it can be concluded that the process is capable of meeting the process specification.

In order to quantify the process capability, the first step is the estimation of variability associated with the analytical response measurement followed by the calculation of capability indices.

Index C_p (Eq. 9.1) is used when the response has a defined maximum allowable amount of variation and has a two-sided specification, with an USL and LSL (relative to a mean result).

$$\widehat{C}_p = \frac{\text{USL} - \text{LSL}}{6s} \tag{9.1}$$

Index C_{pk} represents the measures of the process performance capability and indicates the shifting of the process from the USL and LSL, from the target value or mean value. If the response goal is to minimize, an absolute USL is established

($C_{p,\text{upper}}$). If the response goal is to maximize, an absolute LSL is established ($C_{p,\text{lower}}$). $C_{p,\text{upper}}$ and $C_{p,\text{lower}}$ can be expressed by Eqs. 9.2 and 9.3.

$$\widehat{C}_{p,\text{upper}} = \frac{\overline{x} - \text{USL}}{3s} \tag{9.2}$$

$$\widehat{C}_{p,\text{lower}} = \frac{\overline{x} - \text{LSL}}{3s} \tag{9.3}$$

Index C_{pkm} (Eq. 9.4) is used when the response goal is target and even small deviations of the mean response from target are critical. The specification limits are not necessarily symmetrical.

$$\widehat{C}_{pkm} = \frac{\widehat{C}_{pk}}{\sqrt{1 + \left(\frac{\overline{x} - T}{s}\right)}} \tag{9.4}$$

Index C_{pm} (Eq. 9.5) is used when the response goal is target and the LSL and USL are symmetrical (equal distance from the target).

$$\widehat{C}_{pm} = \frac{\widehat{C}_{pk}}{6\sqrt{s^2 + (\overline{x} - T)}} \tag{9.5}$$

This is followed by the comparison of procedure variability with the attribute specification. Capability index equal 1.00 indicates that the average response is 3 s apart from specification (99.7% of observations meeting specifications) and the process is assumed to be capable and meet specification limits; however, there is a low risk (0.3%) of few observations to fall outside the limit (Fig. 9.6b). Capability index equal 1.33 indicates the distance between mean value and specification limit is equal 4 s. This indicates a very low risk of observations falling outside the specification and shows that the process is capable and meets specification limits (Fig. 9.6c). Capability index equal 2.00 indicates the distance between mean value and specification limit is equal 6 *s* (Fig. 9.6d). This indicates that the process is capable and meets specification limits. In general, any value less than 1.00 can mean that the procedure variation is too wide compared to the specification limits or the performance average is away from the target. Conditions with capability index at or above 1.33 indicate that the process is capable and meets specification limits. The larger the capability index is, the less likely it is that any observation will be outside the specification limits.

A strategy for risk control during optimization studies is the selection of analytical conditions which can reduce the variability and are able to meet specifications (e.g., selecting conditions with capability indices ≥1.00 and if possible ≥1.33).

MU Estimation Using Monte Carlo Simulation and Process Capability Assessment Based on DoE-Derived Models for MODR Design
MU can be estimated at stage 1 from robustness study. Deliberated changes of the different analytical parameters can be explored to allow the estimation of variability derived from variation of parameters around their set point in normal use due to random error. In this example, optimal design was used to derive prediction models for analytical responses for the optimization of a UHPLC-UV procedure (reversed-phase ion-pair LC) for the quantification of impurities in a drug substance using a cubic model to fit the data. Several responses were monitored to design the procedure conditions, such as chromatographic resolution (Rs) between all impurities and adjacent peaks, tailing factor of the basic compound (impurity 6), and retention factor (k) of the least retained impurity (impurity 1). The MU was estimated by Monte Carlo simulation using DoE-derived models for each response, where a number of simulation trials were run under the defined coverage probability of 95% for the estimation of density function of input factors (initially assumed to follow normal distribution) followed by the propagation of input factor distributions for the estimation of the combined uncertainty of the measurand (*y's:* CPQAs) [22]. Fig. 9.7 illustrates the approach.

The desired levels of robustness for each study factor computed for the MU estimation were as follows: (x_1) flow rate ± 0.05 mL/min, (x_2) final % or organic solvent in the mobile phase ±2.0%, (x_3) ion-pairing reagent ±1 mM, and (x_4) column temperature ± 5.0 °C. Process capability assessment was performed to evaluate the ability of the procedure to meet acceptance (risk) criteria. For the response Rs for all

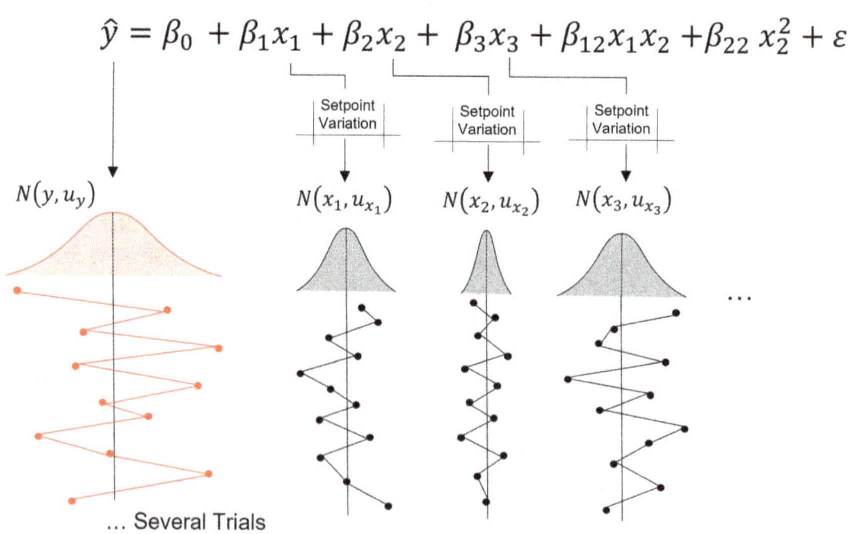

Fig. 9.7 Simplified illustration of MU estimation for predicted response ($\hat{y}$) using Monte Carlo simulation. Legend; β_n, regression coefficients; n, number of coefficients; x,input factors; u, uncertainty (assuming the data is normally distributed)

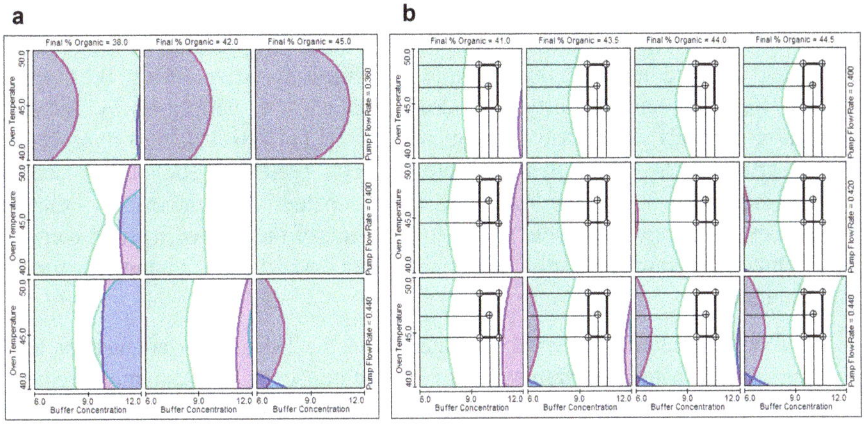

Fig. 9.8 Overlay graphic of acceptable regions for CPQAs defining the knowledge space and MODR (white area) considering (**a**) conditions with wider ranges for factors (flow rate, 0.44–0.36 mL/min, and % organic solvent, 38–45%) and all responses with Cpk ≥ 1.33 (±4 *s* distant from LSL) and (**b**) narrower ranges (flow rate, 0.40–0.44 mL/min, and % organic solvent, 41–44.5%) and conditions with all responses with $C_{pk} \geq 1.33$ with exception of Rs between impurity 5/6 for which $C_{pk} \geq 1.00$ (±3 s distant from LSL – 0.3% probability of failure). Legend: (green) Rs between impurity 5/6 with Rs < 1.5; (purple) Rs between impurity 7/8 with Rs < 1.5, and (blue) tailing factor for impurity 6 > 1.5. Since for Cpk = 1.00, the average of results is 3 s apart from LSL, there is 99.7% of probability that the observations fall within the specification (if assumed the data follow normal distribution). Thus, the risk of 0.3% of observations falling outside the risk criteria (below Rs of 1.5) was accepted. In order to reduce the risk of lack of robustness during routine use or procedure transfer, an additional risk reduction strategy was established by defining system suitability (SST) requirements to increase the detectability of risk throughout the procedure lifecycle. The SST requirements for critical attributes were established as the LSL and USL for each monitored response

impurities, the goal was to maximize Rs, and the $C_{p,\text{lower}}$ was calculated setting up the LSL as the risk criteria for the inaccurate quantification of impurities: not less than Rs of 2.0 for critical impurities (imp. 4/API, and imp. 3/4) and Rs of 1.5 for "non-critical" target impurities. For tailing factor, USL of 1.5 was selected and for retention factor, LSL of 0.5. In this example, the risk control strategy included the selection of conditions that result in Rs with $C_{pk} \geq 1.33$ for all target impurities. For the Rs between impurity 5/6 (green response in Fig. 9.8), conditions with narrower ranges for flow rate and final % of organic solvent were selected as control strategy, additionally to the reduction of C_{pk} from 1.33 to 1.00 in order to enlarge the MODR.

Robustness Assessment and MODR Design Based on Prediction Intervals

An adequate model for TAE can be defined by a statistical model then deriving statistical prediction and/or tolerance intervals for these models. A prediction interval is an interval in which, with a given probability, future observation(s) should fall, given the already observed results [20, 39]. In general, for robustness study, the question to be answered (to allow for proper risk assessment) is how to specify the range in which the response values of future analysis using the procedure are likely to be found. The use of prediction intervals to assess the joint effect (i.e., the simultaneous effect) of the factors has been reported [19, 20], and an application example is detailed below. In this example, robustness assessment for a HPLC assay method (for API and impurity in a drug product) was performed by deriving prediction intervals, from DoE experimental results [11, 16–19]. This approach is also described in the USP General Chapter <1210> Statistical Tools for Procedure Development [9]. These prediction intervals consider the amount of variation induced by the simultaneous effect, the joint effect, of the studied method parameters. Three different prediction intervals were used, each giving a specific information:

- Total error prediction interval: obtained from the DoE total variance for each response, estimates the parameters' joint effect impact on the mean measurement. If the joint effect impact is included in the given acceptance criterion for the response, method robustness is verified for this response.
- Experimental error prediction interval: obtained from the DoE experimental error variance for each response, estimates the experimental error component impact on the parameters' joint effect.
- The prediction interval at parameters' non-significance limits: obtained from the variance at parameters' non-significance limits, estimates the impact of the parameters on the mean effect when their level does not induce anymore any significant effect. When the method robustness is verified at these parameters' non-significance limits, the MODR can be modified according to the corresponding parameters' non-significance levels.

Details of these different prediction intervals' calculation can be found in references [18, 19]. Four parameters were considered in an eight-experiment fractional factorial design including three center points (for estimation of experimental error): ammonium formate concentration (1.2 and 1.4 g/L), pH (3.4 and 3.6), methanol content in the mobile phase (396 and 404 mL/L), and column age (old and new). Three responses were studied: impurity peak symmetry, API peak symmetry, and API retention time. Once the effects were calculated, the robustness assessment was conducted by means of the parameters' effect impact on the mean effect (b_0), using the sum of the effect and its confidence interval, and of the previously described prediction intervals [18], the results of which being presented in Table 9.1.

The results obtained for the three responses clearly demonstrated the interest of the use of prediction intervals for robustness assessment. In the case of the impurity peak symmetry, although some effects were statistically significant (p-value $< 0.1\%$), none of them showed a critical impact on the b_o, and their values remained quite small. Therefore, the variation induced by the parameters' joint effect, estimated by the total error prediction interval, confirmed the procedure robustness. Regarding the API peak symmetry, lack of robustness was observed for the LC column parameter, which showed an unacceptable impact on the response, confirmed by the total error prediction interval. Interestingly, for the response API retention time, although none of the parameters showed to impact to the robustness of the response measurement (p-value $< 0.1\%$), a potential

Table 9.1 DoE and prediction intervals for the considered responses

Impurity peak symmetry (acceptance criterion: 0.8–1.5)					
	b0	Formate	pH	MeOH	LC column
Effect	1.147	0.014	0.030	0.038	−0.065
p-value (%)		4.80%*	0.39%**	0.17%**	0.02%***
Impact on b0 (upper)		1.18	1.19	1.20	1.23
Impact on b0 (lower)		1.12	1.10	1.10	1.07
Total error prediction interval				0.99–1.31/robust	
Experimental error prediction interval				1.12–1.17	
API peak symmetry (acceptance criterion: 0.8–1.5)					
Effect	1.420	−0.001	−0.003	−0.008	−0.083
p-value (%)		82.20%	25.70%	2.04%*	0.00%***
Impact on b0 (upper)		1.43	1.43	1.43	**1.51**
Impact on b0 (lower)		1.41	1.41	1.41	1.33
Total error prediction interval				1.27–**1.57/not robust**	
Experimental error prediction interval				1.41–1.43	
Parameters' non-significance limits' prediction interval				1.41–1.43/robust	
API retention time (acceptance criterion: 6.0–7.0 min)					
Effect	6.581	0.246	−0.159	−0.155	−0.120
p-value (%)		0.45%**	2.03%*	2.22%*	4.86%*
Impact on b0 (upper)		6.95	6.86	6.85	6.82
Impact on b0 (lower)		6.22	6.30	6.31	6.34
Total error prediction interval				**5.79–7.38/not robust**	
Experimental error prediction interval				6.35–6.81	
Parameters' non-significance limits' prediction interval				6.22–6.95/robust	

, moderately significant effect; **, significant effect; *, highly significant effect; non-robust effects' or prediction intervals' limits are marked in bold*

robustness issue was flagged by the total error prediction interval, due to the joint effect of parameters having individual effect estimated between 0.15 and 0.25 min. In this case, the use of the total error interval provided the needed knowledge for the establishment of appropriate risk control and to avoid a misinterpretation of the procedure robustness with regard to this response. Considering these results, the MODR could be designed to ensure procedure robustness, in accordance with the non-significance limits of the tested parameters.

Another advantage of the use of these prediction intervals is that it may provide a framework for the MODR validation with reduced number of experiments, in case a quantitative response is also monitored (content of API in a drug product). If the prediction intervals demonstrate the procedure robustness of the method for this response, this confirms that the quantitative response is stable enough across the defined MODR; therefore, the procedure validation may not need to be done within the entire MODR.

9.2.3 *Considerations for MODR Design and Validation*

In general, the MODR is designed using predictive modeling, using either empirical models (e.g., DOE-derived models), hybrid models, or mechanistic models. A minimum of requirements are expected to be fulfilled when selecting the methodology for MODR design, in terms of experimental design type used to derive the procedure performance modeling and methodology for model building:

- Selection of CPQAs/CAPPs, Factor Ranges, and Levels: If the actual CAPPs (factors and their ranges) are not considered for response modeling, the prediction accuracy may be compromised, and risk controls cannot be designed properly; consequently, the MODR may fail to provide the demonstration of assurance of quality. Missing CPQAs may lead to the same scenario (e.g., monitoring only resolution between critical pairs in a HPLC procedure for impurities' quantification, while one impurity shows poor retention factor, the non-inclusion of retention factor as CPQA for MODR design may not provide enough information for risk assessment). A rationale for designating the CAPPs, CPQAs, and their acceptance limits should be provided. Additionally, a suitable approach should be used to design the MODR containing all CPQAs (e.g., building appropriate model for all CQPAs and finding a set of operating conditions that optimize all responses and/or maintain the responses within the desired ranges).
- Response Modeling and Prediction Model Validation: The selected design needs to provide good statistical properties (e.g., orthogonality and rotatability) and should maintain the number of experiments as low as possible [39].
  - Orthogonality of designs: For empirical and hybrid models, well-designed experiments (e.g., DoE) that consider multivariate interactions between factors should be considered. The selected DoE type should provide accurate estimates of main factor and factor interaction effects on procedure performance (if they have a significant impact on performance). Since the use of very saturated DoE with reduced number of experiments causes aliasing [40], some effects for factors and their interaction cannot be estimated separately from each other (non-orthogonal designs). For designing the MODR, the use of designs with resolution V is highly recommended. Designs with lower resolution may be used in cases where an approach is used to demonstrate that the high-order effects are negligible.
  - Rotatability: In rotatable designs, variance of predicted response is the same at all points x that are the same distance from the design center [40].
  - Allow for high-order models to be built: Usually, complex responses are being monitored, and optimization designs are used to allow for the identification of optimum conditions. These designs allow for fitting high-order models (e.g., quadratic and cubic models) and may be more suitable for the MODR design (unless it is demonstrated good approximation by fitting more simple models to the data). The design should also allow estimation of the experimental error and assessment of the validity of the model tested [39].

- Robustness Assessment: Typically, when the experiments have been performed using a response surface design, a mean acceptable performance region is designed by establishing risk criteria for each monitored CPQA usually interpreted through the overlay of the mean response surfaces or contour plots for all CPQAs. The mean response surface depicts the behavior of the response measured with respect to the range of factors assessed and to the model fitted and does not provide any guarantee that the analytical method can meet the specifications with high probability (the response is observed on average) [39]. For MODR design, other approaches should be applied taking into account the model parameter uncertainty and providing information about how often the specifications will be met. This is essential, since ICH Q14 clearly requires it for a level of assurance guaranteeing that the specifications will be met [39]. Robustness assessment is part of risk assessment which allows for the delimitation of the MODR. Few approaches were discussed under Sect. 4.2. Simple confidence intervals are not the ideal approach to quantify the uncertainty for future observations and consequently may not be suitable to design the MODR. Prediction and tolerance intervals are more suitable approaches for robustness assessment and MODR design.
- MODR Qualification (Validation): The MODR is typically a multidimensional and irregular region imposing a challenge for its experimental validation using traditional validation protocols. The validation of the MODR starts by providing appropriate justification of the methodology used for its design such as the rationale for CPQAs' and CAPPs' selection, appropriate experimental design, and predictive modeling technique used to build the prediction models. Appropriate statistical validation of prediction models should be conducted to ensure suitable prediction accuracy and robustness. The main challenge relies on the experimental validation of the MODR to demonstrate that the conditions within the ranges meet established requirements for performance characteristics, such as accuracy, precision, selectivity, etc. In order to provide additional evidence that the MODR is valid, a portion of the MODR can be selected, and verification runs can be performed to check performance (e.g., SST requirements). DoE can also be used to assess a smaller portion of the MODR, deriving prediction intervals from the models used to predict quantitative responses (e.g., reportable value) to ensure procedure fitness for use across the entire selected portion of the MODR.

9.2.3.1 Changes Within the MODR

As per USP General Chapter <1220> [2], if a suitable approach is used to generate the MODR and this region relates to attributes that are linked to the elimination of bias or minimization of variability, an impact on the final performance of the reportable value is not expected. Thus, the USP <1220> and ICH draft guideline [26] state that analytical condition changes within the MODR range previously qualified (validated) may not require additional experimentation (e.g., validation) before implementation. According to ICH Q14 draft guideline, if appropriately justified and validated, changes within the MODR do not require regulatory notification, and there might not be a need for full validation of new conditions

[26]. Consequently, this may provide regulatory flexibility for post-approval change submission and more flexibility for lifecycle management, such as risk assessment of needed changes during routine application, adjustment of procedure conditions during procedure transfer at the receiving unit, etc. Change of analytical conditions outside the qualified and approved MODR would require a risk assessment, appropriate validation, and regulatory reporting [2, 26]. To sum up, change controls are only required when stepping outside the DS limits [39].

References

1. Guiraldelli A, Weitzel J (2022) Evolution of analytical procedure validation concepts: part I – analytical procedure life cycle and compendial approaches. Pharm Technol
2. USP (2022) USP general chapter <1220> analytical procedure lifecycle
3. Separovic L, Simabukuro RS, Couto AR et al (2021) Measurement uncertainty and conformity assessment applied to drug and medicine analyses – a review. Crit Rev Anal Chem. https://doi.org/10.1080/10408347.2021.1940086
4. Separovic L, Lourenço FR (2020) Frequentist approach for estimation of false decision risks in conformity assessment based on measurement uncertainty of liquid chromatography analytical procedures. J Pharm Biomed Anal
5. Williams A, Magnusson B (eds) (2021) Eurachem/CITAC guide: use of uncertainty information in compliance assessment, 1st edn
6. Separovic L, Saviano AM, Lourenço FR (2018) Using measurement uncertainty to assess the fitness for purpose of an HPLC analytical method in the pharmaceutical industry. Measurement:41–45
7. Bettencourt da Silva R, Williams A (eds) (2015) Eurachem/CITAC guide: setting and using target uncertainty in chemical measurement, 1st edn
8. Rozet E, Marini RD, Ziemons E et al (2011) Total error and uncertainty: friends or foes? TrAC Trends Anal Chem 30:797–806
9. USP general chapter <1210> statistical tools for procedure validation. https://doi.org/10.31003/USPNF_M8646_06_01
10. Feinberg M, Boulanger B, Dewé W, Hubert P (2004) New advances in method validation and measurement uncertainty aimed at improving the quality of chemical data. Anal Bioanal Chem 380:502–514. https://doi.org/10.1007/s00216-004-2791-y
11. Hubert P, Nguyen-Huu JJ, Boulanger B et al (2004) Harmonization of strategies for the validation of quantitative analytical procedures: a SFSTP proposal – part I. J Pharm Biomed Anal 36:579–586. https://doi.org/10.1016/j.jpba.2004.07.027
12. Traple MAL, Saviano AM, Francisco FL, Lourenço FR (2014) Measurement uncertainty in pharmaceutical analysis and its application. J Pharm Anal 4:1–5
13. Moreira C Dos S, Lourenço FR (2020) Development and optimization of a stability-indicating chromatographic method for verapamil hydrochloride and its impurities in tablets using an analytical quality by design (AQbD) approach. Microchem J 154. https://doi.org/10.1016/j.microc.2020.104610
14. Ellison L, Williams A (eds) (2012) Eurachem/CITAC guide CG 4: quantifying uncertainty in analytical measurement, 3rd edn
15. Ramsey M, Ellison S, Rostron P (eds) (2019) Eurachem/EUROLAB/CITAC/Nordtest/AMC guide: measurement uncertainty arising from sampling A guide to methods and approaches, 2nd edn
16. Hubert P, Nguyen-Huu JJ, Boulanger B et al (2007) Harmonization of strategies for the validation of quantitative analytical procedures. A SFSTP proposal – part II. J Pharm Biomed Anal 45:70–81. https://doi.org/10.1016/j.jpba.2007.06.013

17. Hubert P, Nguyen-Huu JJ, Boulanger B et al (2007) Harmonization of strategies for the validation of quantitative analytical procedures. A SFSTP proposal-part III. J Pharm Biomed Anal 45:82–96. https://doi.org/10.1016/j.jpba.2007.06.032
18. Roussel J, Righezza M (2021) Application of prediction intervals to the interpretation of analytical methods robustness studies. J Pharm Biomed Anal 193:113706. https://doi.org/10.1016/j.jpba.2020.113706
19. Roussel J, Schelling C, Righezza M, Veuthey J (2022) Application of prediction intervals to the interpretation of the robustness study of a UHPLC method for the separation of cannabinoids. J Pharm Biomed Anal 220:114977. https://doi.org/10.1016/j.jpba.2022.114977
20. Gryzea S, Langhansb I, Vandebroek M (2007) Using the correct intervals for prediction: a tutorial on tolerance intervals for ordinary least-squares regression. Chemom Intell Lab Syst 87: 147–154. https://doi.org/10.1016/j.chemolab.2007.03.002
21. International Organization of Legal Metrology (2021) International vocabulary of metrology basic and general concepts and associated terms (VIM), 3rd edn
22. JCGM (2008) JCGM 101:2008 evaluation of measurement data – supplement 1 to the "guide to the expression of uncertainty in measurement" – propagation of distributions using a Monte Carlo, 1st edn
23. JCGM (2008) JCGM 100:2008 GUM 1995 with minor corrections. Evaluation of measurement data – guide to the expression of uncertainty in measurement, 1st edn
24. Barnett KL, Mcgregor PL, Martin GP et al (2016) Analytical target profile: structure and application throughout the analytical lifecycle, USP stimuli article. Pharmacopeial Forum 42(5)
25. Burgess C, Curry P, Leblond DJ et al (2016) Fitness for use: decision rules and target measurement uncertainty, USP stimuli article. Pharmacopeial Forum 42(2)
26. ICH (2022) ICH draft guideline Q14 analytical procedure development
27. Weitzel MLJ, Vegge CS, Pane M et al (2021) Improving and comparing probiotic plate count methods by analytical procedure lifecycle management. Front Microbiol 12. https://doi.org/10.3389/fmicb.2021.693066
28. Borman PJ, Chatfield MJ, Damjanov I, Jackson P (2011) Method ruggedness studies incorporating a risk based approach: a tutorial. Anal Chim Acta 703:101–113
29. Swartz ME, Krull IS (2006) Method validation and robustness. LCGC N Am 24:480–490
30. ICH (1994) ICH guideline Q2(R1) validation of analytical procedures
31. ICH (2022) ICH draft guideline Q2(R2) validation of analytical procedures
32. USP general chapter <1225> validation of compendial procedures. https://doi.org/10.31003/USPNF_M99945_04_01
33. ICH (2009) ICH guideline Q8(R2) pharmaceutical development
34. Peterson JJ, Yahyah M (2009) A Bayesian design space approach to robustness and system suitability for pharmaceutical assays and other processes. Stat Biopharm Res 1:441–449. https://doi.org/10.1198/sbr.2009.0037
35. Herrador MÁ, Asuero AG, González AG (2005) Estimation of the uncertainty of indirect measurements from the propagation of distributions by using the Monte-Carlo method: an overview. Chemom Intell Lab Syst 79:115–122. https://doi.org/10.1016/j.chemolab.2005.04.010
36. Carpenter J (1998) Bootstrap methods and their application. In: Davison A, Hinkley DV (eds) Epidemiology and infection, 2nd edn. Cambridge University Press, Cambridge
37. Shrestha DL, Solomatine DP (2006) Machine learning approaches for estimation of prediction interval for the model output. Neural Netw 19:225–235. https://doi.org/10.1016/j.neunet.2006.01.012
38. Perakis M, Xekalaki E (2003) On a process capability index for asymmetric specifications. Commun Stat Theory Methods 32:1459–1492. https://doi.org/10.1081/STA-120021568
39. Rozet E, Lebrun P, Hubert P et al (2013) Design spaces for analytical methods. TrAC Trends Anal Chem 42:157–167
40. Montgomery D (2013) Designs and analysis of experiments, 8th edn. Wiley, New York

Chapter 10
Quality by Design Applications to Development of Analytical Methods for Quality Control of Pharmaceuticals

Yamile Soledad Caro, María Silvia Cámara, and María Mercedes De Zan

Abstract Analytical quality by design (AQbD) strategy is presented and discussed from a practical point of view. In a case study, several important AQbD tools are applied to the development of a new chromatographic method for the quality control of a pharmaceutical product. The purpose and required performance of the method are defined in the analytical target profile (ATP). The critical method attributes (CMAs) are identified in a scouting phase. A quality risk management (QRM) process is applied to classify the method variables as a function of their effects on CMAs. Design of experiments (DoE) and response surface methodology (RSM) are used to assess the critical method parameters (CMPs) and to define the method operable design region (MODR). An optimal working point (OWP) is determined, and a control strategy is set to ensure the required quality of the reportable result.

10.1 Introduction

The pharmaceutical industry is constantly developing new active pharmaceutical ingredients (APIs), excipients, and pharmaceutical products. In parallel, the quality control laboratory must develop efficient methods to guarantee that the attributes of the new pharmaceutical item fall within the range that has been demonstrated to be efficient and safe for patient treatment. This requires the development of analytical methods to be employed in quality control for that particular intended use; in addition, those methods should be robust and under control during their entire lifecycle, which in turn should be linked to the pharmaceutical item lifecycle.

Y. S. Caro · M. S. Cámara · M. M. De Zan (✉)
Laboratorio de Control de Calidad de Medicamentos (LCCM), Cátedra de Control de Calidad, Facultad de Bioquímica y Ciencias Biológicas, Universidad Nacional del, Litoral, Santa Fe, Argentina
e-mail: ycaro@fbcb.unl.edu.ar; mcamara@fbcb.unl.edu.ar; mmdezan@fbcb.unl.edu.ar

M. C. Breitkreitz, H. C. Goicoechea (eds.), *Introduction to Quality by Design in Pharmaceutical Manufacturing and Analytical Development*, AAPS Introductions in the Pharmaceutical Sciences 10, https://doi.org/10.1007/978-3-031-31505-3_10

In some cases, those requirements can be easily met. However, for complex formulations involving more than one API, each one showing different physical and chemical properties, the development of a single analytical method for their characterization presents a great challenge to any analytical scientist. A similar situation occurs when it is necessary to develop a method to simultaneously determine assay and impurities in a drug substance or drug product, as in stability-indicating methods.

In every case, a multifactorial approach is required because, due to interaction effects, a change in one method parameter may modify the effect of another parameter on method performance. Identifying these variables is the first task, and sometimes the result is surprising: a list of more than 40 or perhaps 50 parameters, which together define, for example, a high-performance liquid chromatographic (HPLC) method with ultraviolet detection.

The sample preparation stage includes several factors related to the analytical balance, glassware, solvents, dissolution and dilution procedures, centrifugation, and/or filtration of solutions, in addition to environmental conditions and analyst training. On the other hand, for the chromatographic system, both the instrument modules (autosampler, column oven, pump, and detector) as well as the analytical column and the mobile phase contribute a large number of variables.

The quality indicators of an analytical method are known as *analytical properties* and were traditionally classified into three groups: capital, basic, and productivity-related properties. Capital analytical properties are representativeness, accuracy (for quantitative results), and reliability (for qualitative analysis). Basic analytical properties are precision, sensitivity, selectivity, and robustness, and they support the capital properties, so that accuracy or reliability cannot be achieved if the analytical procedure is not precise, sensitive, and selective enough. Finally, productivity-related properties are the attributes of an analytical procedure with very high practical and/or social relevance, namely, expeditiousness (capacity to develop the analytical process from raw sample to results over a short time), cost-effectiveness (price of equipment, reagents, consumables, and cost of maintenance and analyst training), greenness, and personnel safety [1].

Risk assessment, statistical design of experiments, multiple linear regression models, multiple response optimization techniques, and control strategies are valuable tools of the *analytical quality by design* (AQbD) methodology, which allows the efficient management of a large number of variables in the development of a new analytical method that meets the requirements of the analytical properties.

The previous chapters have outlined the fundamentals and regulatory perspectives of AQbD in pharmaceutical development. In this chapter, we detail the application of AQbD concepts through an example in the framework of the analytical method lifecycle.

10.2 Analytical Problem Statement

As already mentioned, the requirements for the development of a new analytical method are related to the drug substance or drug product attributes involved, as well as the type of analysis to be performed (e.g., identification, impurities, assay, content, dissolution). In this sense, before starting with the development, it is necessary to gather all relevant information about the drug substance or the composition of the drug product and well-defined analytical objectives. The relevant information that must be evaluated and taken into account includes the physical and chemical properties of the analytes, information from pharmacopeias and in the scientific literature, as well as any existing procedures for the same type of sample.

In this chapter, we describe a case study that consists of the development of an analytical method to be employed in the quality control of a topical solution (TS) for dermal use, containing two APIs (API_1, which is expected in the concentration of 2.0% w/v, and API_2, expected in the concentration of 0.5% w/v) and a preservative (EX) in an alcoholic solution. Two related compounds for API_1 ($IMP_{1.1}$ and $IMP_{1.2}$) and two for API_2 ($IMP_{2.1}$ and $IMP_{2.2}$) are likely to be present in the commercial product.

Both APIs are pharmacopeial drug substances, consisting of small organic molecules with significant absorptivity in the ultraviolet spectrum. API_1 has anti-inflammatory activity. It is a nonpolar molecule with a Log P value of 4.1 and poor water solubility and is very soluble in ethanol and acetone. API_2 has fungicidal activity. It is a compound with a Log P value of 6.3 and a pKa = 4.7; it is partially soluble in water and very soluble in ethanol.

The related compounds are described in the respective drug substance monographs, but limited information about their physical-chemical properties is available. The EX is also a small molecule with bacteriostatic activity in the pH range of 5.0–8.0 and has high solubility in ethanol and methanol. Ethanol is used as API and EX solvent and as formulation diluent in the TS.

The formulation must meet the following quality acceptance criteria:

- Assay: 90.0–110% of the label claim for each API.
- Organic impurities:
  - $IMP_{1.1}$ and $IMP_{1.2}$: no more than 50 μg/L.
  - $IMP_{2.1}$ and $IMP_{2.2}$: no more than 10 μg/L.

10.3 Analytical Target Profile

The first important step of the AQbD methodology consists of defining the goals by means of an *analytical target profile* (ATP). The ATP should be based on the intended use of the method and serves as the basis for procedure development, qualification, and monitoring during the method lifecycle. It is a prospective description of the desired performance of the analytical procedure that defines the required quality of the reportable result. In turn, the ATP must be in accordance with the quality target product profile. Identifying the quality of the procedure in terms of the reportable results provides a framework for development and helps to ensure that validation acceptance criteria will be met without last-minute unexpected results.

The ATP can be defined by the method developer in several ways, as long as attention is focused on achieving acceptable levels of bias and precision for the reportable result across the analytical range. Separate or combined acceptance criteria for trueness and precision may be provided. In the latter case, a total allowable analytical error is indicated [2].

Notice that in the pharmaceutical environment, the term *accuracy* is used instead of *trueness*, often leading to confusion. According to the International Organization for Standardization (ISO) [3], accuracy is a combination of trueness and precision and is therefore evaluated by assessing both systematic and random error. In addition, an increasingly common expression of accuracy is *measurement uncertainty* [4]. It is interesting to highlight that the ATP can be defined independent of an analytical technique.

In our case study, the method is intended to be applied in the routine quality control of the TS; hence, the criteria for the ATP were related to ICH Q2 (R1) guideline validation parameters [5]. Typical acceptance criteria were adopted, such as mean bias ≤ 2.0% and mean relative standard deviation (RSD) ≤ 2.0% for API quantification within the range of 80.0 to 110.0% of the label claim. For impurity quantitation, higher bias (≤ 10%) and RSD (≤ 10%) were accepted within the range of quantitation limits (LOQs) to 150% of the specification limit. The criterion for the LOQs was established to be ≤ 50% of the specification limits. The expeditiousness and greenness of the procedure, as well as the cost-effectiveness, were also taken into account in the design of the new method.

The ATP was defined as follows: *The analytical procedure must be able to simultaneously quantify* API_1*,* API_2*,* $IMP_{1.1}$*,* $IMP_{1.2}$*,* $IMP_{2.1}$*, and* $IMP_{2.2}$ *in the topical solution in the presence of EX in a short analysis time. APIs must be accurately quantified within a range from 80.0 to 120.0% of the expected concentration with a trueness = 100.0% ± 2.0% and a precision ≤ 2.0% for the reportable value. IMPs must be accurately quantified within a range from LOQs to 150% of the maximum allowed concentration with a trueness = 100% ± 10% and a precision ≤ 10% for the reportable value.*

Other interesting ways to define the ATP can be found elsewhere [6–8].

10.4 Technique Selection

Once ATP is defined, the analytical technique capable of fulfilling the ATP requirements should be selected. There is a wide diversity of analytical technologies, and their applicability overlaps, with each technique having strengths and weaknesses. The selected technique must be available at both the development and transfer sites; in addition, staff expertise in its routine operation and maintenance is an important issue to be taken into account.

Methods based on chromatographic techniques are widely used in pharmaceutical laboratories because their scope can be extended to simultaneously assess identification, potency/API content, and determination of organic impurities. Thus, for the case study, when we defined the ATP, as indicated in the previous section, we considered a chromatographic instrumental analytical technique. Specifically, a reversed-phase high-performance liquid chromatography with diode array detection (RP-HPLC-DAD) was selected as the most suitable technique to meet the ATP requirements in a safe, efficient, and cost-effective way.

The pre-analytical stage has an enormous impact on method performance; therefore, the most appropriate analytical technique or operation for the sample pre-treatment must be carefully selected, taking into account the characteristics of the sample and analytes. Suitable homogenization, dissolution, dilution, extraction, derivatization, filtration, and/or centrifugation procedures should be adopted in order to ensure that the ATP requirements will be met.

For the TS analysis, the only pre-treatment required was a dilution of the original homogenized sample in the chromatographic mobile phase using proper glassware.

10.5 Scouting Phase: Initial Procedure and Preliminary Experiments

Considering the ATP and the selected analytical technique, the next step is to propose a preliminary procedure and to perform a few experimental tests to obtain initial information about the system. For this, it is necessary to gather relevant information about the sample matrix, the analytes (see Sect. 10.2), and the analytical technique performance. Thus, variables like sample solvent and stationary phase and/or mobile phase components in HPLC methods are usually selected early in the scouting phase. Moreover, both a calibration strategy and a replication strategy for the routine analysis should be proposed at this phase. Certainly, as method development process progresses, and even during the validation stage, any of these parameters may be subjected to changes as necessary.

As previously mentioned, in the current case study, the APIs in the commercial product are pharmacopeial substances. Both monographs propose the use of C18, a 3.5-μm-diameter porous particle as a stationary phase, but they differ in the lengths of the columns to achieve an adequate separation of APIs and their respective related substances. The separation mode is isocratic in both monographs. In the case of API_1, the mobile phase is a mixture of water and acetonitrile 40:60 v/v at a flow rate of 1.0 mL/min resulting in a runtime of 15 min. In API_2 monograph, the mobile phase is composed of a 40 mM phosphate solution and methanol 50:50 v/v and is delivered at 1.0 mL/min during a runtime of 22 min. For API_1, the column oven temperature is 30 °C, whereas for API_2, no control of this parameter is established. In both monographs, the detection wavelength to quantify the API and the related substances is 254 nm.

Taking into account this information, preliminary tests were carried out under the following initial conditions:

- Instrument: Shimadzu system equipped with LC-20AT pumps, a SIL-30AC Nexera autosampler, a CTO-10ASVP column oven, an SPD-M20A UV–Vis detector, and a DGU-20A5R degasser.
- Column: Zorbax Eclipse XDB C18 (100 × 4.6 mm, 3.5 μm).
- Column oven temperature: 30.0 °C.
- Mobile phase: phosphate buffer 30 mM at pH 6.00: acetonitrile 40:60 v/v in isocratic mode.
- Flow rate: 1.00 mL/min.
- Autosampler temperature: 25.0 °C.
- Injection volume: 20 μL.
- Detector: UV at 254 nm.

A few univariate changes in some parameters were also carried out to initially explore the system variability.

Combined standard solutions, sample solutions, spiked sample solutions, and placebo solutions (as a blank) were used to evaluate this preliminary system. Individual stock standard solutions were prepared by weighing and dissolving pharmacopeial standards in ethanol, whereas sample solutions were obtained by diluting TS 1/20 (v/v) in mobile phase. Spiked sample solutions were prepared with the impurities at their maximum desirable LOQ.

Figure 10.1 shows a chromatogram obtained with a spiked sample solution in the preliminary conditions. As can be observed, proper signals are achieved for all the analytes in a short runtime; however, the resolution between some of the peak pairs is not adequate.

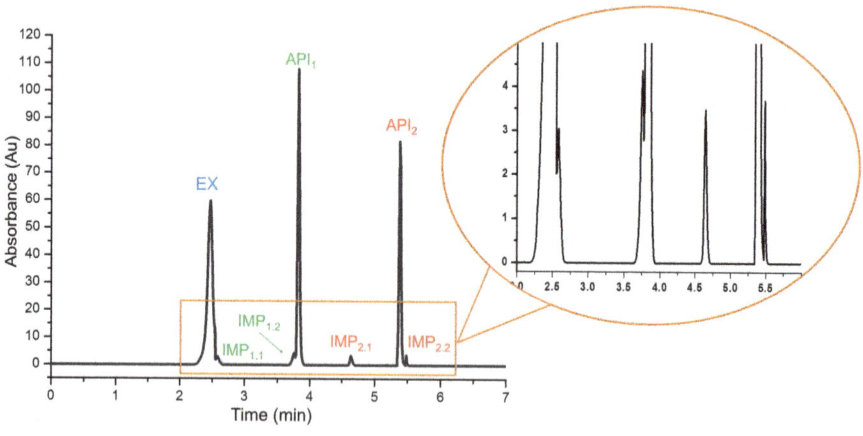

Fig. 10.1 Representative chromatogram of a spiked sample solution under the preliminary chromatographic conditions

10.6 Critical Method Attributes (CMAs)

As a general rule, the ICH analytical performance characteristics are necessary to meet the quality level declared in the ATP of a pharmaceutical quality control method. That is, if the procedure lacks linearity or selectivity or does not have suitable precision or sensitivity, the ATP will not be met.

The next step is to determine which key parameters of the analytical responses can be evaluated as indicators of the analytical property performance. These key variables are known as *critical method attributes* (CMAs) and are directly correlated to a mathematical representation of the quality of method performance [9, 10]. In other words, the CMAs translate the ATP requirements into procedure response parameters.

CMAs depend on the analytical technique, working principle, and purpose of the method. They are usually related to the selectivity, precision, sensitivity, or trueness (recovery and/or bias) of the method.

For example, in chromatographic methods, sensitivity requirements may be translated to the magnitude of the parameters related to the analytical signal, such as height, area, or signal-to-noise ratio of the analyte peaks. Selectivity demands may be monitored through variables that evaluate adequate separation between adjacent peaks such as resolutions (*Rs*) or separation factor (α), whereas tailing factor (T), retention factor (k), and number of theoretical plates (N) can be evaluated to ensure a good peak shape, which is related to precision and selectivity. Finally, the runtime can be adopted as a CMA that is correlated to the productivity-related analytical properties.

Once the CMAs are selected, target value or acceptable ranges have to be established for the CMAs to support the following procedure development activities and to define the control strategy. In the current case study, the information gathered during the preliminary experimental phase allowed us to determine that accuracy was the capital analytical property that was affected in the initial procedure, due to a lack of selectivity given by the incomplete separation of the chromatographic peaks. As for the sensitivity required for the quantification of the impurities, it does not represent a difficulty. Despite the lack of resolution, peaks with appropriate signal-to-noise ratios were recorded for the four impurities in the spiked sample solutions.

Based on this analysis, to continue with method development, we defined the CMAs as the peak pairs *Rs* between APIs, their respective impurities, and the EX. Thus, the CMAs were identified as the *Rs* between EX-$IMP_{1.1}$ (R_1), $IMP_{1.1}$-$IMP_{1.2}$ (R_2), $IMP_{1.2}$-API_1 (R_3), API_1-$IMP_{2.1}$ (R_4), $IMP_{2.1}$-API_2 (R_5), and API_2-$IMP_{2.2}$ (R_6). An acceptable value of $Rs \geq 1.5$ was established as a criterion for all the CMAs.

10.7 Critical Method Parameters (CMPs)

Once the CMAs are defined, the next step in the development of the new method consists of searching for the operating conditions that ensure acceptable values for the CMAs and, consequently, meet the requirements of the ATP. This task must be done on a scientific and risk management basis to deeply understand and control the influence of the variability of the method parameters on the quality of the reportable result, not only during the validation stage but also in routine use and/or transfer of the method.

As a rule, method parameters may be classified as non-critical, critical, or unclassified [11]. An important task involved in developing a new analytical method based on the AQbD concept is to determine if the unclassified variables are truly critical or non-critical for the quality performance of the procedure.

Non-critical variables are those for which relatively large input ranges may be accepted. Critical variables are those parameters whose levels affect, to a greater or lesser extent, one or more CMAs and, therefore, should be monitored and/or controlled in a target value or in a bounded range. These critical variables are known as the *critical method parameters* (CMPs) [10] or critical procedure parameters (CPPs).

10.8 Quality Risk Management

Risk is defined by ICH Q9(R1) [12] as the combination of the probability of occurrence of harm and the severity of that harm. In the framework of pharmaceutical quality control, risk has to be considered from two points of view. The first and most important one is focused on patient safety. Hence, the risk to cause harm to the patient due to an incorrect analytical result leading to the release of products lacking quality must be carefully considered. The second point of view is an economic one, since economic losses cause harm to the business; indeed, incorrect analytical results may lead to the rejection of good-quality batches.

The general *quality risk management* (QRM) process, as described in the ICH guide, is a formal and documented process that includes several stages and sub-stages, namely, risk assessment (identification, analysis, and evaluation), risk control, and risk review, which may be implemented in AQbD. QRM leads to a deep understanding of the procedure variability, so the risk can be monitored, controlled, and reduced when possible. Thus, QRM helps to ensure, with a certain degree of confidence, that the analytical method will generate reportable results that meet the ATP requirements across the method lifecycle.

Risk assessment in method development can be more or less demanding, depending on the initial procedure performance. Likewise, the effort to control risk during routine use of the method will depend on the success achieved in reducing risk during method development. In order to achieve the purposes of the

QRM process, a variety of tools can be applied, either alone or in combination. It is also possible to apply them more than once during the analytical method development process.

10.8.1 Risk Assessment

10.8.1.1 Risk Identification

The first sub-stage in risk assessment consists of systematically identifying all the variables associated with the analytical procedure that may have an influence on the CMAs. This task must include all aspects of the complete analytical procedure, i.e., sampling, reagents, preparation of the sample and reference standards, facilities and equipment operating conditions, calibration and replication strategy, and reportable result calculation, among others. A process map of the preliminary procedure can be used to sort the identified variables in a list, a table, or an Ishikawa diagram (also known as fishbone diagram). For complex systems, it can be useful to break the process down into steps.

In the case study, the method parameters were summarized systematically using fishbone diagrams to formalize the risk identification exercise in each of the analytical steps. As an example, Fig. 10.2 shows the fishbone diagram constructed to identify the factors of the chromatographic system with possible effects on the CMAs. An interesting example of how to identify risks in a pre-analytical stage can be found in [13].

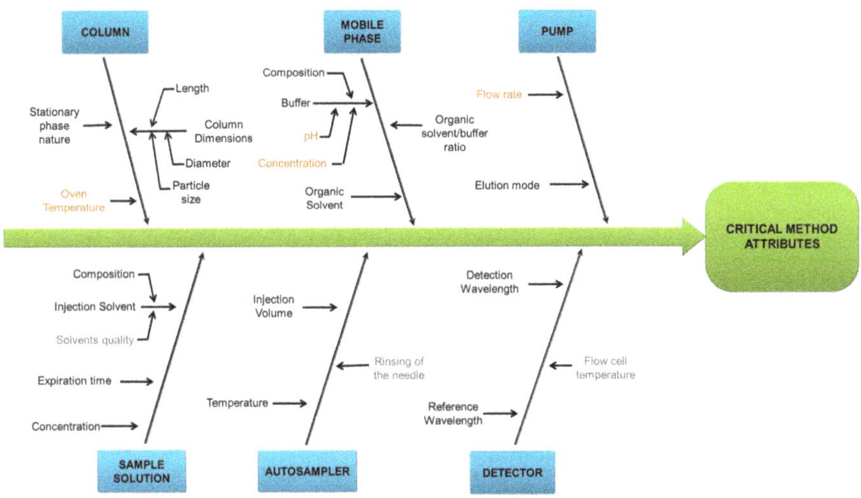

Fig. 10.2 Fishbone diagram elaborated to identify potential risk factors in critical method attributes for the chromatographic system

10.8.1.2 Risk Analysis

Once risk variables have been identified, the associated hazard should be estimated by analyzing the probability of failure occurrence and the severity of the consequence on CMAs. This process may be performed in a qualitative or quantitative manner using a variety of risk analysis tools, such as CNX (Control, Noise, eXperimental) method; heat maps; failure modes and effects analysis (FMEA); failure mode, effects, and criticality analysis (FMECA); and fault tree analysis (FTA), among others [12]. At this stage, based on theory, previous knowledge, and evaluation of the preliminary procedure, decisions can be made as to which of the identified variables pose or do not pose a real risk. Thus, the risk analysis step aims to assess, of the many variables in the output from the risk identification step, those whose impact on the CMAs needs to be understood in order to establish appropriate levels and/or control limits.

The CNX risk analysis tool classifies the variables according to certain criteria. Variables that are difficult, expensive, or impractical to control or are known to be of low risk are called noise variables (N, depicted in gray in Fig. 10.2). Other variables that, from prior knowledge or operative convenience, should be set in a target value or within a certain range are called control variables (C, depicted in black in Fig. 10.2). Finally, variables that could potentially affect one or more CMAs and, therefore, have to be experimentally evaluated are called experimental variables (X, depicted in orange in Fig. 10.2).

Environmental variables, such as ambient temperature and humidity, can be considered N variables as long as the analytes are not hygroscopic or thermolabile. Otherwise, these parameters should be considered C variables, and appropriate controls and operational limits have to be provided. Batches of reagents or columns are normally considered N variables. However, these uncontrolled variables may result in unacceptable procedure performance. Therefore, system suitability test (SSTs) should be included in the control strategy to monitor their incidence during routine use of the method.

Another qualitative tool to support risk analysis is the heat map, also called risk estimation matrix (REM) [14]. This approach is complementary to CNX and provides a visual indication of which variables are considered to have potentially strong, medium, or minor impact on procedure performance. By using heat maps, low-risk variables can be left to chance (i.e., classified as N variables) or can be fixed at convenient levels (i.e., assumed as C variables), whereas medium or strong impact variables must be controlled early and/or experimentally evaluated to determine their effects.

In the current case study, a combination of CNX and heat map tools was used to do risk analysis. For this, the knowledge source (theoretical, from the literature or from the scouting phase) and the knowledge degree (low, medium, or high) about the impact of the variables on the CMAs were considered. Then, a risk level was assigned according to the following scale: strong impact, medium impact, or minor impact. Finally, the variables to be controlled at suitable operation levels

Table 10.1 Risk analysis matrix. Combination of heat map and CNX tools

Analytical unit of operation	Variable	Potential hazard on the CMAs	Knowledge		Risk on CMAs	
			Source	Degree	Impact	CNX
Column	Stationary phase	Separation efficiency	T – L	High	Strong	C
	Column dimensions	Separation efficiency	T – L	High	Strong	C
	Oven temperature	Separation efficiency and peak shape	L – SP	Low	Strong	X
Mobile phase	Buffer composition	Peak shape and detectability	L	Medium	Medium	C
	Buffer concentration	Peak shape	L	Low	Strong	X
	Buffer pH	Retention time and peak shape	T – L	Low	Strong	X
	Organic solvent	Retention time and peak shape	L	High	Medium	C
	Organic solvent and buffer ratio	Separation efficiency	L – SP	Medium	Minor	C
Pump	Flow rate	Separation efficiency	L – SP	Low	Strong	X
	Elution mode	Separation efficiency	L	Medium	Strong	C
Sample solution	Solution concentration	Peak shape and size	L – SP	High	Minor	C
	Injection solvent	Peak shape	L	Medium	Minor	C
	Expiration time	Stability of test samples	L – SP	High	Minor	C
Autosampler	Injection volume	Peak shape and size	L – SP	High	Minor	C
	Temperature	Stability of test samples	L	High	Minor	C
Detector	Detection wavelength	Peak size	L – SP	High	Minor	C

C control variable, ***L*** literature, ***SP*** scouting phase, ***T*** theoretical, ***X*** experimental variable

were selected. The results from this analysis are shown in Table 10.1. Note that, even though three of the C variables were marked as having strong impact (the stationary phase, the column dimensions, and the elution mode), they were not studied experimentally since information from pharmacopeial methods and the scouting phase results allowed to fix these variables, avoiding over-dimensioning the experimental work.

Given their level of criticality, as well as the lack of knowledge about their effects on the separation efficiency, four variables were considered to be in-depth studied through experimental work: mobile phase buffer concentration and pH, flow rate, and column oven temperature.

10.8.1.3 Risk Evaluation

The risk evaluation continues with the critical (C in the CNX classification) and those variables (X in the CNX classification) for which we do not know a priori their most convenient set point or the real effect of their variability on the performance of the method.

This is an experimental stage that allows to classify the variables as a function of the magnitude of their effects on the CMAs. This new classification can be achieved through the response surface methodology (RSM), by applying designs of experiments (DoE), as will be seen in Sect. 10.9. Thus, the strength of evidence for CMPs' management is considered in the risk evaluation stage.

10.8.1.4 Risk Control and Risk Review

The risk control stage involves decision-making to reduce or to accept the known risk. Risk reduction can include actions to reduce severity and probability and/or improve detectability. Through DoE, the risk evaluation step provides the knowledge and understanding about which of the experimental procedure variables (X) impact the ATP. Thus, those variables can be assigned as CMPs, and their most convenient level can be found, and the appropriate controls can be established (see Sect. 10.14).

The risk management process is sequential and iterative; therefore, if adequate control of a CMP under certain conditions is not possible, the analytical procedure may need to be re-designed, or adequate changes may be implemented.

However, no matter how much effort is made during the method optimization, risk cannot be completely eliminated. Therefore, a certain degree of residual risk must be accepted. The control strategy defined during method development and improved during the validation stage has the objective to reduce and maintain the residual risk at acceptable levels, whereas risk review on a regular basis should be a part of the routine procedure monitoring process [15].

10.9 Design of Experiments. Response Surface Methodology

The RSM is a very useful collection of statistical and mathematical techniques for developing, improving, and optimizing different types of products or processes [16], which has been adopted in the AQbD concept. This methodology allows us to analyze both mathematically and graphically the relationship between a response variable yield (the CMA performance) and several input variables: critical variables

(C in the CNX classification) and unclassified method parameters (X in the CNX classification). The unknown true response function is modeled using experimental data obtained by applying DoE, as mentioned in previous chapters.

First-order models, constructed from data acquired with first-order experimental designs, are appropriate to explore the response surfaces over a relatively small experimental region and to determine the effects of the method parameters on the CMAs. With suitable designs, the effects of the interaction between the variables may be also modeled. This screening allows to investigate which method parameters are the most important in explaining the CMAs' performance and the level of risk associated with each of them. Thus, it is possible to reduce the number of candidate CMPs, so that in a subsequent stage, optimization experiments will be more efficient and require fewer runs.

Another important use of the information gathered with first-order DoE is to determine whether the initial levels of the CMPs are within the region of experimentation where conditions that simultaneously meet all the CMA requirements can be found. If this is not the case, it will be necessary to move the region of experimentation, i.e., to change the level of the CMPs toward the zone that the screening experiments have indicated as leading to best results, and re-evaluate them. This can be considered a risk reduction activity.

Graphical tools like Pareto chart and normal and half-normal plot may be used in addition to the regression models to assess variable effect significance on CMAs and to decide how to move the experimental domain in an attempt to find the best analytical conditions. In some cases, when sufficient prior knowledge about the CMP effects is available, the work involved in the screening phase may be minimal or even avoided, provided that a rational planning of the optimization phase is possible.

The ultimate objective of RSM applied to AQbD is the definition of the *method operable design region* (MODR) and of the *optimal working point* (OWP), i.e., of operational conditions in which the risk is known, reduced, and controlled. Then, once the favorable region of experimentation is established, a second-order experimental design, also known as response surface design, should be run in the optimization phase to obtain second-order models that describe as closely as possible the true functions explaining the CMAs. When these models are well adjusted, they can be used as predictive models to construct two-dimensional graphs (the contour plots) or three-dimensional graphs (the response surfaces).

In this optimization phase, the investigation is generally focused on two or three continuous CMPs. A greater number, or the introduction of a categorical variable in this stage, leads to a number of experimental runs that might be too high.

Appropriate software packages are available to carry out DoE studies, such as Statistica, Minitab, Modde, Design-Expert, MATLAB, and R, among others. In the current case study, Design-Expert 11.0 software (Stat-Ease Inc., Minneapolis, USA) was used to generate the experimental plans, construct and evaluate the regression models, and build the response surface graphs.

10.9.1 First-Order Designs

Among first-order designs, Plackett and Burman, 2^k full factorial, or 2^{k-1} fractional factorial designs are traditionally used for an initial experimental evaluation of those method parameters identified as critical or unclassified in the scouting phase.

In the case study, for the first screening of the X variables, we used a 2^{k-1} fractional factorial design (fFD1) with three replicates of the center points. Each variable was evaluated at low (−1) and high (+1) levels assuming as center point (0) the level of the variable in the initial chromatographic system. Therefore, factor variables and their experimental ranges were established as follows: **A** = pH (5.00–7.00), **B** = buffer concentration (10.0–50.0) mM, **C** = flow rate (0.80–1.20) mL/min, and **D** = temperature (25.0–35.0) °C.

Experimental runs were carried out using a spiked sample solution as described in Sect. 10.5 and in a randomized order as determined by the software to ensure the independency of results, thereby minimizing the effects of uncontrolled factors. The responses (CMAs) were measured in each resulting chromatogram using the instrument control software and were recorded in the experimental matrix (Table 10.2). As observed, the least favored CMAs in the experimental domain of fFD1 were R_1 and R_3, i.e., no satisfactory *Rs* were achieved for Ex-IMP$_{1.1}$ or for IMP$_{1.2}$-API$_1$ under the evaluated conditions.

Pareto charts were used to assess the significance of the main effects and the binary interaction effects of the variables. As can be seen in Fig. 10.3a, in four of the six CMAs, buffer pH and the interaction between buffer pH and column oven temperature (**AD**) have shown significant effects. The flow rate shows a clear significant effect on R_1, R_3, and R_5, whereas it appears to have a potentially significant effect on R_4. The buffer concentration has a lower effect on the responses compared to the other factors, but it has a clear significant effect especially on R_3 (positive) and R_6 (negative).

Table 10.2 Conditions of the two-level fractional factorial screening design (fFD1) and results of the measured CMAs

	CMPs				CMAs					
Run	A	B (mM)	C (mL/min)	D (°C)	R_1	R_2	R_3	R_4	R_5	R_6
1	6.00	30.00	1.00	30.00	0.80	4.31	0.92	5.41	5.80	1.81
2	7.00	50.00	1.20	35.00	1.66	4.71	1.65	2.67	6.07	1.87
3	7.00	10.00	0.80	35.00	0.00	4.44	0.00	1.47	4.62	2.02
4	5.00	10.00	0.80	25.00	0.00	2.36	0.00	3.99	6.35	2.09
5	5.00	10.00	1.20	35.00	1.12	4.18	0.00	6.68	6.18	1.82
6	5.00	50.00	0.80	35.00	0.00	4.28	0.00	4.48	4.21	1.71
7	6.00	30.00	1.00	30.00	0.68	4.14	0.66	4.52	6.11	1.84
8	7.00	50.00	0.80	25.00	0.00	6.57	1.75	6.83	2.30	1.65
9	7.00	10.00	1.20	25.00	1.28	5.95	1.58	7.01	3.78	1.74
10	5.00	50.00	1.20	25.00	1.86	2.83	1.41	3.99	8.19	1.98
11	6.00	30.00	1.00	30.00	0.90	4.00	0.38	4.43	6.00	1.86

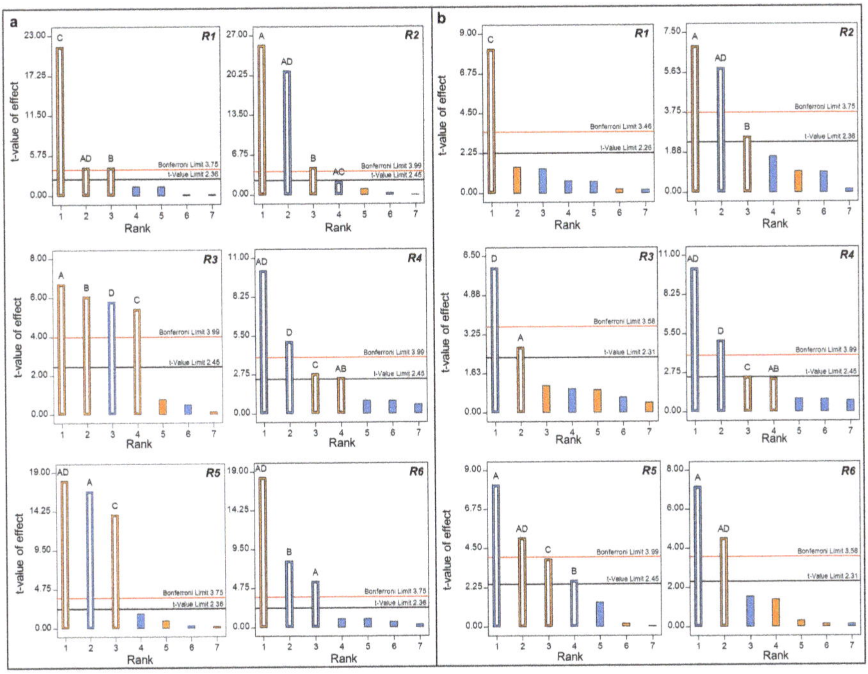

Fig. 10.3 Pareto charts showing the main and interaction effects of factors on the CMAs assessed by fFD1 (**a**) and by fFD2 (**b**) experiments

The confirmation of the significant effect of all the four analyzed variables on some of the responses allowed us to define them as CMPs. Therefore, these four variables had to be further studied to find their appropriate operational levels.

With this acquired knowledge and according to the sequential methodology of RSM, a new experimental domain was evaluated to search for more favorable conditions for R_1 and R_3. A second 2^{k-1} fractional factorial design (fFD2) with three center points was constructed to evaluate the four CMPs in a new experimental space (Table 10.3). In this case, the buffer concentration (**B**, 10.0–50.0) mM and column oven temperature (**D**, 25.0–35.0) °C, levels were kept as in the first experimental domain, whereas the ranges of both buffer pH (**A**, 6.50–7.50) and flow rate (**C**, 1.00–1.50) mL/min, were moved to a new experimental region. The latter decision was made because both compromised CMAs (R_1 and R_3) were favored by an increase in the mobile phase flow rate and buffer pH.

The results obtained from fFD2 indicated that, in the new experimental domain, R_1 and R_3 have considerably been improved; however, R_6 becomes more critical. As can be seen in Fig. 10.3b, the increase in flow rate had a significant positive effect on R_1 and R_5, but not on the other responses. In general, on the latter (R_2, R_3, R_4, and R_6), the most significant variables were buffer pH and oven column temperature as well as their interaction, with conflicting effects on the CMAs.

Table 10.3 Two-level fractional factorial screening design (fFD2). Experimental design and its results

	CMPs				CMAs					
Run	A	B (mM)	C (mL/min)	D (°C)	R_1	R_2	R_3	R_4	R_5	R_6
1	7.00	30.00	1.25	30.00	3.99	4.73	1.88	5.63	5.23	1.69
2	7.50	50.00	1.50	35.00	5.12	5.29	1.93	2.70	4.24	1.50
3	7.50	10.00	1.00	35.00	3.07	4.99	1.53	1.69	3.89	1.49
4	6.50	10.00	1.00	25.00	2.35	2.69	2.26	4.11	6.23	1.74
5	6.50	10.00	1.50	35.00	5.23	4.68	1.33	6.79	5.56	1.63
6	6.50	50.00	1.00	35.00	2.49	4.96	1.00	4.69	4.03	1.53
7	7.00	30.00	1.25	30.00	3.98	4.24	2.84	4.73	5.89	1.65
8	7.50	50.00	1.00	25.00	3.49	7.97	3.74	7.03	2.38	1.31
9	7.50	10.00	1.50	25.00	5.52	6.03	3.31	7.10	3.98	1.40
10	6.50	50.00	1.50	25.00	5.35	3.42	2.97	4.12	6.47	1.72
11	7.00	30.00	1.25	30.00	3.10	4.07	2.21	4.70	5.99	1.59

10.9.2 Second-Order Designs

Several designs are available to carry out the optimization stage, as described in Chap. 3. The most widely used symmetrical ones are the three-level full factorial, Box-Behnken, central composite, Doehlert matrix, and Taguchi designs. Optimal designs may meet the requirements of asymmetrical experimental domain when experimental restrictions are present, such as when certain combinations of pHs and temperature levels lead to the degradation of the API. On the other hand, mixture designs and mixture-process variables can be very useful to investigate the effect of the proportion of the ingredients in a mixture, as a sample diluent or the mobile phase in a chromatographic method.

In the case study, buffer pH and column oven temperature were selected as factors to drive the optimization stage, because they showed to be CMPs with significant effect on most of the CMAs, with their interaction effect being also significant to some responses. The buffer pH, hereafter named factor **A**, was investigated within the range of 6.50 to 7.50, whereas column oven temperature, hereafter named factor **B**, varied from 25.0 to 35.0 °C.

As for the buffer concentration, its effect was less significant for the overall CMAs; hence, we decided to fix a 30.0 mM concentration, since it is the central point of the evaluated range and, therefore, it provides a good degree of robustness. The flow rate was fixed at 1.50 mL/min based on its known favorable effect on analysis time reduction. Further increase of this parameter was considered to be inconvenient from an operational point of view, since very high pressures can be generated in the HPLC mode.

A central composite design (CCD) with coded distance for α equal to 1.41 was selected to achieve the model prediction isovariance criterion by rotation. Factor

Table 10.4 Experimental runs and results of the central composite design (CCD)

		CMPs		CMAs					
Run	Block	A	B (°C)	R_1	R_2	R_3	R_4	R_5	R_6
1	1	7.35	33.60	5.16	6.32	3.12	7.25	3.07	1.65
2	1	7.35	26.40	4.23	6.39	3.66	8.88	1.79	0.85
3	1	6.65	33.60	5.22	4.59	1.14	1.96	7.83	3.16
4	1	7.00	30.00	4.85	5.02	2.28	5.93	5.30	2.76
5	1	7.00	30.00	4.81	5.12	2.55	5.80	5.06	2.89
6	1	7.00	30.00	4.64	5.13	2.67	5.73	5.03	2.50
7	1	6.65	26.40	4.36	4.53	1.49	4.12	6.48	2.04
8	2	7.00	30.00	4.42	5.98	2.36	6.17	5.22	2.37
9	2	7.00	35.00	4.78	5.23	1.62	4.61	6.44	1.28
10	2	7.00	25.00	3.76	5.26	2.61	7.36	3.82	0.00
11	2	7.00	30.00	4.71	5.31	2.28	6.11	5.15	1.94
12	2	7.00	30.00	4.64	5.19	2.46	6.38	4.87	2.20
13	2	7.50	30.00	4.93	7.21	3.98	9.18	1.66	0.98
14	2	6.50	30.00	4.76	4.02	1.18	1.90	8.51	3.03

ranges were established in terms of α; therefore, no extreme experimental points were necessary. The experimental matrix was divided into two blocks; thus, it was possible to carry out the runs in two different days. Within each block, 3 replicates of the center points were included, resulting in a total of 14 experimental runs.

As seen in the experimental matrix shown in Table 10.4, the peak pair resolutions between $IMP_{1.2}$-API_1 and API_2-$IMP_{2.2}$ were, as expected, the least favored CMAs in this experimental design. Nevertheless, since all the CMAs attain acceptable values in some of the runs, it was possible to model and analyze the responses to find out an acceptable solution.

Second-order polynomial models were obtained by means of least squares regression technique using the software tools. In each model, the estimated coefficients were evaluated by ANOVA, and a backward regression procedure was applied to eliminate the non-significant terms (p-value > 0.10). Thus, parsimonious models including only the significant terms and those necessary to maintain the hierarchy were constructed and evaluated again by ANOVA to determine the model significance (p-value < 0.05) and the lack of fit (p-value > 0.05). Moreover, residuals were carefully analyzed to verify that ANOVA assumptions were met.

Later, the response surface plots for each of the fitted models were obtained and evaluated. In this example, an antagonistic behavior of the CMAs R_3 and R_6 is clearly observed (Fig. 10.4a); therefore, a compromise solution was required for these CMAs to define adequate operational conditions.

Other interesting approaches to risk assessment and reduction by using DoE in analytical method developments can be found in the literature [7, 14, 17, 18].

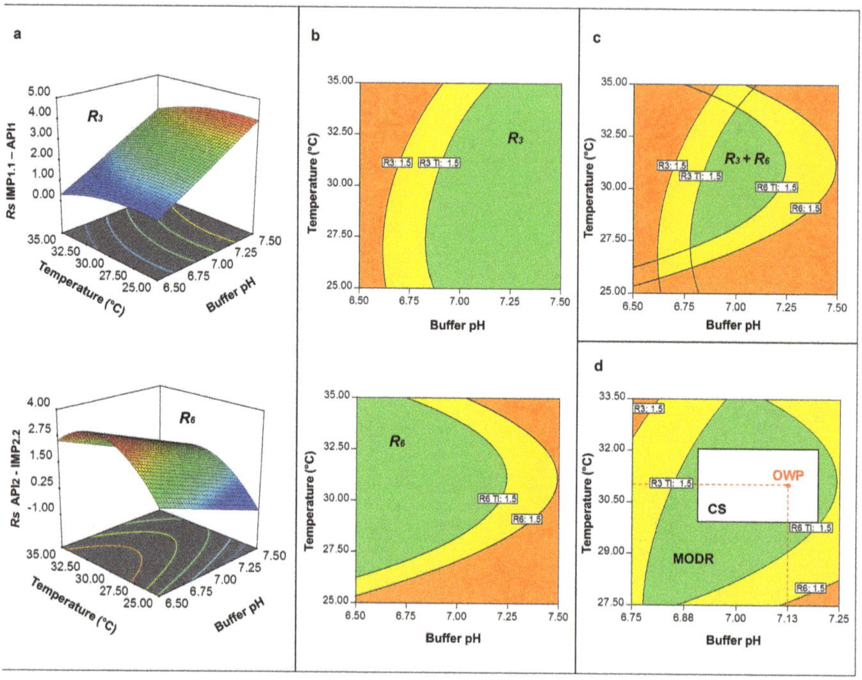

Fig. 10.4 (**a**) Response surface plots for R_3 and R_6 fitted models. (**b**) Contour plots. (**c**) Overlay plot showing the MODR (green area). (**d**) The CS is indicated with a white rectangle and the OWP is represented by a red point

10.10 Method Operable Design Region and Control Space

The MODR is a multidimensional combination of procedure parameters, where all combinations have been demonstrated to provide required performance and robustness with an acceptable level of confidence. In other words, it is a multivariate domain of the CMPs that ensures that all the CMAs achieve acceptance values included within predefined limits with proven assurance of quality [19].

Therefore, in AQbD concept, the analytical method is no longer defined by a single point in the space of its operating parameters, but as a range of operating conditions. Thus, the larger the MODR, the more robust is the method. The MODR is established using the adjusted second-order models built for the CMAs. It is usually defined graphically as a multi-response surface plot overlapping the contour plots of each CMA as a function of the CMPs. However, it cannot be defined using the overlay plot of the mean response surfaces; indeed, it only represents a region where the average response is observed. To provide assurance of quality in the MODR, the contour plots of the predicted responses should be constructed taking into account the model parameter uncertainties. Therefore, the confidence intervals (CI) of the responses, constructed with a certain degree of confidence (usually 95%), or, even better, the tolerance intervals (TI), which include the proportion of new measurements (usually 90–99%), are used to depict the overlapped plot defining the MODR in a consistent and robust manner.

To visualize the operability region properly, only the appropriate bound of the CI or TI should be depicted in the graph. So, if a maximum acceptable value for the CMA is set in the specification, then the upper bound of the interval will delimit the region where new measurements will be no larger than the acceptable value.

Moreover, the uncertainty bounds of the MODR can be validated by means of Monte Carlo simulations, Bayesian modeling, or bootstrapping. These are statistical tools that allow to estimate the probability of meeting the CMA specification inside the MODR, using the uncertainty parameters of the models.

One of the major benefits of building a MODR is to ensure the procedure robustness during development, instead of evaluating it in the validation stage, as traditionally done [20]. Another important aid of MODR is the fact that changes in method parameters can be safely made during the lifecycle, without the need for further validation tests. Thus, MODR offers greater flexibility for the method during its routine use.

In the hypothetical TS case study, the contour plots of the adjusted models, with the 95%-TI bound with a confidence level of 95%, were depicted to define the MODR, as shown for R_3 and R_6 in Fig. 10.4b. The contour plots show three different delimited areas. The orange areas represent the region where the CMA acceptance criterion has a high risk of not being met. The yellow areas correspond to the 95%-TI for $Rs = 1.5$, i.e., the region where 50–95% of new measurements of these CMAs will reach Rs values $\geq$1.5 with 95% of confidence. Finally, the green areas indicate the regions where the 95% of new measurements of the Rs will be $\geq$1.5 with 95% of confidence. The latter areas are then areas of reduced risk for the CMAs.

In Fig. 10.4c, the overlay plot shows the common region defined by the lowest tolerance bound of the responses, where both CMAs simultaneously met the predefined acceptance criteria. Note that particularly in this example, only overlapped contour plots of R_3 and R_6 are shown because they are the most critical attributes. For the other CMAs, the 95%-TI for $Rs = 1.5$ was far from the compromise zone, so it was not necessary to overlap them. However, in other situations, three, four, or even more contour plots may be overlapped to define the MODR.

Finally, to simplify operability, a *control space* (CS) inside the MODR was established by the following ranges, buffer pH 6.90–7.20 and column oven temperature 30.0–32.0 °C, as illustrated in Fig. 10.4d.

Another interesting strategy to define and validate the MODR can be found in [21].

10.11 Optimal Working Point

The OWP is a specific point within the CS that is selected as operating point for the validation stage and the routine use of the method. It can be chosen based on operative convenience or on statistical criteria such as the desirability value, as long as it is in a robust zone within the CS.

The maximization of the *Derringer and Suich desirability function* (D) allows us to find the experimental conditions to reach the best values of compromise for the CMAs within the CS. This is achieved by converting the multiple regression models into a single function that collects the analyst's expectations and priorities. Details about this powerful tool applied to analytical developments can be found in the literature [22].

In the current case study, the desirability function was constructed with the aim to maximize the CMAs and establish for R_3 and R_6 a lower boundary in $Rs = 1.5$. Thus, the OWP resulted for buffer pH = 7.13 and column oven temperature = 31.0 °C, with a D = 0.456 (Fig. 10.4d). The predicted responses for CMAs in the OWP were as follows: $R_1 = 4.80$, $R_2 = 5.75$, $R_3 = 2.76$, $R_4 = 6.60$, $R_5 = 4.37$, and $R_6 = 2.22$, all of them exceeding the required optimization criteria.

Other strategies to define the working point can be found in [8, 14, 23].

10.12 Experimental Check

Once the OWP is selected, some experimental tests under those operating conditions must be done to verify, before formal validation, whether the ATP requirements are met.

A first objective of this stage is to evaluate if the responses predicted for the OWP by the regression models are experimentally accomplished. Well-adjusted models and proper optimization criteria should ensure the success of this stage. Moreover, these studies are intended to provide supporting evidence for the absence of significant bias and to confirm that the precision of the reportable result is at an appropriate level. At this point, it may be necessary to define, review, or deepen controls for the CMPs as well as for calibration or replication strategies.

For the case study, Fig. 10.5 shows the chromatogram achieved under the OWP conditions. As can be seen, all analyte peaks have now baseline resolution, confirming the models' predicted values for the CMAs under this operating condition. Regarding the recoveries of the impurities in the spiked samples, appropriate results were obtained by using an external standard calibration strategy. Finally, triplicates of sample preparation provided adequate precision for the reportable results; therefore, these calibration and replication strategies were established for the final procedure.

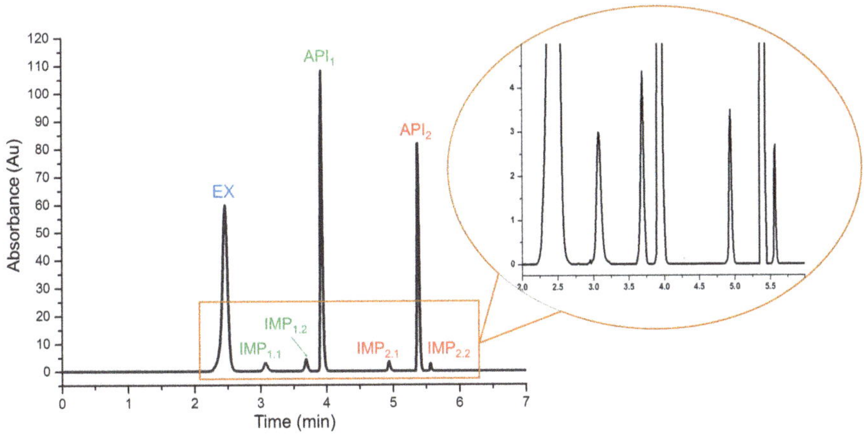

Fig. 10.5 Representative chromatogram of a spiked sample solution in the optimal working point

10.13 Method Validation

Once design activities are complete, the gathered knowledge is documented, preliminary controls are set, and analytical properties in the OWP are checked, the performance of the method can be formally qualified. The purpose of qualification is to confirm that the procedure meets the ATP criteria and that it will be appropriate for testing the pharmaceutical item in the environment where used routinely [24].

Normally, a draft of the standard operating procedure (SOP) that includes the preliminary control strategy is elaborated. Then, a validation study is conducted following these instructions. The main acceptance criteria, described in the ATP, are incorporated into the validation protocol by providing a detailed description of the experiments and statistical approach to be used to analyze the data in order to demonstrate requirement compliance.

As already stated, one of the greatest advantages of the AQbD strategy is that chances of failure to meet the validation criteria are minimized, since all the development activities were focused on the ATP. However, some minor modifications in procedure and/or in the control strategy may be still necessary through qualification results. These modifications can be made in a simple way and may be supported by the knowledge acquired during method development.

As a result of the validation stage, a final SOP, including the final control strategy, must be written, revised, and approved to be implemented in the routine work.

10.14 Control Strategy and Routine Monitoring

The control strategy is defined by a set of controls derived from the understanding of the procedure performance and is implemented with the objective of mitigating and monitoring the known risk to guarantee that the ATP requirements will be met.

Controls should be established on both CMPs and CMAs, so that while CMP controls are aimed at decreasing the probability of hazard occurrence, CMA controls are aimed at increasing the detectability of hazards. Together, they minimize the risk of harm to the quality of the reportable results.

Controls on CMPs must be designed with the objective to reduce the negative impact of parameters on CMA quality. For that, their optimal value and their acceptable ranges, established from risk assessment, should be explicitly stated in the final SOP. For example, in the case study, mobile phase buffer pH of 7.13 ± 0.05 and column oven temperature of 31.0 ± 0.1 °C were specified.

As already mentioned, CMPs that are not experimentally evaluated should be also controlled. For the pre-analytical stage, detailed instructions for the preparation of test solutions (e.g., standard and sample mass, uncertainty of mass measurement, glassware size and calibration control, solvent quality, sonication/extraction time/temperature) must be provided.

On the other hand, controls on CMAs are implemented by establishing SSTs and through quality control sample (QC) used to evaluate and ensure the required CMAs' performance during the routine use of the method. The function of SSTs and QC is to detect failures in method analytical properties before the reportable result is generated. Parameters such as injection precision, signal-to-noise ratio, *N*, *T*, and *Rs* are typically controlled in SSTs. In the case study, a minimum *Rs* of 1.5 between the critical peak pair ($IMP_{1.2}$- API_1) and (API_2- $IMP_{2.2}$) was an SST criterion.

If the controls show out-of-specification outcomes in the routine use of the method, the analysis must be stopped, and the procedure should be appropriately adjusted or modified to ensure an adequate performance. Thus, the risk of wrong results is minimized and controlled, leading to assurance of quality that is consistent with the method ATP during its whole lifecycle.

References

1. Valcárcel M (2012) Analytical chemistry today and tomorrow. In: Krull IS (ed) Analytical chemistry. Intech, London, pp 93–114
2. USP-NF (2022) Analytical procedure life cycle<1220>. https://www.usp.org. Accessed 7 Mar 2022
3. Technical Committee ISO/TC 69/SC 6 Measurement methods and results (1994) ISO 5725 accuracy (trueness and precision) of measurement methods and results – parts 1–6. International Organization for Standardization, Geneva

4. Eurachem A focus for analytical chemistry in Europe (2014) The fitness for purpose of analytical methods. A laboratory guide to method validation and related topics. https://www.eurachem.org. Accessed 9 Mar 2022
5. ICH International Council on Harmonization of Technical Requirements for Pharmaceuticals for Human Use (2005) Q2(R1) Validation of analytical procedures: text and methodology. https://www.ich.org. Accessed 10 Mar 2022
6. Jackson P, Borman P, Campa C, Chatfield M et al (2019) Using the analytical target profile to drive the analytical method lifecycle. Anal Chem 91:277–2585
7. Mitrović M, Protić A, Malenović A et al (2020) Analytical quality by design development of an ecologically acceptable enantioselective HPLC method for timolol maleate enantiomeric purity testing on ovomucoid chiral stationary phase. J Pharm Biomed Anal. https://doi.org/10.1016/j.jpba.2019.113034
8. Teng J, Zhu C, Lyu J et al (2022) Analytical lifecycle management (ALM) and analytical quality by design (AQbD) for analytical procedure development of related substances in tenofovir alafenamide fumarate tablets. J Pharm Biomed Anal. https://doi.org/10.1016/j.jpba.2021.114417
9. Dispas A, Avohou HT, Lebrun P et al (2018) 'Quality by design' approach for the analysis of impurities in pharmaceutical drug products and drug substances. Trends Anal Chem 101:24–33. https://doi.org/10.1016/j.trac.2017.10.028
10. Deidda R, Orlandini S, Hubert P et al (2018) Risk-based approach for method development in pharmaceutical quality control context: a critical review. J Pharm Biomed Anal 161:110–121. https://doi.org/10.1016/j.jpba.2018.07.050
11. Lionberger RA, Lee SL, Lee LM et al (2008) Quality by design: concepts for ANDAs. AAPS J 10:268–276. https://doi.org/10.1208/s12248-008-9026-7
12. ICH International Council on Harmonization of Technical Requirements for Pharmaceuticals for Human Use (2021) Q9(R1) quality risk management. https://www.ich.org. Accessed 10 Mar 2022
13. Kochling J, Wu W, Hua Y et al (2016) A platform analytical quality by design (AQbD) approach for multiple UHPLC-UV and UHPLC-MS methods development for protein analysis. J Pharm Biomed Anal 125:130–139. https://doi.org/10.1016/j.jpba.2016.03.031
14. Saini S, Sharma T, Patel A et al (2020) QbD-steered development and validation of an RP-HPLC method for quantification of ferulic acid: rational application of chemometric tools. J Chromatogr B Biomed Appl. https://doi.org/10.1016/j.jchromb.2020.122300
15. Kovacs E, Ermer J, McGregor PL et al (2016) Analytical control strategy. Pharmacopeial Forum 42
16. Myers RH, Montgomery DC, Anderson-Cook CM (eds) (2016) Response surface methodology: process and product optimization using designed experiments. Wiley, Hoboken
17. Moreira C, Lourenço FR (2020) Development and optimization of a stability-indicating chromatographic method for verapamil hydrochloride and its impurities in tablets using an analytical quality by design (AQbD) approach. Microchem J. https://doi.org/10.1016/j.microc.2020.104610
18. Pasquini B, Orlandini S, Furlanetto S et al (2020) Quality by design as a risk-based strategy in pharmaceutical analysis: development of a liquid chromatography-tandem mass spectrometry method for the determination of nintedanib and its impurities. J Chromatogr A. https://doi.org/10.1016/j.chroma.2019.460615
19. Rozet E, Lebrun P, Hubert P et al (2013) Design spaces for analytical methods. Trends Anal Chem 42:157–167. https://doi.org/10.1016/j.trac.2012.09.007
20. Breitkreitz MC (2021) Analytical quality by design. Braz J Anal Chem 8:1–5. https://doi.org/10.30744/brjac.2179-3425.editorial.mcbreitkreitz.N32

21. Abreu JC, Mahr AG, do Lago CL (2021) Stability-indicating method development for quantification of bromopride, its impurities, and degradation products by ultra-high performance liquid chromatography applying analytical quality by design principles. J Pharm Biomed Anal. https://doi.org/10.1016/j.jpba.2021.114306
22. Vera Candioti L, De Zan MM, Cámara MS et al (2014) Experimental design and multiple response optimization. Using the desirability function in analytical methods development. Talanta 124:123–138. https://doi.org/10.1016/j.talanta.2014.01.034
23. Deidda R, Avohou HT, Baronti R et al (2019) Analytical quality by design: development and control strategy for a LC method to evaluate the cannabinoids content in cannabis olive oil extracts. J Pharm Biomed Anal 166:326–335. https://doi.org/10.1016/j.jpba.2019.01.032
24. USP-NF (2016) Stimuli to the revision process: analytical target profile: structure and application throughout the analytical life life-cycle. https://www.usp.org. Accessed 15 Mar 2022

Chapter 11
Design of Experiments Applied to the Extraction of Active Compounds

Ieda Spacino Scarminio and Roy Edward Bruns

Abstract Design of experiments (DOE) strategies for the extraction of active compounds from diverse plant matrices can consist of process variable factorial designs, mixture variable designs, as well as process-mixture variable combinatorial designs. Linear, quadratic, and cubic mixture models are described in detail and equations given for combined process-mixture variable models. Analysis of variance (ANOVA) simultaneous component analysis (ASCA) for mixed process variable-mixture variable models are useful for extraction optimization that is monitored by multivariate spectral data. Principal component analysis (PCA) is often applied to spectral data matrices for exploratory investigations and classification problems. ASCA applications recommend performing PCA on effect matrices and can provide detailed information about spectral regions containing characteristic bands for the identification of extracted compounds. A five-solvent mixture design on metabolite yields extracted from *Annona muricata* L. leaves is described. Two different strategies for exploring the effect of solvent interactions on the mobile phase separation of metabolites in extracts of *Erythrina speciosa* Andrew leaves are summarized. A crossed simplex centroid design with mixture axial points to study the influence of different solvents on the extraction media and RP-HPLC mobile phase composition of green tea are discussed.

I. S. Scarminio (✉)
Laboratory of Chemometrics in Natural Sciences, Chemistry Department, State University of Londrina, Londrina, PR, Brazil
e-mail: ieda@uel.br

R. E. Bruns
Institute of Chemistry, University of Campinas (UNICAMP), Campinas, SP, Brazil
e-mail: bruns@unicamp.br

M. C. Breitkreitz, H. C. Goicoechea (eds.), *Introduction to Quality by Design in Pharmaceutical Manufacturing and Analytical Development*, AAPS Introductions in the Pharmaceutical Sciences 10, https://doi.org/10.1007/978-3-031-31505-3_11

11.1 Introduction

The great increase in the use of plant extracts with potential use in medicine, research, and industrial applications, as well as their economical impact, has provoked a renewed interest in the knowledge of the chemical compositions of different matrices of plant materials in order to develop new active ingredients for the food, cosmetic, and pharmaceutical industries. Plants produce a great variety of organic compounds that can be divided into primary and secondary metabolites based on their chemical structures. Primary metabolites are involved in processes of growth and development of the plants, while the secondary metabolites (active compounds) have features that can be expressed in terms of ecological, taxonomical, and biochemical differentiations [1]. These active compounds are often used in the preparation of dietary supplements, nutraceuticals, functional food ingredients, food additives, pharmaceuticals, and cosmetics [2]. They can be classified into three major groups, terpenoids, alkaloids, and phenolic compounds. In many cases, they are unique to individual plant species usually occurring in low concentrations [3] and are contained inside specific parts of the plants such as the roots, fruits, flowers, leaves, barks, and seeds.

There are two main types of methods used for active compound extraction: traditional or classical methods and modern or advanced methods. Figure 11.1

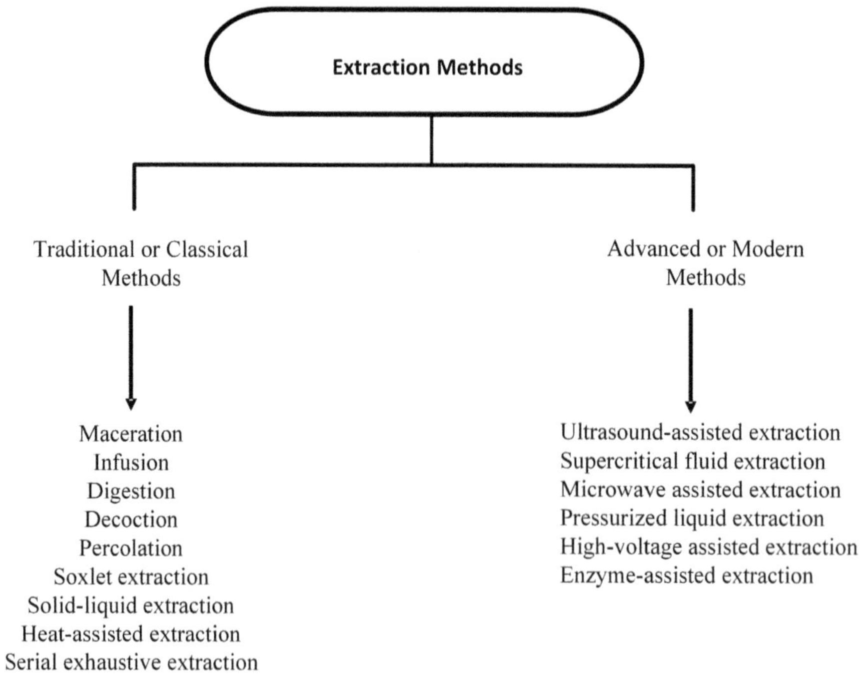

Fig. 11.1 Classification of the extraction methods

Table 11.1 Extraction methods and variables more frequently used in optimization

Extraction methods	Variables
Classical extraction methods	Solvent, extraction time, pH, temperature, stirring during extraction, liquid/solid ratio
Soxhlet extraction	Temperature, number of cycles
Supercritical fluid extraction	Temperature, extraction time, pressure, co-solvents, solvent flow rate
Pressurized liquid extraction	Temperature, extraction time, pressure
Ultrasound-assisted extraction	Solvent, pressure, temperature, extraction time, power, ultrasound frequency, amplitude, intensity
Microwave-assisted extraction	Solvent, temperature, extraction time, microwave power
Enzyme-assisted extraction	Solvent, temperature, time, enzyme

shows the methods commonly used for extraction. Moreover, modifications of these methods are continuously being developed as an alternative to conventional extraction methods to shorten extraction time, decrease solvent consumption, and increase extraction yield. These methods limit degradation, exposure to excess heat, light, and time during the extraction process. Therefore, appropriate extraction methods are needed to extract these active compounds because their functional groups and polarities differ. Combinations of extraction methods or several extraction steps may be needed to efficiently extract the chemical compositions of plant materials.

The use of different extraction methods or the same method under different experimental conditions results in extracts with different compositions and properties. Extract quality and the magnitude of qualitative and quantitative losses of active compounds of an extract are influenced by several variables like the extraction procedure, temperature and pressure used, time of extraction, plant material/solvent ratio, etc. The methods more frequently used and the main variables that influence the yields of extracts are shown in Table 11.1. Although the choice of the extraction method may have a significant effect on the quality of the extract, the solvent used provides the most obvious means of affecting the qualitative composition and the number of active compounds of the extract.

11.2 Choosing the Appropriate Solvent

The choice of appropriate solvent or solvent mixture is one of the most critical decisions for efficient extraction. However, selection of the appropriate solvent is not always easy because various properties of the solvent must be considered such as dipole-dipole interactions, π-electron interaction, and hydrogen bonding. Several classifications of organic solvents exist being divided into different classes as

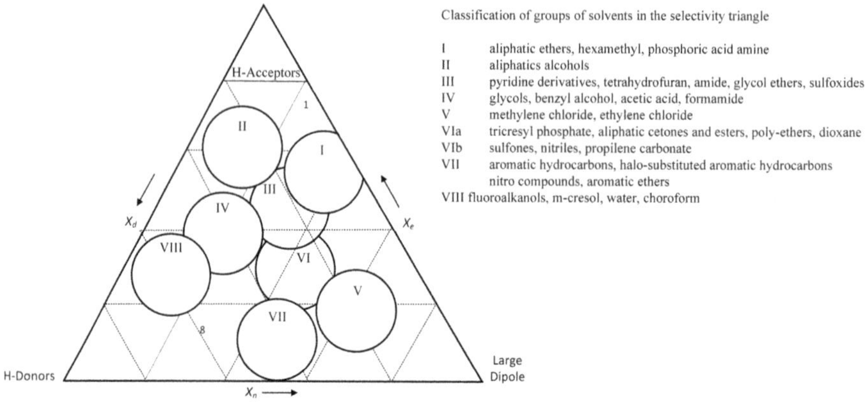

Fig. 11.2 Solvent selectivity triangle

follows [4]: (1) liquids capable of forming three-dimensional networks of strong hydrogen bonds; (2) liquids that have both active hydrogen atoms and donor atoms, but do not form three-dimensional networks; (3) liquids composed of molecules containing donor atoms, but no active hydrogen atoms; (4) liquids composed of molecules containing active hydrogen atoms but no donor atoms; and (5) liquids with no hydrogen bond forming capability and no donor atoms.

Snyder [5] proposed a selectivity triangle as an alternative for solvent choice (Fig. 11.2). In addition to the polarity index (P′), Snyder used different solutes to study three interactions, acidity measured with dioxane (*xd*), basicity measured with ethanol (*xe*), and dipolarity which was measured with nitromethane (*xn*). In this way, Snyder separated more than 80 solvents into 8 groups. The advantage of the Snyder scheme is the grouping of solvents according to the type of interaction in which they participate. Chemically similar solvents that have similar selectivity parameters (*xe*, *xd*, and *xn*) are part of the same group, although they may have different polarity values. To create the biggest change in selectivity, the solvents must be in diverse parts of the selectivity triangle.

In many extraction systems, the solvents are employed in substantially pure form as so-called neat solvents, although pure solvents can fail to make an optimal extraction. In this case, it is important to use solvent mixtures. Mixtures of two or three solvents from different groups shown in Fig. 11.2, in different proportions, allow a continuous variation of the solvent selectivity that can increase extraction efficiency.

In the literature, different solvent systems have been used for the extraction of active compounds. For example, methanol, acetone, and their water mixtures are most widely used for the extraction of phenolic compounds [6]. Other solvent mixtures with different polarities such as methanol, water, ethyl acetate, and petroleum ether have also been reported for phenolic compound extractions [7, 8]. On the other hand, alkaloids can be properly extracted with hydrocarbons moistened with water or water-rendered alkaline. Flavonoids and terpenes can be extracted at neutral pH with ethyl acetate or aliphatic ketones [9]. Chlorophylls are normally extracted with methanol, acetone, or other solvents miscible with water, while carotenoids can be extracted with acetone, methanol, or ethanol and solvents miscible with water. Non-polar solvents such as hexane are also used for the extraction of pigments [10, 11].

For classical extraction methods, solvent temperature is also an important variable, as higher temperatures can cause losses of active compounds by degradation and reduce the extraction yield. pH is also an important variable because it converts active acidic or basic compounds into non-ionized species, modifying the solubility in different solvents. Agitation during extraction is another important variable because it can promote mass transfer in the extraction process. Some of these variables are also crucial in advanced extraction methods.

The next step is to determine how many detectable components are present in the sample. To ensure analytical standardization of the plant extract, qualitative tests should not be limited to a single substance or to a few groups of substances. Qualitative determination based on a greater number of substances will make it possible to obtain the chemical composition of the extract more safely. Among the analysis methods, high-performance liquid chromatography (HPLC) has become an indispensable routine and effective method due to its ease in performing the separation, identification, and quantification of chemical species, by itself or in conjunction with other instrumental analysis techniques. In the chromatographic field, both the strength and the nature of the solute-solvent interactions must be considered jointly to permit an understanding of the overall solvent effect on the process studied.

In the field of liquid chromatography, there are a wide variety of solvents at our disposal for defining the mobile phase components and sample solvent. The selection of these solvents is generally determined by solvent characteristics related to detection, separation, flow resistance (viscosity), and miscibility. Once again, an alternative for choosing the best extracting solvent is the use of the solvent selectivity triangle proposed by Snyder [5]. This approach leads to a description of different solvents in terms of their polarities or chromatographic strengths and according to their selectivities or relative abilities to engage in hydrogen bonding or be affected by dipole interactions (proton donor (xd), proton acceptor (xe), and strong dipole (xn)) [12]. Even so in HPLC, the complete separation of substances from complex mixtures obtained from plant extracts presents several problems owing to the great variation of polarities, as well as the tendency to group substances with similar polarities.

In short, the quality of an extract, separation, identification, and quantification of chemical species is influenced by several factors such as solvent, extraction time, pH, temperature, stirring during extraction, liquid/solid ratio, co-solvents, solvent flow rate, and methods, among others. All these can be investigated simultaneously with statistical mixture analysis. In earlier times, the choice of the most appropriated solvent, time, temperature, etc. was purely empirical.

11.3 Experimentation Strategy

Let's assume that the researcher wants to obtain the highest yield of an active compound from a plant material. Looking at the methods and variables in Table 1, the following question arises. Which experimentation strategies should be used to get the largest yield of an active compound? In any experiment, the results and conclusions largely depend on how the data was collected.

The experimentation strategy is the general way of planning or conducting experiments. One of the strategies used is to vary the value of one factor or mixture variable one at a time, leaving the others constant. This strategy is not efficient when there are interactions between the solvent variables, that is, the optimized value of a factor or solvent variable depends simultaneously on the levels of the others being optimized. Experimental designs based on statistical principles are more efficient strategies, as they can extract the most useful information from the system under study, performing a minimum number of experiments. To obtain a successful outcome of an experiment, the following steps must be followed: well-defined objectives, choice of factor variables and, when applicable, choice of levels, selection of the response variables, choice of the appropriate experimental design, and statistical analysis of the data.

There are different experimental designs that can be applied to optimize one or more responses, including full factorial design, fractional factorial design, central composite design, and mixture design, among others. The use of one of these types will depend on the choice of method. If the researcher wants to investigate how the yield of an active compound is influenced by increasing temperature, pressure, pH, etc., factorial designs are the most suitable (Fig. 11.3a), while if the objective is to investigate how the different proportions of solvent influence the yield of an active compound, mixture designs are the most efficient strategy for applications (Fig. 11.3b). When process and mixture variables are included in the design, the interactions between these variables can also be investigated by combining the two experimental designs (Fig. 11.4a–c).

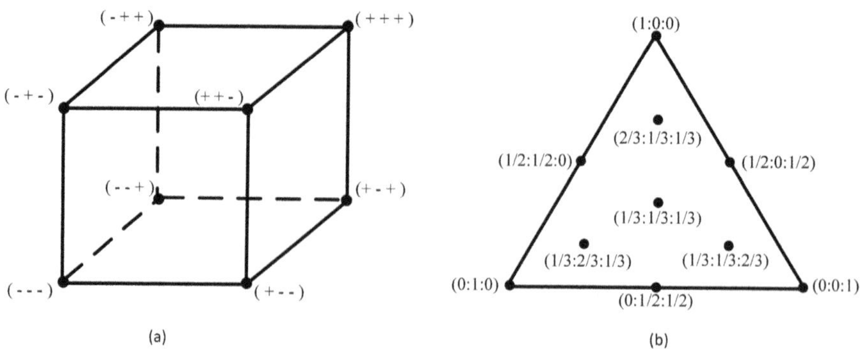

Fig. 11.3 (**a**) Three factors investigated at two levels indicated by the + and – signs. (**b**) A simplex centroid mixture design augmented with three axial points

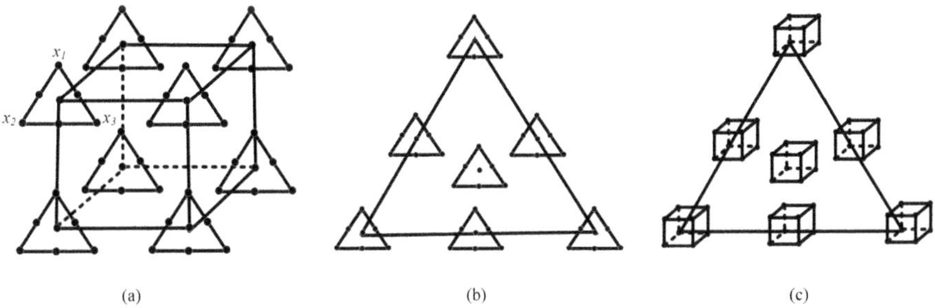

Fig. 11.4 (**a**) Three mixture component simplex centroid designs embedded in a 3^2 factorial design. (**b**) Seven simplex centroid designs embedded in another three-component simplex centroid design. (**c**) Seven 2^3 factorial designs embedded in a simplex centroid design

11.4 Mixture Models

Mixture models [13, 14] differ from the models discussed in the preceding chapters in that their results do not depend on the absolute values of the factors being optimized but on their *proportions*. For a mixture with three components, this can be expressed as

$$x_1 + x_2 + x_3 = 1 \tag{11.1}$$

where x_i represents the proportion of the ith mixture component. As such, there are only two independent variables. Knowing the percentages of two components permits determining the percentage of the remaining component. As these component variables are not independent, ordinary least squares regressions cannot be carried out. The singularity can be removed using the so-called Scheffé models that are derived from the equations of models for process variables. The expected response value, $\widehat{y}$, for a linear model for three process variables can be expressed as

$$\widehat{y} = b_0 + b_1 z_1 + b_2 z_2 + b_3 z_3 \tag{11.2}$$

where the letter z is used to represent the levels of the process variables. Expressing Eq. (11.2) in terms of mixture variables and knowing that the sum of the proportions of the three components is one, Eq. 11.1 can be used as a multiplier of the constant term, b_0, without changing the equality.

$$\widehat{y} = b_0(x_1 + x_2 + x_3) + b_1 x_1 + b_2 x_2 + b_3 x_3 \tag{11.3}$$

Rearranging and consolidating constants, one has an additive mixture model

$$\widehat{y} = (b_0 + b_1)x_1 + (b_0 + b_2)x_2 + (b_0 + b_3)x_3 = b_1^* x_1 + b_2^* x_2 + b_3^* x_3 \tag{11.4}$$

for which $b_i^* = b_0 + b_i, i = 1, 2, 3$. The three starred constants are the response values of their pure mixture components. An additive mixture model can be determined without actually mixing the ingredients, but this is not recommended.

Figure 11.3a shows a three-process variable factorial design. Performing experiments at each vertex, representing different combinations of factor levels, allows the determination of not only a linear model but also higher-order interaction effects.

Most process factors or mixed solvents exhibit synergic or antagonistic effects. This can most easily be taken into consideration using a quadratic model. For process variables

$$\widehat{y} = b_0 + b_1z_1 + b_2z_2 + b_3z_3 + b_{11}z_1^2 + b_{22}z_2^2 + b_{33}z_3^2 + b_{12}z_1z_2 + b_{13}z_1z_3 + b_{23}z_2z_3. \quad (11.5)$$

These models demand experimentation at the vertices of the factorial design as well as those of a star design. Again, to treat solvent proportions, the relation for mixture proportions can be introduced into the above process variable equation as given below:

$$\widehat{y} = b_0(x_1 + x_2 + x_3) + b_1x_1 + b_2x_2 + b_3x_3 + b_{11}x_1(1 - x_2 - x_3) + b_{22}x_2(1 - x_1 - x_3) + b_{33}x_3(1 - x_1 - x_2) + b_{12}x_1x_2 + b_{13}x_1x_3 + b_{23}x_2x_3.$$

On rearrangement and consolidation of constants as performed for the linear model, one has

$$\widehat{y} = b_1^*x_1 + b_2^*x_2 + b_3^*x_3 + b_{12}^*x_1x_2 + b_{13}^*x_1x_3 + b_{23}^*x_2x_3 \quad (11.6)$$

where $b_i^* = b_0 + b_i + b_{ii}$ and $b_{ij}^* = b_{ij} - b_{ii} - b_{jj}$, with $i \neq j$.

The first three terms of this quadratic model represent additive mixture effects, whereas the last three terms correspond to synergic or antagonistic interactions between the mixture components. Positive values of b_{ij}^* indicate synergic effects and negative ones antagonistic interactions. Mixture response surfaces can be represented as contour lines within a triangle representing the mixture proportions as will be seen later in solvent extraction applications. To determine the model coefficients, a minimum of six experiments must be performed with at least three pure solvents and three 1:1 binary mixtures being recommended. Figure 11.3b shows points at the vertices and the centers of the sides of the triangle. This design is called a simplex lattice. Replicates and points in the interior of the triangle are necessary to determine lack of fit and experimental error for model validation.

Cubic models have also been used to describe mixture response surfaces and are given by the equation

$$\widehat{y} = b_1x_1 + b_2x_2 + b_3x_3 + b_{12}x_1x_2 + b_{13}x_1x_3 + b_{23}x_2x_3 + d_{12}x_1x_2(x_1 - x_2) + d_{13}x_1x_3(x_1 - x_3) + d_{23}x_2x_3(x_2 - x_3) + b_{123}x_1x_2x_3 \quad (11.7)$$

The first six terms are the same as those of the quadratic mixture model above. The last term represents possible synergic or antagonistic three-component interactions. The three terms with the d_{ij}^{*} coefficients describe two-component curvature effects that do not have simple interpretations. In practice, for extractions with solvent mixtures, Eq. 11.7 is simplified by neglecting these two-component curvature effects. This results in the special cubic model that is used in many solvent extraction applications

$$\widehat{y} = b_1^* x_1 + b_2^* x_2 + b_3^* x_3 + b_{12}^* x_1 x_2 + b_{13}^* x_1 x_3 + b_{23}^* x_2 x_3 + b_{123}^* x_1 x_2 x_3. \qquad (11.8)$$

Mixture designs are used to determine the coefficients in the above models. A minimum design to determine the quadratic model involving three solvents consists of performing experiments using the three pure solvents and three (1:1) binary mixtures from each combination of solvents. To determine the special cubic model, one must add a (1:1:1) ternary mixture of the three solvents. However, these minimum designs do not allow validation of the models as the number of mixture experiments is equal to the number of regression constants to be determined. For a three-component system, one can add three ternary solvent mixtures with proportions of (2/3, 1/6, 1/6), (1/6, 2/3, 1/6), and (1/6, 1/6, 2/3) and a center point while also performing replicate determinations for the (1:1:1) ternary mixture. This augmented simplex centroid design is shown in Fig. 11.3b. This allows the determination of the variance explained by the model using seven degrees of freedom with three degrees of freedom remaining for the quantification of the residual variance. The residual variance is a sum of lack of fit and experimental error variance, often called pure error variance. Replicates of the center point are recommended to measure pure error. A four-component simplex lattice design requires ten mixtures, four pure solvents, and six 1:1 binary mixtures. The simplex centroid requires the addition of four 1:1:1 ternary and a mixture with each component at 25%. The simplex centroid design can be represented by a tetrahedron with points at the four vertices, six points at the centers of the edges, four points in the centers of the triangles forming the sides, and one point at the center of the tetrahedron. It should be noted that the vertices of these designs do not necessarily have to represent pure solvents. They can be mixtures as it is possible to model mixtures of mixtures.

The recommended strategy for solvent extractions is to test the linear, quadratic, and special cubic models successively determining the most precise model using the same ANOVA as used for process variable models. The ANOVA table contains the sums of squares and variances explained by the model and the residuals that are split into two variances, lack of fit and pure error. All these variances permit model validation.

Mixture models are not restricted to investigations for which all the ingredients vary from 0% to 100%. For these cases, pseudocomponents can be used to study subspaces, triangular or irregular, within the experimental mixture domain.

Pseudocomponents are used when it is desired to study problems for which the mixture components have lower limits above 0% and upper limits less than 100%. For example, a study might be made varying solvents between 20% and 80%. However, solvent extractions are conveniently performed with all ingredients varying between 0 and 100%. This helps the researcher compare results of pure solvents with those of mixtures permitting assessments of binary and ternary mixture effects relative to those for pure solvents. Most of the applications discussed later in this chapter will be of this type.

Solvent extraction optimization normally depends on both the levels of the process variables and the mixture proportions. These studies can be carried out combining process and mixture models. For simplicity, we show how these models can be constructed for two process variables and two mixture proportions. One combined model can be obtained multiplying process and mixture models. For a linear model for two process variables and quadratic mixture model for two solvents, we have

$$\hat{y} = (b_0 + b_1 z_1 + b_2 z_2)\left(b_1^* x_1 + b_2^* x_2 + b_{12}^* x_1 x_2\right) \tag{11.9}$$

$$y = c_1 x_1 + c_2 x_2 + c_{12} x_1 x_2 + d_{11} z_1 x_1 + d_{12} z_1 x_2 + d_{112} z_1 x_1 x_2 + d_{21} z_2 x_1 + d_{22} z_2 x_2 + d_{212} z_2 x_1 x_2. \tag{11.10}$$

The model coefficients can be determined performing three 2^2 factorial designs one for each pure solvent and one for their binary 50%:50% mixture. It is always recommended that replicate points be performed at least at the center of the factorial design. In this way, the quantity of variance described by the model and variances for lack of fit and pure error can be determined for model validation. The number of model terms increases very rapidly as the number of variables and model orders increase, for example, a three-process variable - quadratic model (ten terms), Eq. (11.5), combined with a three solvent – quadratic mixture model (six terms), Eq. (11.6), would involve a minimum of 60 experiments. However, replicate experiments need to be performed for model validation. However, replicates increase the number of experiments, and this may not be practical from an operational point of view. As an alternative, cumulative normal probability plots can be used to obtain an estimate of the significant effects.

The diagram on the left of Fig. 11.4a presents a combined design for a three-process variable factorial and 3 mixture components with 56 design points. The one on the right (Fig. 11.4c) shows an alternative to this design. The center diagram (Fig. 11.4b) represents the combination of two mixture solvent models.

As these combined designs contain many experiments, complete randomization is not practical to carry out. Experimental work can be eased using a split-plot approach [15]. For the figure on the left, randomization is carried out only for the solvent points within each triangle with the center point located at a vertex. The order of treating the vertices is also performed randomly. But split-plot methods have a cost in that the error analysis is much more complicated than if all experiments are performed randomly. If error analysis is carried out as for a completely randomized data set, experimental error could be underestimated. Cumulative probability plots can also be used for obtaining approximate estimates of significant effects. Split-plots have been useful in chromatographic studies of solvent extractions carried out under different process variable conditions with different mobile phases. Diagrams like those in the figure above can be used in this kind of study.

Chromatographic analysis of solvent extractions could demand the combination of two mixture models, one for the extraction and the other for the chromatographic mobile phase employing the center solvent-solvent design in Fig. 11.4b. If both are three-solvent quadratic models, their multiplication necessitates the performance of at least 36 experiments.

Many applications in chemistry involve the optimization of more than one response. In these situations, a separate mixture model can be determined for each response. Simultaneous optimization of more than one response can be carried out using the method of Derringer and Suich [16]. However, fingerprinting methods for natural products and metabolomics demand examination of an entire spectra or chromatogram that consists of hundreds or even thousands of response values necessitating a reduction in the number of variables to be carried out while losing a minimum of useful information. Principal component analysis (PCA) is the most widely used method in chemometrics for variable reduction. ASCA (ANOVA simultaneous component analysis) combines analysis of variance (ANOVA) , factorial and mixture designs, and PCA that can be applied to data sets with thousands of variables to quantify and determine the significance of design variable effects [17, 18].

11.5 ANOVA Simultaneous Component Analysis

ANOVA [19] is a widely used method for univariate analysis and determines and quantifies the effects of each factor and interaction effects involving more than one factor. Our discussion will not include the statistical foundation of ANOVA including the basic assumptions but will focus on how to do the calculations. The reader should consult references providing an in-depth treatment of ANOVA.

For an ANOVA model with a single factor investigated at two levels, the model equation is

$$x_i = \mu + \alpha + \epsilon_i \tag{11.11}$$

where x_i is the result of the ith experiment, μ is the populational value of the average of the experimental results, and α is the populational value of the difference of the average results for the factorial level at which the experiment was performed relative to the total average. The populational residual or error is given by ϵ_i that is assumed to follow a normal distribution. The estimated values of these parameters are determined by the equation

$$x_{ti} = \bar{x} + (\bar{x}_t - \bar{x}) + \quad (x_{ti} - \bar{x}_t) \tag{11.12}$$

where x_{ti} stands for the result of the ith experiment performed at level t, $\bar{x}_t$ is the average of the experiments executed at level t, and $\bar{x}$ is the mean of all experimental results.

One can express Eq. 11.12 in the form of vectors

$$\boldsymbol{X} = \mathbf{1}\boldsymbol{m} + \boldsymbol{X}_{\mathbf{effect}} + \boldsymbol{\epsilon} \tag{11.13}$$

where $\boldsymbol{X}$ is the vector of a number of experimental results, $\mathbf{1}$ is a column vector of ones, $\boldsymbol{m}$ is a row vector with all elements being the average of all results, $\boldsymbol{X}_{\mathbf{effect}}$ contains differences between the factorial level average and the total average, and $\boldsymbol{\epsilon}$ contains the residual values. These vectors are given below for simulated results of reaction yield for one factor of a two-level design. This will help understand how ASCA works for spectra or chromatograms with many variables as each one is treated the same way.

Each column below contains elements of the vectors in Eq. (11.13) that have been obtained from a factorial design.

$$\begin{matrix} \text{Level} & & \text{Yields} & & \text{Average} & & \text{Effect} & & \text{Residual} \\ \begin{bmatrix} A \\ B \\ A \\ B \\ A \\ B \\ A \\ B \end{bmatrix} & ; & \begin{bmatrix} 29.80 \\ 84.25 \\ 20.70 \\ 96.40 \\ 18.54 \\ 95.73 \\ 23.75 \\ 85.45 \end{bmatrix} & = & \begin{bmatrix} 56.83 \\ 56.83 \\ 56.83 \\ 56.83 \\ 56.83 \\ 56.83 \\ 56.83 \\ 56.83 \end{bmatrix} & + & \begin{bmatrix} -33.63 \\ 33.63 \\ -33.63 \\ 33.63 \\ -33.63 \\ 33.63 \\ -33.63 \\ 33.63 \end{bmatrix} & + & \begin{bmatrix} 6.60 \\ -6.21 \\ -2.50 \\ 5.94 \\ -4.66 \\ 5.27 \\ 0.55 \\ -5.01 \end{bmatrix} \end{matrix}$$

The first column identifies the level of each experiment, A or B, and the remaining columns present the experimental yields, the total average, the effect values, and the residuals.

As the vectors are mutually orthogonal, as can be verified for the columns above, we can write an equation analogous to Eq. (11.13) for the sum of squares of the elements in each column

$$||\boldsymbol{X}||^2 = ||\mathbf{1}\boldsymbol{m}||^2 + ||X_{\text{effect}}||^2 + ||\boldsymbol{\epsilon}||^2. \tag{11.14}$$

The averages of the A and B factor levels, the total average, and the sum of squares of each column are $\bar{x}_A = 23.26$, $\bar{x}_B = 90.4$, $\bar{x} = 56.83$, $||\boldsymbol{X}||^2 = 35{,}081$, $||\mathbf{1}\,\boldsymbol{m}'||^2 = 25{,}837$, $||\boldsymbol{X}_{\mathbf{effect}}||^2 = 9048$, and $||\,\boldsymbol{\epsilon}\,||^2 = 199$.

Assuming a normal distribution, an F test can be performed where the effect has one and the residuals have six degrees of freedom.

$$F = \frac{9048/1}{199/6} = 272.9 \; F_{1,6,0.05} = 5.99$$

As the ratio of the sums of squares has a value more than the F value of a normal distribution, the treatment effect is significant well above the 95% confidence level. The p significance value is 3.14×10^{-6}.

A randomization test that does not depend on whether the data are normally distributed can also be made. This is accomplished randomizing the design results.

The values in the second column below have been randomized independent of the levels at which the experiments were performed in the above application.

$$\begin{matrix} \text{Level} & & \text{Yields} & & \text{Average} & & \text{Effect} & & \text{Residual} \\ \begin{bmatrix} A \\ B \\ A \\ B \\ A \\ B \\ A \\ B \end{bmatrix} & ; & \begin{bmatrix} 84.25 \\ 18.54 \\ 96.40 \\ 95.73 \\ 29.80 \\ 23.75 \\ 85.45 \\ 20.70 \end{bmatrix} & = & \begin{bmatrix} 56.83 \\ 56.83 \\ 56.83 \\ 56.83 \\ 56.83 \\ 56.83 \\ 56.83 \\ 56.83 \end{bmatrix} & + & \begin{bmatrix} 17.15 \\ -17.15 \\ 17.15 \\ -17.15 \\ 17.15 \\ -17.15 \\ 17.15 \\ -17.15 \end{bmatrix} & + & \begin{bmatrix} 10.27 \\ -21.14 \\ 22.42 \\ 56.05 \\ -44.18 \\ -15.93 \\ 11.47 \\ -18.98 \end{bmatrix} \end{matrix}$$

The values necessary to perform an F test are $\bar{x}_A = 73.98$, $\bar{x}_B = 39.68$, $\bar{x} = 56.83$, $\|\boldsymbol{X}\|^2 = 35{,}081$, $\|\mathbf{1}\,\boldsymbol{m}\|^2 = 25{,}837$, $\|\boldsymbol{X}_{\mathbf{effect}}\|^2 = 2353$, and $\|\boldsymbol{\epsilon}\|^2 = 6894$.

The F test ratio is $\frac{2353/1}{6894/6} = 2.05$ that is much less than the tabled F value above with a p-value of 0.726, clearly not significant at the 95% confidence level.

Thousands of randomly permuted sets of experimental values can be carried out. Smilde and coworkers provide an application for which 100,000 randomizations were carried out [18]. If the sum of squares of the estimated treatment effect from the original data is larger than those of many permuted values, the treatment effect is significant. If the original treatment effect sum of squares is larger than those of all the permuted values, the p-value is smaller than 1 divided by the number of randomizations. The concepts behind randomization test for significance are discussed in the classic book by Box, Hunter, and Hunter [19]. If appropriate reference sets are used, assumptions about data normality and independence (data are not autocorrelated) are not necessary.

11.6 Principal Component Analysis

Spectral and chromatographic responses can be represented digitally with hundreds or more variables. It is not possible to perform ANOVA as described above for each variable. For this reason, data reduction methods must be employed. PCA [20] is the most widespread method for doing this. The data matrix, $\boldsymbol{X}$, has dimensions (N × M) where N indicates the number of spectra and M the total number of variables. Each variable contains digitalized signal values such as transmittance or absorbance at a particular wavelength or frequency. The first principal component is along the direction explaining the largest data variance in the M-dimensional space. The second component is perpendicular to the first one and explains the second largest variance in the data. Additional principal components can be found in the same way. Ideally, this process is continued until all useful information in the data matrix has been explained. Graphs of the accumulated quantities of variance explained as a function of the number of principal components are often used to estimate the number of principal components containing the most useful information for analysis. The remaining components are expected to contain mostly signal noise. Only a few significant principal components are subjected to ANOVA. The equation for a two-principal component model is

$$\boldsymbol{X} = \boldsymbol{T}_1\boldsymbol{P}_1^t + \boldsymbol{T}_2\boldsymbol{P}_2^t + \boldsymbol{\epsilon} \quad (11.15)$$

with $\boldsymbol{T}$ being the (N × 2) score matrix and $\boldsymbol{P}^t$ the (2 × M) loading matrix. Analysis with more than two principal components will have more $\boldsymbol{TP}'$ terms.

An ANOVA model for two principal components can be expressed as

$$x_{cdij} = \mu_j + \alpha_{cj} + \beta_{dj} + (\alpha\beta)_{cdj} + \varepsilon_{cdij} \quad (11.16)$$

where c and d are the levels of the two factors, i the replicate number and the variable j. Here, α and β refer to the first two principal components and $(\alpha\beta)$ refers to the interaction between them. The corresponding matrix equation is

$$\boldsymbol{X} = \mathbf{1}\boldsymbol{m}' + \boldsymbol{X}_a + \boldsymbol{X}_b + \boldsymbol{X}_{ab} + \boldsymbol{\epsilon}. \quad (11.17)$$

Each column of the $\mathbf{1}\boldsymbol{m}'$ matrix contains the averages of all the measurements obtained for the j variable (frequency or retention time as examples). The $\boldsymbol{X}_a$ and $\boldsymbol{X}_b$ contain estimates of the effect values of the scores of the first two principal components. The columns of these matrices have the same structure as the data in the ASCA discussion. The $\boldsymbol{X}_{ab}$ matrix contains interaction effects and $\boldsymbol{\epsilon}$ the residuals. As in the one-variable ANOVA discussion, the analogous sums of square matrices can be written as

$$\|\boldsymbol{X}\|^2 = \|\mathbf{1}\boldsymbol{m}\|^2 + \|\boldsymbol{X}_a\|^2 + \|\boldsymbol{X}_b\|^2 + \|\boldsymbol{X}_{ab}\|^2 + \|\boldsymbol{\varepsilon}\|^2. \quad (11.18)$$

The corresponding sums of squares of the matrices in this equation can be subjected to F and randomization tests for model validation for both the original and permuted data sets as carried out for the one-variable ANOVA.

PCA being a variable reduction method often permits the projection of higher-dimensional space into two- or three-dimensional space in which discrimination between different classes of samples can be visualized. Usually, the degree of separation is estimated by visual inspection of score graphs. This is a highly subjective procedure. ASCA can be used for separation problems even if factorial designs are not used to generate the experimental data. A separation of two classes can follow a procedure like that in the ASCA discussion. ANOVA and PCA can be applied to problems with more than two classes. The ASCA method provides an objective way to determine the statistical significance of class separation. There are freeware [21] and commercial computer programs [22] readily available for executing the ASCA calculations.

11.7 Application Section

Scarminio and collaborators [23] performed a factorial design fingerprint discrimination of *Coffea arabica* beans under elevated atmospheric carbon dioxide and limited water conditions involving a 2^2 factorial design and a simplex centroid mixture design with the PCA and ASCA methods. The levels of the two factors were an elevated CO_2 abundance between 540 and 590 $\mu L\ L^{-1}$ and current CO_2 atmospheric concentration of about 390 $\mu L\ L^{-1}$ for irrigated and unirrigated conditions. First, a statistical mixture design was investigated with four solvents: ethanol, dichloromethane, hexane, and ethyl ether. Four pure solvents and six 1:1 binary, three 1:1:1 ternary, and three replicated mixtures of all four solvents in equal parts at the center of the simplex centroid design were used. The 17-mixture design extraction procedure was carried out on coffee beans collected from high, low, and intermediate tree heights for each point of the factorial design. UV spectra between 190 and 400 nm were obtained for the extract fingerprints. PCA resulted in the selection of pure ethanol and a 1:1 ethanol-dichloromethane mixture as the best solvents to discriminate beans from enriched and normal CO_2 environments. This choice was based on visual inspection of PCA score graphs and a statistical t-test carried out on the factor scores of each class.

Both ethanol and ethanol-dichloromethane extracts provided similar results. No PCA score plot showed discrimination between elevated and current CO_2 levels for irrigated or unirrigated trees although partial separations of clusters could be observed. The loading plots indicated changes in CGA and kahweol abundances with changes in the factor levels. Unirrigated trees under current CO_2 level beans had lower CGA and kahweol abundances than those for CO_2 at the high level. The opposite behavior was found for irrigated trees. This suggests an interaction effect occurs between these factors.

ASCA calculations were made on both pure ethanol and ethanol-dichloromethane data sets. Applying permutation tests to the ethanol extracts resulted in p-values of 0.263 and 0.379 for CO_2 and water separations. The values for the binary solution were 0.074 and 0.11 values indicating that this solution results in a better class discrimination than pure ethanol. Although ASCA results led to the conclusion that no main effect was significant at the 95% confidence level for both solvent extracts, a significant interaction effect between water availability and CO_2 conditions was found at the 99% confidence level. The existence of this effect would not have been discovered by visual inspection of PCA score plots and performing t-tests on the factor scores of each class.

A mixture design was proposed by Scarminio and collaborators [3] to investigate the effects of five extraction solvents and their mixtures on the yields of metabolites in crude and fractioned extracts of *Annona muricata* L. leaves. Extraction media were prepared using simplex centroid mixtures of ethanol, ethyl acetate, dichloromethane, acetone, and chloroform. Solvent selection was made considering the solvent selectivity triangle, where solvents from different groups in the triangle have different selectivity characteristics. Solvent proportions used in the extraction resulted in 33 extractions consisting of 31 different mixtures and 2 extra replicate runs that were performed at the center point. These extracts were fractionated by liquid-liquid extraction resulting in fibers and four more fractions, neutral (fat and grease), organic (terpenoids and phenolic compounds), basic (alkaloids), and polar (quaternary alkaloids and n-oxide). Except for the polar fraction, gravimetric measurements were performed on all the other fractions. On comparing the crude extract yields, the results show that ethanol-dichloromethane, ethanol-chloroform, and ternary ethanol-dichloromethane-chloroform mixtures are more efficient extractors, while ethanol-ethyl acetate-dichloromethane, ethanol-acetone-chloroform, ethanol-ethyl acetate-dichloromethane-chloroform, and quinary mixtures are more efficient extractors for the neutral fraction. For organic fractions, pure ethanol and the ethanol-dichloromethane-acetone mixture are the most efficient extractors. Quaternary mixtures of ethanol-ethyl acetate-acetone-chloroform, ethanol-ethyl acetate-dichloromethane-chloroform, and quinary mixtures significantly affect the yields of the basic fraction. Ethyl acetate-dichloromethane, dichloromethane-chloroform, ethyl acetate-dichloromethane-acetone-chloroform, and ethyl acetate-acetone-chloroform mixtures are more efficient extractors of the fiber fraction. Ethanol-dichloromethane, ethanol-chloroform, ethanol-dichloromethane-acetone, and ethanol-dichloromethane-chloroform mixtures are recommended for residual material. The results indicated that the mobile phase composed of methanol/acetonitrile/water (26:27:47 v/v/v) was more suitable for the analysis of basic fraction at 254 nm, whereas the mobile phase composed of methanol/acetonitrile/water (35:35:30 v/v/v) was more adequate for the organic fraction analysis at this same wavelength. The results indicated that the number of peaks and the chromatographic profiles were significantly affected by the mobile phase strength and analysis wavelength.

Two different strategies with the aim of exploring the effect of solvent interactions of mobile phase separation metabolites in extracts from *Erythrina speciosa* Andrews leaves were investigated by Scarminio and collaborators [1]. These were applied to open column chromatography. In the first procedure, 29 mixtures were prepared for the statistical mixture design involving the 4 solvents. These were used to investigate the influence of solvent proportions on chromatographic fractionation. The second strategy employed these same four solvents but only as pure solvents and some of their binary mixtures. The eluting solutions were passed through the column in increasing polarity order, as done in the classical chemical procedure for chromatographic fractionation. Principal component analysis was used to compare the

UV spectra obtained from RP-HPLC-DAD and to estimate the number of independent factors contained in the chromatographic data of the extracts. When pure solvents are used, larger groups of similar spectra are observed in the factor analysis score graphs indicating the same or a limited number of metabolite classes. In contrast, solvent mixtures produced score graphs with a larger number of clusters indicating greater metabolic diversity. Besides resulting in more peaks than the pure solvents, the chromatographic data of the design mixtures resulted in larger numbers of significant principal components confirming the greater chemical diversity of their extracts. With these results, it was possible to determine that solvent mixture play an important role in separation. In fact, greater selectivity and more diverse classes of metabolites can be seen when the mobile phases are mixtures of two or more solvents having synergistic effects that do not occur in their pure solvents. Mixture designs are very convenient to study these effects since they contain ternary mixtures that are absent in classical separation techniques. Thus, if the objective of an analysis is to obtain metabolites of the same class, one should use pure solvents; otherwise, binary or ternary mixtures are more efficient for studying class diversity.

Green tea has been frequently studied by chemists and others owing to its important antioxidant and beneficial effects. Scarminio and Almeida [9] studied the influence of different solvents on the extraction medium and RP-HPLC mobile phase composition. Statistical mixture designs were used to optimize solvent proportions and to prepare fingerprints of this medicinal herbal extract. A crossed simplex centroid design with mixture axial points, as illustrated in the center in Fig. 11.5, was used in which the mobile phase designs were embedded within the extraction solvent simplex centroid. Complete randomization for carrying out laboratory procedures on so many experiments was not practical, suggesting use of a

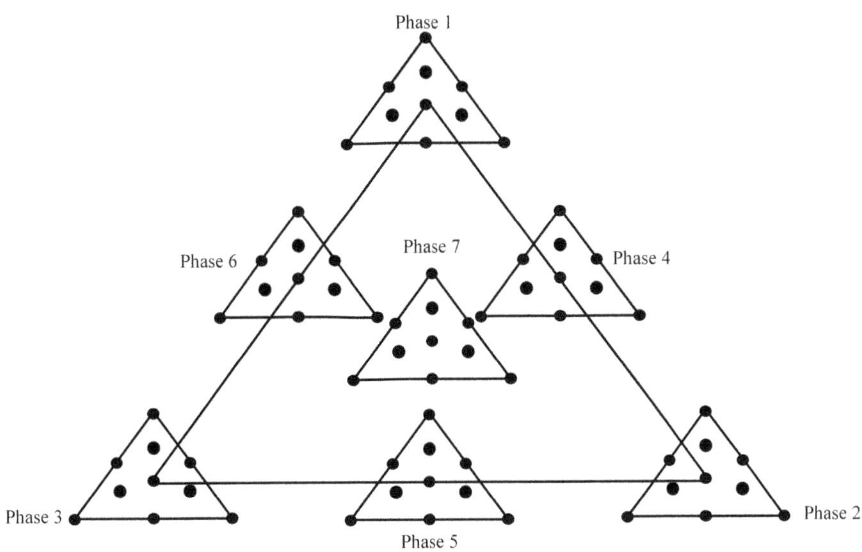

Fig. 11.5 Crossed simplex centroid mixture design for mobile phase and simplex centroid with axial point for solvent extraction media

split-plot strategy. Mobile phase solvent changes were randomly made for each solvent extract. The measure of fingerprint information was the number of peaks in each chromatogram.

Response surface contour diagrams can be analyzed for each one of the seven mobile phases in Fig. 11.5. As an example, the response contour map for the number of peaks for the acetonitrile mobile phase, phase 2 in Fig. 11.5, was used to estimate linear, quadratic, and special cubic model coefficients that can satisfactorily determine the best extraction medium composition. The extraction solvents were ethyl acetate (ea), ethanol (et), and dichloromethane (d). The regression model obtained for the quadratic model with significant coefficients in bold face is

$$\hat{y} = \underset{(\pm 2.079)}{\mathbf{7.12}\,\text{ea}} + \underset{(\pm 2.079)}{4.76\,\text{et}} + \underset{(\pm 2.079)}{4.20\,\text{d}} + \underset{(\pm 9.54)}{4.67\text{ea-et}} + \underset{(\pm 9.54)}{3.60\,\text{ea-d}} + \underset{(\pm 9.54)}{\mathbf{42.83}\text{et-d}}. \quad (11.19)$$

The first three coefficients in Eq. (11.19) are in excellent agreement with the number of peaks observed in each pure component extraction medium (7, 5, and 4). There is only one significant synergic interaction between ethanol-dichloromethane, so the model predicts an optimum solvent extraction composition of 50% ethanol and 50% dichloromethane resulting in the elution of more than 14 peaks, confirmed by inspection of the response surface contour plot in Fig. 11.6.

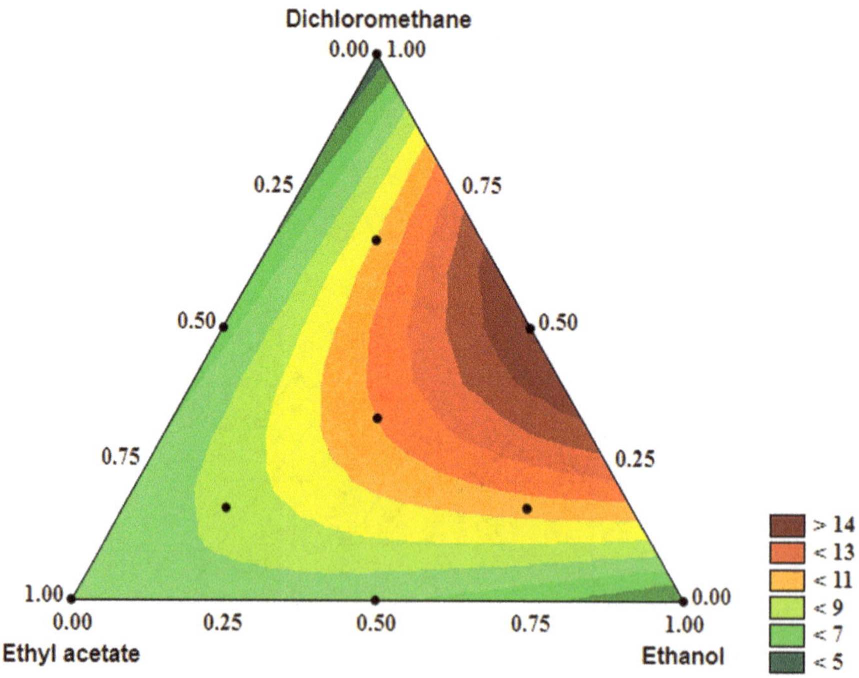

Fig. 11.6 Response surface contour diagram of the number of peaks predicted for the acetonitrile mobile phase with varying extraction solvent proportions

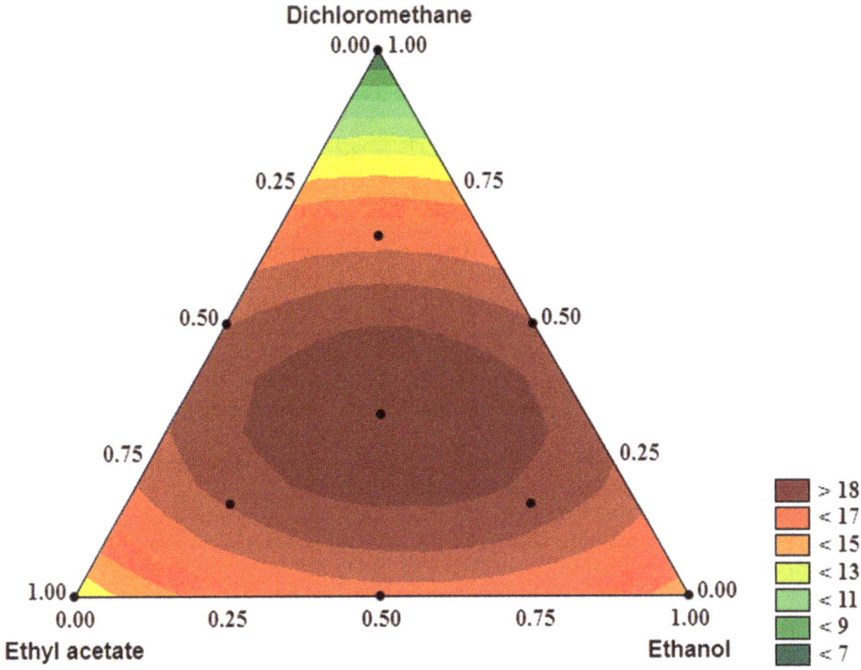

Fig. 11.7 Response surface contour diagram of the number of peaks predicted for the fourth mobile phase with varying extraction solvent proportions

The maximum number of chromatographic peaks was found for mobile phase 3 in Fig. 11.5. Its extraction solvent response surface contour plot is presented in [9].

The optimum compositions of solvents to chemical extract substances from green tea for fourth mobile phase chromatographic analysis (Fig. 11.7) were ethyl acetate/ethanol/dichloromethane (33.3:33.4:33.3 v/v/v) and methanol/acetonitrile (50:50 v/v/v), respectively. This system results in 20 peaks in the chromatographic fingerprint. These results show that an incorrect choice of modifiers for mobile phase composition and solvent extraction hampers the detection of a maximum number of peaks and produces a poor fingerprint.

References

1. Xavier D, Scarminio IS, Bruns RE (2011) Mixture designs for exploring class diversity and metabolite fingerprinting: an efficient column chromatographic strategy. Anal Chim Acta 702: 288–294
2. Chavan UD, Shahidi M, Naczk M (2001) Extraction of condensed tannins from beach pea (Lathyrus maritimus L.) as affected by different solvents. Food Chem 75:509–512

3. de Souza EBR, da Silva RR, Afonso S, Scarminio IS (2009) Enhanced extraction yields and mobile phase separations by solvent mixtures for the analysis of metabolites in *Annona muricata* L. leaves. J Sep Sci 32:4176–4185
4. Rydberg J, Cox M, Musikas C, Choppin GR (2004) Solvent extraction, principles and practices, 2nd edn. Marcel Dekker, New York
5. Snyder LR, Kirkland JJ, Glajch JL (1997) Practical HPLC method development, 2nd edn. Wiley, New York, p 35
6. Čanadanović-Brunet JM, Djilas SM, Ćetković GS, Tumbas VT, Mandić AI, Čanadanović VM (2006) Int J Food Sci Technol 41:667–675
7. Cheung LM, Cheung PCK, Ooi VE (2003) Antioxidant activity and total phenolics of edible mushroom extracts. Food Chem 81:249–255
8. Singh RP, Murthy KNC, Jayaprakasha GK (2002) Studies on the antioxidant activity of pomegranate (*Punica granatum*) peel and seed extracts using in vitro models. J Agric Food Chem 50:81–86
9. de Almeida AA, Scarminio IS (2007) Statistical mixture design optimization of extraction media and mobile phase compositions for the characterization of green tea. J Sep Sci 30:414–420
10. Rebecca LJ, Sharmila S, Das MP, Seshiah C (2014) Extraction and purification of carotenoids from vegetables. J Chem Pharm Res 6:594–598
11. Lichtenthaler HK (1987) Chlorophylls and carotenoids: pigments of photosynthetic biomembranes. In: Packer L, Douce R (eds) Methods in enzymology, vol 148. Academic Press, London, pp 350–381
12. Snyder LR (1978) Classification of the solvent properties of common liquids. J Chromatogr Sci 16:223–234
13. Cornell JA (1990) Experiments with mixtures: designs, models and the analysis of mixture data, 2nd edn. Wiley, New York
14. Bruns RE, Scarminio IS, Neto BB (2006) Statistical design – chemometrics, vol 25. Elsevier, Amsterdam, pp 313–340
15. Cornell JA (1988) Analyzing data from mixture experiments containing process variables: a split-plot approach. J Qual Technol 20:2–23
16. Derringer G, Suich R (1980) Simultaneous optimization of several response variables. J Qual Technol 12:214–219
17. Smilde AK, Jansen JJ, Hoefsloot, Lamers RJAN, van der Greef J, Timmerman ME (2005) ANOVA-simultaneous component analysis (ASCA): a new tool for analyzing designed metabolomics data. Bioinformatics 21:3043–3048
18. Hoefsloot HCJ, Vis DJ, Westerhuis JA, Smilde AK, Jansen JJ (2020) Multiset data analysis: ANOVA simultaneous component analysis and related methods. In: Brown SRT, Walczak B (eds) Comprehensive chemometrics: chemical and biochemical data analysis, 2nd edn. Elsevier, Amsterdam, pp 453–472
19. Box GEP, Hunter WG, Hunter JS (1978) Statistics for experimenters: an introduction to design, data analysis and model building. Wiley, New York
20. Brereton RG (2003) Chemometrics: data analysis for the laboratory and chemical plant. Wiley, London, pp 184–205
21. Kokova DA, Kostidis S, Morello J, Dementeva N, Perina EA, Ivanov VV, Ogorodova LM, Sazonov AE, Saltykova IV, Mayboroda OA (2017) Exploratory metabolomics study of the experimental opisthorchiasis in a laboratory animal model (golden hamster, *Mesocricetus auratus*). PLoS Negl Trop Dis 11(10):e0006044. https://doi.org/10.1371/journal.pntd.0006044
22. PLS_Toolbox software-https://eigenvector.com/software/pls-toolbox/ PLS_Toolbox. Eigenvector Research, Manson, 98831; software available at http://www.eigenvector.com
23. Marcheafave G, Pauli ED, Tormena CD, Ortiz MCV, de Almeida AA, Rakocevic M, Bruns RE, Scarminio IS (2020) Factorial design fingerprint discrimination of Coffea arabica beans under elevated carbon dioxide and limited water conditions. Talanta 209:120591

Index

M. C. Breitkreitz, H. C. Goicoechea (eds.), *Introduction to Quality by Design in Pharmaceutical Manufacturing and Analytical Development*, AAPS Introductions in the Pharmaceutical Sciences 10, https://doi.org/10.1007/978-3-031-31505-3

Printed in the USA
CPSIA information can be obtained
at www.ICGtesting.com
LVHW021111200823
755749LV00005B/192

9 783031 315046